AF588010

PHYSICS RESEARCH AND TECHNOLOGY

HELIUM: CHARACTERISTICS, COMPOUNDS, AND APPLICATIONS

PHYSICS RESEARCH AND TECHNOLOGY

Additional books in this series can be found on Nova's website under the Series tab.

Additional E-books in this series can be found on Nova's website under the E-book tab.

CHEMISTRY RESEARCH AND APPLICATIONS

Additional books in this series can be found on Nova's website under the Series tab.

Additional E-books in this series can be found on Nova's website under the E-book tab.

PHYSICS RESEARCH AND TECHNOLOGY

HELIUM: CHARACTERISTICS, COMPOUNDS, AND APPLICATIONS

LUCAS A. BECKER
EDITOR

Nova Science Publishers, Inc.
New York

For permission to use material from this book please contact us:
Telephone 631-231-7269; Fax 631-231-8175
Web Site: http://www.novapublishers.com

LIBRARY OF CONGRESS CATALOGING-IN-PUBLICATION DATA

Helium : characteristics, compounds, and applications / [edited by] Lucas A. Becker.
p. cm.
Includes index.
ISBN 978-1-61761-213-8 (hardcover)
1. Helium. I. Becker, Lucas A.
QD181.H4H455 2010
546'.751--dc22

2010031193

Published by Nova Science Publishers, Inc. † New York

CONTENTS

PREFACE

Helium is the chemical element with atomic number 2 and an atomic weight of 4.002602, which is represented by the symbol He. It is a colorless, odorless, tasteless, non-toxic, inert monatomic gas that heads the noble gas group in the periodic table. Its boiling and melting points are the lowest among the elements and it exists only as a gas except in extreme conditions. Next to hydrogen, it is the second most abundant element in the universe, and accounts for 24% of the elemental mass of our galaxy. This book presents current research data in the study of helium, including helium in the terrestrial upper atmosphere; an experimental and analytical investigation of helium in metals diffusion; the release of mantle helium and its tectonic implications; and the study of helium behavior in nuclear fuel materials.

Chapter 1 - Helium, one of light components of the near-Earth space environment, determines atmospheric density and temperature at heights of 300-2000 km. These characteristics are studied by ground-based methods using infra-red helium emission 1083 nm with intensity of approximately 1 kilo-Rayleigh. Measurements of this emission are carried out by spectrophotometric method during twilight time at solar zenith angle $\chi_{\odot} = 100 \div 150^{\circ}$. This emission has been for the first time discovered on the geophysical station Zvenigorod (Russia) during intensive low-latitude aurorae on February 11, 1958. Later, the theory of its occurrence has been developed for the stable and disturbed geomagnetic conditions. The 1083 nm emission is caused by fluorescence (in a solar radiation) of metastable helium atoms formed due to excitation by electrons with energy of 25 eV. For stable geomagnetic conditions, such electrons originate from photoionization of the atmospheric components by solar ultraviolet radiation of 30.4 nm emmited by helium ions. During auroras, the electron fluxes with energy nearly 10 keV create the electron fluxes with energy of 25 eV due to interaction with atmospheric atoms and molecules.

Detection of the 1083 nm helium emission was the first proof of photoelectrons with 25eV energies to exist in the upper atmosphere. Characteristics of the fluorescence mechanism of the excited metastable helium atoms in the upper atmosphere gives preference to the 1083 nm emission in comparison to other emissions as the scattering coefficient of photons for 1083 nm considerably exceeds the coefficients for other emissions. Ground-based measurements of the spatial and temporal variations of intensity of the helium 1083 nm emission allow us to obtain the geophysical data on structure of upper layers of the terrestrial atmosphere and characteristics of solar ultraviolet radiation. Long-term ground-based measurements of variations of the helium emission intensity have allowed us to reveal the regularities of the variations during twilight periods depending on a zenith solar angle, phase age of the Moon, season, levels of solar and geomagnetic activities. The analysis of the observational data has revealed latitudinal asymmetry determined by seasonal position of the Sun. The accumulated data provided a basis for construction of empirical model of variations of characteristics of infrared radiation and, hence, contents of helium atoms in the near-Earth space environment depending on the different helio and geophysical conditions. Besides, a comparison of results of the measurements with the calculated dependence of the intensity changes during twilight on zenith solar angle allowed us to determine temperature of the upper layers of the Earth's atmosphere.

Chapter 2 - In the first part of this paper experimental and analytical investigation of helium diffusion and bubble formation and growth in aluminum is presented [1;2]. A theoretical model for equation of state of aluminum with helium bubbles and its validation in shock wave experiments are presented in the second part [3].

A pure aluminum with 0.15% wt of 10B was neutron-irradiated in a nuclear reactor to get homogeneous helium atoms in the metal according to the reaction $^{10}B+n \rightarrow {}^{7}Li+{}^{4}He$. Formation and growth of helium bubbles was observed in situ by heating the post-irradiated metal to 470ºC in TEM with a hot stage holder. It was found that above 400ºC the time scale for bubble's shape change is seconds. In other experiments the Al-10B was first heated in its bulk shape and then observed in TEM at room temperature. In this case the helium bubble formation takes hours.

Analytical evaluation of the diffusion processes in both cases was done to explain the experimental results. The number of helium atoms in a bubble was calculated from the electron energy loss spectrum (EELS) measurements. These measurements confirmed the hard sphere equation of state (EOS) for inert gases that was used in the analytical diffusion calculations.

It was also found that the helium-rich area expands due to helium migration. Electron beam diffraction revealed that the preferred orientation of the helium atoms migration is normal to plane. The results are consistent with models for helium atoms migration between interstitial sites for an fcc metal.

At the second part of this paper a theoretical model for equation of state of aluminum with helium bubbles is presented. Based on this equation of state, the influence of helium bubbles on shock loading is examined. The Hugoniot curve (temperature vs. pressure as well as shock velocity vs. particle velocity) for aluminum containing bubbles is calculated for various bubbles mass, bubbles percentage and helium equation of state models. The bubble mass and concentration seem to affect measurably the Hugoniot curve. The equation of state model implied for the helium in the bubbles has minor significance, which means the model is not sensitive to the details of the helium EOS. These findings were confirmed in shock wave experiments.

Chapter 3 - Helium is the lightest noble gas and both stable isotopes, 3He and 4He, are produced in the crust in a ratio of ~ 0.02 Ra, with Ra being the atmospheric 3He/4He ratio of 1.4×10-6. Higher values are an indication of helium from the mantle where 3He captured during planetary accretion has been stored. It has been suspected for some time that degassing of the planet does not occur homogeneously over the Earth's surface, but is rather concentrated along plate boundaries, where the dynamics of the lithosphere are more intense and mantle helium from the Earth's interior can be more easily transported to the surface. The authors indicate that the spatial distribution of 3He/4He ratios in gas samples from crustal fluids are considered to provide potentially useful information for determining not only latent magmatic activity but also potential pathways for mantle volatiles, such as in tectonically active zones.

Chapter 4 - At the interaction between the spin-polarized excited atom and paramagnetic ground state atom or molecule in gas discharge, elastic and inelastic processes can take place simultaneously. It means that besides the chemo-ionization of the atom or molecule at the expense of atom's excitation energy (inelastic process), an exchange of electrons is possible without a great depolarization (elastic process, or spin exchange). In such a case these two processes give rise to a remarkable spin polarization transfer between colliding particles. Influencing each other, these two processes result in a change in the spin exchange and frequency shift cross section values.

The helium metastable atoms (He*), having a large store of internal energy (19.8 eV), are capable to ionize the molecule or atom (A) even at thermal energies of relative motion

Chapter 5 - Helium in the state of plasma has the properties strikingly different from those of atomic helium. Using helium plasma, one may obtain VUV (vacuum ultraviolet) radiation. At present, He VUV radiation is widely used to modify porous layers based on silicon dioxide. In the present work, the authors demonstrate the experimental data on the action of He plasma treatment on methyl-doped silicon dioxide layers. It is shown that the He VUV radiation affects the materials with different pore sizes in principally different ways. The authors propose a model based on the experimental data. The reasonable character of the model was confirmed by quantum chemical calculations. According to the authors' model, the action of He plasma is considered mainly as the VUV radiation (21 eV) causing rupture of chemical bonds. Detachment of hydrogen atom from the methyl groups causes the appearance of the positive charge on silicon atoms, which leads to an increase in dπ-pπ overlapping, which, in turn, decreases the reactivity of the layers. This causes passivation of the layers.

Chapter 6 - From its discovery, helium had a leading role in many scientific developments. In this chapter, the authors are interested by its use as a lifting agent for lighter than air vehicles, more specifically unmanned airships. An important application, bridge monitoring, is highlighted in this chapter. The disaster caused by the collapse of one of Minneapolis (Minnesota) highway bridges points to the need for better technologies to inspect bridges. A flight simulator software is first developed then kinematic and dynamic models of this lighter than air vehicle are developed. Finally, the trim trajectories, used in the flight simulator, are detailed and a robust control method introduced.

Chapter 7 - The α-decay product helium (He) is an important safety-limiting factor of nuclear fuel materials due to its low solubility in the fuel matrix, especially in the actinide dioxide lattice with fluorite structure. When the concentration of He exceeds approximately 1%, He contributes to diminishing the mechanical strength of the fuel by initiating the precipitation. The authors present here the results of a first-principles study to understand the diffusion mechanism as well as clustering behavior of He in primarily UO2. Performing energetic calculations, the favored locations of a He atom are calculated as an octahedral interstitial site (OIS) in a defect-free UO2 lattice, and as a uranium vacancy when radiation induced vacancy defects are present. Helium has a strong agglomerating tendency, resulting in the creation of point defects in the fuel lattice. The He behavior in other oxide fuels, ThO2 and PuO2, is investigated and compared with that obtained for UO2.

Chapter 8 - The thermodynamic properties of helium at wide range of temperature and pressure are analytically derived from the knowledge of the pair spherical potentials to which quantum corrections are superposed.

The double Yukawa potential model is considered to describe the intermolecular ttraction and repulsion energies. Low temperature quantum effects are incorporated by using the first order quantum correction of the Wigner-Kirkwood expansion. A fundamental equation of state is formulated including Helmholtz energy as an explicit function of temperature and density. The thermodynamic properties are expressed as an explicit combination of the Helmholtz energy and its derivatives. The obtained values are compared to the thermodynamic data and the Molecular Dynamic calculations, a satisfactory correspondence with simulation results realized. The feature of helium thermodynamics in the critical region discussed. Contrary to most previous similar works, the present theory retrieves the main features of the helium at wide temperature and pressure from analytical formulation.

Chapter 9 -Charged particles such as pions, kaons or antiprotons (produced either by nature or by man in High Energy Physics laboratories under ultrahigh vacuum conditions) are soon decelerated when brought into a medium made of ordinary matter, where their fate essentially depends on their charge. Positive particles (as π+ or K+) just hang around in the space between different atoms, until they spontaneously decay with their characteristic lifetime. Negative particles (as π–, K– or $\bar{p}$), are instead quickly captured by the atomic nuclei in the medium, leading to their annihilation on nanosecond or picoseconds timescale.

This is actually what happens most of the times yet with one exception: helium. In fact, when a negative particle such as π–, K– or $\bar{p}$is captured by an helium atom, ejecting one of the two electrons, an exotic atom π–He+, K–He+, or $\bar{p}$He+respectively is formed. A small yet measurable fraction of these atoms, about 2-3 percent, shows a much longer lifetime, which is anyway limited by the intrinsic decay of the particle in the first two cases ($\tau_\pi \simeq 26$ ns, $\tau_k \simeq 12.4$ ns), though it can be as long as a few microseconds in the case of antiprotonic helium. Although quite puzzling at the beginning, this behaviour was realized to stem from the fact that a small fraction of negative particles were trapped in long-lived states, giving rise to the formation of "metastable" atoms with lifetimes of the order of 1 μs. These veryspecial atoms must be very resilient to the effects usually shortening their lifetime, mainly Stark decay caused by the collisions with the surrounding helium atoms and Auger decay transforming $\bar{p}$He+into $\bar{p}$He++. This fact leaves them affected by only the radiative decay, which is usually much slower and has timescales comparable to what is observed.

Subsequent studies have actually revealed that only a few states of the antiprotonic atom have these features, enabling one to identify them and eventually to measure, with very high accuracy, the energy involved in the transitions between one of these states and one of the normal, short-lived ones.

In: Helium: Characteristics, Compounds…
Editors: Lucas A. Becker

ISBN: 978-1-61761-213-8

Chapter 1

HELIUM IN THE TERRESTRIAL UPPER ATMOSPHERE: SPATIAL AND TEMPORAL DISTRIBUTION OF ITS EMISSION IN THE INFRARED SPECTRAL REGION

N. N. Shefov* and A. I. Semenov
A.M. Obukhov Institute of Atmospheric Physics, Russian
Academy of Sciences, Moscow, Russia

ABSTRACT

Helium, one of light components of the near-Earth space environment, determines atmospheric density and temperature at heights of 300-2000 km. These characteristics are studied by ground-based methods using infra-red helium emission 1083 nm with intensity of approximately 1 kilo-Rayleigh. Measurements of this emission are carried out by spectrophotometric method during twilight time at solar zenith angle $\chi_{\odot} = 100 \div 150^{\circ}$. This emission has been for the first time discovered on the geophysical station Zvenigorod (Russia) during intensive low-latitude aurorae on February 11, 1958. Later, the theory of its occurrence has been developed for the stable and disturbed geomagnetic conditions. The 1083 nm emission is caused by fluorescence (in a solar radiation) of metastable helium atoms formed due to excitation by electrons with energy of 25 eV. For stable geomagnetic conditions, such electrons originate from photoionization of the atmospheric components by solar ultraviolet radiation of 30.4 nm emmited by helium ions. During auroras, the electron fluxes with energy nearly 10 keV create the electron fluxes with energy of 25 eV due to interaction with atmospheric atoms and molecules.

Detection of the 1083 nm helium emission was the first proof of photoelectrons with 25eV energies to exist in the upper atmosphere. Characteristics of the fluorescence mechanism of the excited metastable helium atoms in the upper atmosphere gives preference to the 1083 nm emission in comparison to other emissions as the scattering coefficient of photons for 1083 nm considerably exceeds the coefficients for other emissions. Ground-based measurements of the spatial and temporal variations of intensity of the helium 1083 nm emission allow us to obtain the geophysical data on structure of

* Corresponding author: nikoshefov@yandex.ru, anasemenov@yandex.ru

upper layers of the terrestrial atmosphere and characteristics of solar ultraviolet radiation. Long-term ground-based measurements of variations of the helium emission intensity have allowed us to reveal the regularities of the variations during twilight periods depending on a zenith solar angle, phase age of the Moon, season, levels of solar and geomagnetic activities. The analysis of the observational data has revealed latitudinal asymmetry determined by seasonal position of the Sun. The accumulated data provided a basis for construction of empirical model of variations of characteristics of infrared radiation and, hence, contents of helium atoms in the near-Earth space environment depending on the different helio and geophysical conditions. Besides, a comparison of results of the measurements with the calculated dependence of the intensity changes during twilight on zenith solar angle allowed us to determine temperature of the upper layers of the Earth's atmosphere.

1. INTRODUCTION

Helium is one of the most important components of the upper atmosphere at heights above 300 km. Originally, Helium was detected in the solar atmosphere and then in the Earth's lower atmosphere. However, before the International Geophysical Year (1957-1959, IGY), the helium content of the upper atmosphere was estimated only based on the theoretical models of its composition. Helium emissions have high excitation energies [Telegin and Yatsenko 2000], which require a presence of electrons with energies of about 25 eV (Figure 1). Such electrons are formed in the upper atmosphere of the Earth basically due to photoionization of atoms and molecules by solar ultraviolet radiation of $\lambda < 30.4$ HM. Unfortunately, many helium emissions obtained by ground-based observations are overlapped by the spectral lines of the hydroxyl molecule bands. Therefore, it is very difficult to register them even during auroras.

The detection of the orthohelium 1083 nm fluorescent emission as a result of its sharp enhancement during the intensive sunlit aurora on February 11th in 1958, was one of the new important results obtained at Zvenigorod Measurement Station ($\varphi = 55.7°$N, $\lambda = 36.8°$E; $\Phi = 51.2°$N, $\Lambda = 120.0°$E) when the studies of the upper atmosphere glow were intensified during IGY [Krassovsky and Galperin 1960; Mironov et al 1959; Khomich et al 2008]. In twilight conditions for solar zenith angle $\chi_{\odot} < 105°$ and under stable geomagnetic conditions the helium 1083 nm emission has intensity of ~ 1 kilo-Rayleighs (kR) (10^9 photon··cm^{-2}·s^{-1}). The intensity of 1083.0 nm exponentially decreases when the solar zenith angle increases. It is necessary to notice, that at spectral devices' resolution of ~0.3 nm, this emission is blended by the $Q_1(1)$ line of the (5-2) band of hydroxyl molecule with an intensity of about 2 kR (Figure 2, Table 1). This problem has not allowed to detect the 1083 nm helium emission during regular twilight observations prior to 1958.

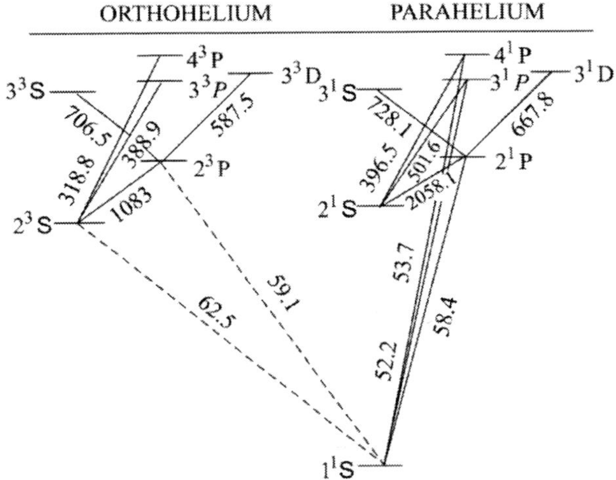

Figure 1. The energy level structure of the helium atom and the wavelengths of the basic transitions [Khomich et al 2008]

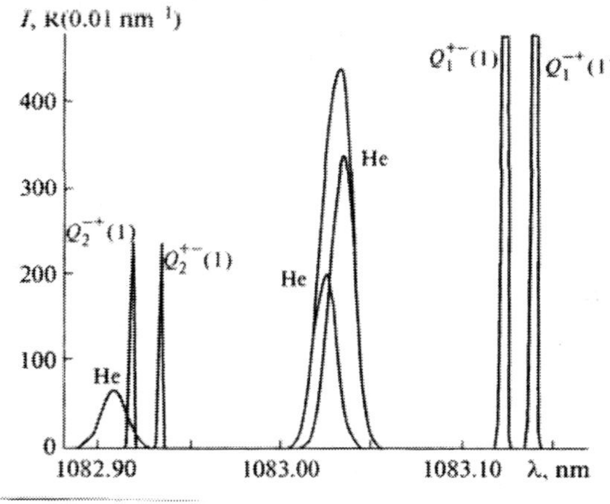

Figure 2. Spectral structure of the helium and hydroxyl lines in the 1082.88 – 1083.70 nm [Shefov et al 2009 a]

Table 1. Parameters of the helium and hydroxyl emission lines [Shefov et al 2009 a]

λ, nm	Symbol	Intensity I, Rayleighs	Sum of lines, Rayleighs	I_{max}, R·(0.01 нм)$^{-1}$
1082.908	He	110	110	67
1082.918	$Q_2^{-+}(1)$ OH(5−2)	60	120	237
1082.933	$Q_2^{+-}(1)$ OH(5−2)	60		237
1083.025	He	330	890	200
1083.034	He	560		340
1083.123	$Q_1^{+-}(1)$ OH(5−2)	500	1000	1980
1083.139	$Q_1^{-+}(1)$ OH(5−2)	500		1980

However, during the sunlit intense aurora observed on February 11th in 1958, the emission of 1083 nm had a much higher intensity (~ 60 kR) and could be detected. Such a specific feature was confirmed when the aurora of March 31st in 1960, was observed [Fedorova 1961a, 1962]. The special studies made it possible to detect the presence of this emission with an intensity of ~1 kR in the sunlit upper atmosphere based on the measurements at high [Fedorova 1961a,b, 1967] and middle latitudes [Shefov 196lb, 1967; Taranova 1967; Toroshelidze 1970, 1976, 1984. 1991], by Fabry-Perot interferometer [Shcheglov1962a, b] (Figure 3).

Further measurements were conducted under usual twilight conditions at low and middle geographic latitudes at Zvenigorod (55.7°N, 36.8°E) [Shefov 1963a,b, 1967, 1968; Taranova 1967; Shefov and Yurchenko 1970], Socorro (34.0°N, 252.8°E) [Christensen et al 1971, 1972; Tinsley 1968; Tinsley and Christensen 1976], Saskatoon (52.1°N, 253.3°E) [Neo and Rundle 1969], Calgary (51.0° N, 245.9°E) [Harrison 1969; Harrison and Cairns 1969], Abastumani (41.8°N, 42.8°E) [Toroshelidze 1970, 1971, 1976, 1984, 1991], Mount Agulhas Negras (22.4°S, 315.3°E) [Teixeira et al 1975], Huancayo (12.0°S, 286.7°E) [Tinsley and Christensen 1976], Kakioka (36.2°N, 140.2°E) [Suzuki 1983], Millstone Hill (42.6°N, 288.5°E), and Arecibo (18.3°N, 293.2°E) [Kerr et al 1996; Noto et al 1998] stations. The observations were conducted onboard an airplane during the solar eclipse near Rostov-on-Don (47°N, 39°E) [Shouyskaya 1963], and at Loparskaya station (68.6°N, 33.4°E) during sunlit auroras [Fedorova 1967; Sukhoivanenko and Fedorova 1976].

During auroras, the emission intensity was usually about 10–12 kR but did not reach its maximum observed value of 60 kR, registered in the red aurora on February 11th in 1958.

According to the high-latitude observations in Tixie Bay (Φ = 65.6°N, Λ = 195.2°E), at Spitzbergen (Φ = 73.1°N, Λ= 129.4°E), and in Norway (Φ = 64.3°N, Λ = 104.9° E), the helium 587.6 nm emission with the shifted profile, caused by precipitating helium ions, was regularly observed during intense auroras [Stoffregen 1969; Henriksen and Sukhoivanenko 1982]. Its intensity reached 120 Rayleighs (R).

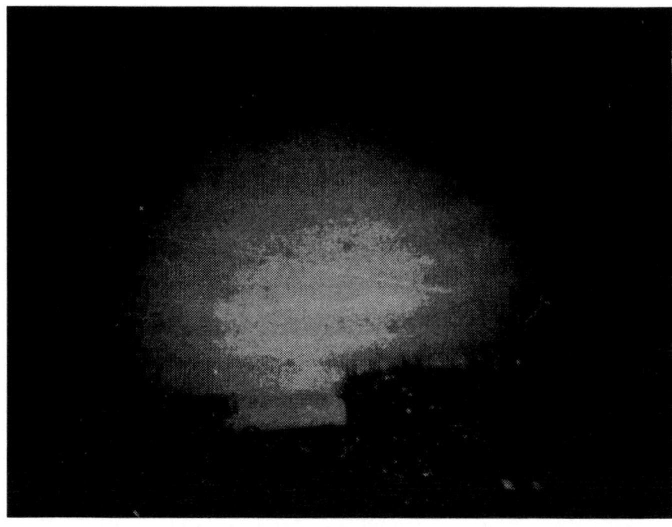

Figure 3. Interferogram of the 1083.0-nm helium emission taken in twilight [Shcheglov 1962 a,b; Khomich et al 2008]

On the base of available data of the intensity measurements of infrared orthohelium 1083 nm emission and developed theory of excitation processes, the statistical analysis of this material is made and analytical approximations of different types of emission variations are obtained for various geophysical conditions.

2. THE MAIN MECHANISM OF EXCITATION OF METASTABLE ORTHOHELIUM ATOMS

The scheme of the excited levels of the helium atoms and the emission lines are presented in Figure 1. Transitions between various electronic states unequivocally confirm that an appearance of the excited atoms originate probably both from impacts of neutral atoms with electrons and from helium ion recombinations. All these processes were studied in details in laboratory conditions. Data on cross sections of excitation and deactivation of various states and recombination coefficients have been obtained (Tables 1,2,3,4,5) [Massey and Barhop 1952; Wiese et al 1966; Allen 1973; Banks and Kockarts 1973 a, b; Lindinger et al 1974; Egorov et al 1994]. All these data have been analyzed after the detection of helium infrared 1083 nm emission during sunlit aurora on February 11th in 1958 [Shefov 1961 a, b, c]. The analysis has led to a conclusion that the only possible process for excitation of 1083 nm emission is the fluorescence of metastable orthohelium atoms caused by a sunlight.

As follows from the structure of the helium atom energetic levels [Telegin and Yatsenko 2000], the 1083 nm line corresponds to the 2^3P-2^3S resonant transition of orthohelium, the lower level of which is metastable with a high excitation energy (19.3 eV) relative to the ground state 1^1S, and the emission process proceeds according to the reaction

$$He(2^3P) \rightarrow He(2^3S) + h\nu(1083\ \text{nm}).$$

This transition is permitted and has a probability of $A_{1083} = 1.05 \cdot 10^7\ s^{-1}$ [Allen 1973].

An analysis of the mechanism of helium emission excitation, caused by the resonant fluorescence in the solar radiation of metastable orthohelium atoms, made it possible to show that this emission should be present in the twilight and daytime airglow [Shefov 1961 a,b, 1962 b]. This means that at this time, electrons with energies of about 25 eV are produced in the upper atmosphere at heights above 250–300 km due to the processes of photoionization of helium atoms and molecules of oxygen and nitrogen. Under quiet geomagnetic conditions, electrons with energies of 25 eV are produced in the atmosphere as a result of photo-ionization of atoms and molecules in the upper atmosphere by solar radiation, which is chatacterized by the highest intensity of the solar emission in the ionized helium He^+ 30.4 nm line [Shefov 1962a, b]. The detection of the helium 1083 nm emissions for the first time directly indicated that substantial photoelectron fluxes exist in the Earth's upper atmosphere [Shefov 1962b], which was previously suggested by Shklovsky [1951].

Table 2. Wavelengths λ, level energy E, transition probabilities A, photon scattering factors g for the emission from orthohelium atoms [Shefov 1961 b; Khomich et al. 2008]

λ, nm	Transition	Level energy E, eV		A, s^{-1}	g, s^{-1}
1083.0341	$2^3P–2^3S$	20.97	19.82	1.05(7)	16.8
1083.0250	$2^3P–2^3S$	20.97	19.82		
1082.9081	$2^3P–2^3S$	20.97	19.82		
388.8648	$3^3P–2^3S$	23.01	19.82	9.4(6)	0.0485
318.7747	$4^3P–2^3S$	23.71	19.82	6.1(6)	0.0287
706.5719	$3^3S–2^3P$	22.72	20.97	1.6(7)	0.89
706.5188	$3^3S–2^3P$	22.72	20.97		
587.5989	$3^3D–2^3P$	23.08	20.97	7.2(7)	9.6
587.5650	$3^3D–2^3P$	23.08	20.97		
587.5618	$3^3D–2^3P$	23.08	20.97		
471.3373	$4^3S–2^3P$	23.60	20.97	3.3(6)	0.0335
471.3143	$4^3S–2^3P$	23.60	20.97		
447.1688	$4^3D–2^3P$	23.74	20.97	2.0(7)	0.715
447.1477	$4^3D–2^3P$	23.74	20.97		
412.0993	$5^3S–2^3P$	23.97	20.97	2.0(7)	0.008
412.0812	$5^3S–2^3P$	23.97	20.97		
402.6362	$5^3D–2^3P$	24.05	20.97	1.2(7)	0.143
402.6189	$5^3D–2^3P$	24.05	20.97		
62.5585	$2^3S–1^1S$	19.82	0.00	2.2(–5)	
59.1406	$2^3P–1^1S$	20.97	0.00	<1(3)	

During auroras, photoelectrons are produced by penetration of fast electrons with energies of ~10 keV [Shefov 196la,b,c,d, 1962b, 1963b]. The performed computations made it possible to obtain the content of helium above 300 km altitude. The measurements of the intensity of the twilight helium emission gave a possibility to detect the global asymmetry in the helium concentration distribution in the Earth's atmosphere and an increased helium concentration in the winter hemisphere [Shefov 1967, 1970,1973].

3. Processes of Excitation and Deactivation of the Metastable Helium Atoms

An analysis of the excitation processes of the helium emission made it possible to show that the presence of only 1083 nm emission and the absence of the well-known helium emissions in the visible region of the spectrum (Tables 2,3) indicate that this emission is caused by fluorescence of metastable orthohelium atoms, produced during collisions with electrons, in the solar radiation [Shefov 1961 b,c]..

Below, we present the main reactions and their rate constants of excitation and deactivation. The rate of excitation by photoelectrons is governed by the process [Khomich et al., 2008]

$$He(1^1S) + e(E > 25\ eV) \rightarrow He(2^3S) + e \qquad q = 4 \cdot 10^{-9}\ cm^3 \cdot s^{-1},$$

where the coefficient rate q is given for the moderate level of solar activity:

Here, the production rate of metastable orthohelium atoms is

$$q = \int_{E_o}^{\infty} \Phi(E) \cdot \sigma(E) \cdot dE,$$

where $\Phi(E)$ is the photoelectron flux, and $\sigma(E)$ is the excitation cross section of metastable helium atoms.

Table 3. Wavelengths λ, level energy E, transition probabilities A, photon scattering factors g for the emission from parahelium atoms [Shefov 1961 b; Khomich et al 2008]

λ, nm	Transition	Level energy E, eV		A, s^{-1}	g, s^{-1}
2058.130	$2^1P - 2^1S$	21.22	20.62	2.0(6)	13.5
501.5675	$3^1P - 2^1S$	23.09	20.62	1.4(7)	1.73
396.4727	$4^1P - 2^1S$	23.74	20.62	8.7(6)	0.0463
361.3641	$5^1S - 2^1P$	24.05	20.62	4.3(6)	0.0255
728.1349	$3^1S - 2^1P$	22.92	21.22	1.9(7)	1.13
667.8149	$3^1D - 2^1P$	23.08	21.22	6.6(7)	13.0
504.7736	$4^1S - 2^1P$	23.68	21.22	5.5(6)	0.079
492.1929	$4^1D - 2^1P$	23.74	22.22	2.0(7)	1.25
443.7549	$5^1S - 2^1P$	24.01	21.22	3.4(6)	0.0186
438.7928	$5^1D - 2^1P$	24.05	21.22	8.6(6)	0.12

Table 4. Wavelengths λ [Striganov and Odintsova 1982], level energy E, transition probabilities A [Allen 1973; Wiese et al 1966], statistical weights g, absorption cross sections σ for the helium ion [Khomich et al 2008]

λ, nm	Transition	E_u, eV	E_l, eV	g_u	g_l	A, s^{-1}	σ, cm^2
23.73307	$5^2P^o - 1^2S$	52.24	0.0	6	2	5.50(8)	2.42(-15)
24.30266	$4^2P^o - 1^2S$	51.02	0.0	6	2	1.09(9)	5.14(-15)
25.63170	$3^2P^o - 1^2S$	48.37	0.0	6	2	2.68(9)	1.48(-14)
30.37822	$2^2P^o - 1^2S$	40.81	0.0	6	2	1.00(10)	9.22(-14)
99.2364	$7^2D - 2^2P^o$	53.31	40.81	10	6	8.23(7)	1.47(-14)
102.5273	$5^2D - 2^2P^o$	52.90	40.81	10	6	1.51(8)	2.97(-14)
108.4944	$5^2D - 2^2P^o$	52.24	40.81	10	6	3.30(8)	7.7)(-14)
121.5088	$4^2D - 2^2P^o$	51.02	40.81	4	2	1.55(8)	6.10(-14)
121.5171	$4^2D - 2^2P^o$	51.02	40.81	10	4	1.07(9)	5.26(-13)
164.0332	$3^2D - 2^2P^o$	48.37	40.81	4	2	3.59(8)	3.47(-13)
164.0474	$3^2D - 2^2P^o$	48.37	40.81	6	4	1.03(9)	7.48(-13)
164.0490	$3^2D - 2^2P^o$	48.37	40.81	4	4	1.00(8)	4.84(-14)

The production of photoelectrons with energies of ~ 25 eV is caused by the influence of the solar radiation mainly with a wavelength of 30.4 nm. This radiation is generated in the solar corona and is proportional to the solar activity index F10.7. According to the empirical relationship [Krinberg 1978], the photoelectron flux depends on solar activity

$$\Phi = \Phi_0 \cdot \exp\left(\frac{F10.7 - 144}{72}\right),$$

where Φ_0 is the photoelectron spectrum at heights above 250 km, corresponding to the radioemission flux F10.7 = 144.

In its turn, the thermospheric temperature also depends on the solar activity level. This dependence $\Psi(F10.7)$ can be taken into account based on the data of the corresponding models (e.g., [Hedin 1991]). Then, the rate constant of excitation of metastable helium atoms will have the form

$$q = 4 \cdot 10^{-9} \cdot \Psi(F10.7)\ \text{cm}^3 \cdot \text{s}^{-1}$$

Formation of the $He(2^3S)$ atoms is possible due to the recombination of helium ions too

$$He^+(1^2S) + e \rightarrow He(2^3S) + h\nu(\lambda \le 267.0\ \text{nm}) \qquad \alpha = 1 \cdot 10^{-12} \cdot \left(\frac{1000}{T_e}\right)^{0.5}\ \text{cm}^3 \cdot \text{s}^{-1} \cdot$$

However, the contribution of this process becomes significant only at altitudes above 1000 km. During twilight measurements of this height region, the corresponding solar zenith angles $\chi_\odot$ are larger than 120°, when the emission intensity becomes lower than 100-150 R. This is also confirmed by the fact [Waldrop et al., 2005] that the intensity of emission due to recombination is about 40 R (Figure 4) [Shefov 1961 b; Ferguson and Schlüter 1962; Shefov et al. 2009 a,b].

Naturally, there is a series of additional processes, including the processes resulting from the intercombination transitions of metastable atoms $He(2^1S)$ during collisions with thermal electrons [Shefov 1962b, 1963b, Khomich et al. 2008].

$$He(2^1S) + e \rightarrow He(2^3S) + e\,.$$

The cross section for this process is $\sigma_{2^1S,2^3S} = (1.5 \div 8) \cdot 10^{-16}\ \text{cm}^2$ [Ivanov et al. 1991], but the contribution of these processes is small.

Fluorescence proceeds as a result of the reactions:

$$He(2^3S) + h\nu(\lambda 1083\ \text{nm}) \rightarrow He(2^3P) \qquad g_{2^3S} = 16.8\ \text{s}^{-1},$$

$$He(2^3P) \rightarrow He(2^3S) + h\nu(\lambda 1083\ \text{nm}) \qquad A_{1083} = 1.05 \cdot 10^7\ \text{s}^{-1}.$$

The scattering coefficient of photons is

$$g = 1.33 \cdot 10^{-21} \cdot S_{\lambda} \cdot r_{\lambda} \cdot \lambda^4 \cdot \frac{g_k}{g_i} \cdot A_{ki} \cdot \frac{A_{ki}}{\sum_j A_{kj}},$$

where S_λ is the solar radiation flux at the boundary of the Earth's atmosphere ($\mathrm{photon} \cdot \mathrm{cm}^{-2} \cdot \mathrm{s}^{-1} \cdot \mathrm{nm}^{-1}$) ; λ is the wavelength (mcm) of the exciting radiation; r_λ is the residual intensity of the Fraunhofer line in the solar spectrum at the wavelength λ.; g_k and g_i are the statistical weights of the upper and lower states, respectively; and A_{ki} is the transition probability.

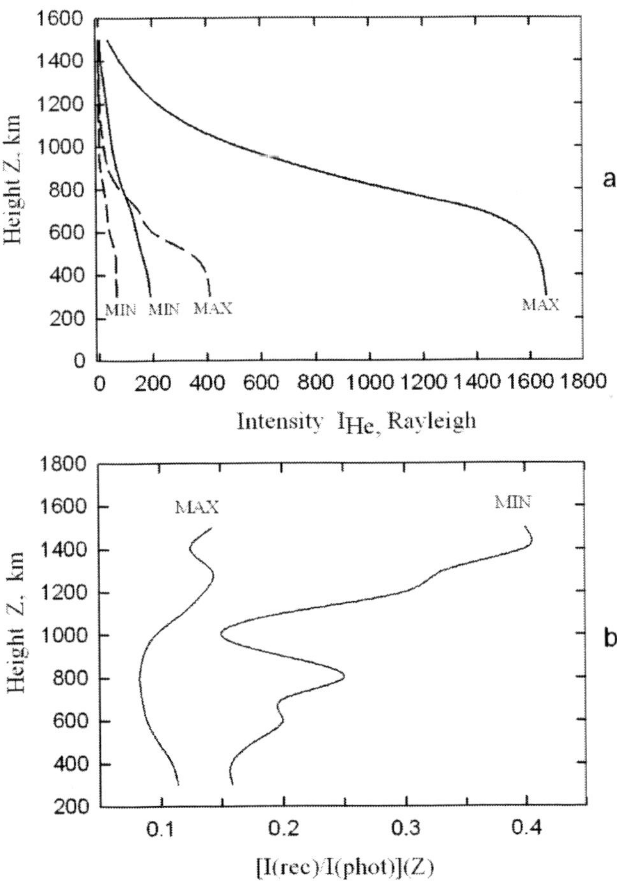

Figure 4. Dependences on the Earth's shadow height for the conditions of solar maximum and minimum of: (a) the helium 1083 nm emission intensity and (b) the ratio of the intensity components caused by the excitation due to recombination and by photoelectrons. Solid and dashed curves show the intensities due to the excitation of neutral atoms by photoelectrons and due to the production of metastable atoms as a result of helium ion recombination [Shefov et al. 2009 b]

It is necessary to note, that the difference of value $g = 17.7\ c^{-1}$ presented in [Bishop and Link, 1993], from $g = 16.8\ c^{-1}$ [Shefov 1961 b, c] is, presumably, connected to the fact that it does not consider influence of remnant intensity in the Fraunhofer profile of the solar spectrum. Actually, the ratio $16.8/17.7 = 0.95$ almost exactly corresponds to value of the residual intensity on wavelength of helium emission (Figure 5) according to [Minnaert et al. 1940; Molher et al. 1950].

Besides the excitation of the 2^3S metastable state, there occurs a deactivation of this state in collisions with atoms, molecules, and electrons, a photoionization, and a transition to the ground state. The level 2^3S is purely metastable. According to the available estimates, the two-photon radiation transition to the state 1^1S is possible. As this takes place, the transition energy, which corresponds to the wavelength 62.6 nm, is divided continuously between two photons:

$$He(2^3S) \rightarrow He(1^1S) + h\nu + h\nu \quad A_{62.6} \approx 2.2 \cdot 10^{-5}\ s^{-1}.$$

The transition probability is taken from the work by Mathis [1957]

$$A_{62.6} \approx 2.2 \cdot 10^{-5}\ s^{-1}$$

According to [Bely 1968]

$$A_{62.6} \leq 10^{-10}\ s^{-1}.$$

Later studies also considered one-photon transitions [Woodworth and Moos 1975]

$$He(2^3S) \rightarrow He(1^1S) + h\nu \qquad A_{62.6} \approx 1.10 \cdot 10^{-4}\ s^{-1}.$$

Deactivation of excited atoms due to photoionization by solar UV proceeds according to the reaction [Rundle 1960; McElroy 1965; Patterson 1967]:

$$He(2^3S) + h\nu(\lambda \leq 164\ nm) \rightarrow He^+ + e \qquad j_{2^3S} = 1.6 \cdot 10^{-3}\ s^{-1}.$$

Woodworth and Moos [1975], based on the data on photoionization cross sections [Stebbings et al. 1973], have obtained $j_{2^3S} = 1.9 \cdot 10^{-3}\ s^{-1}$.

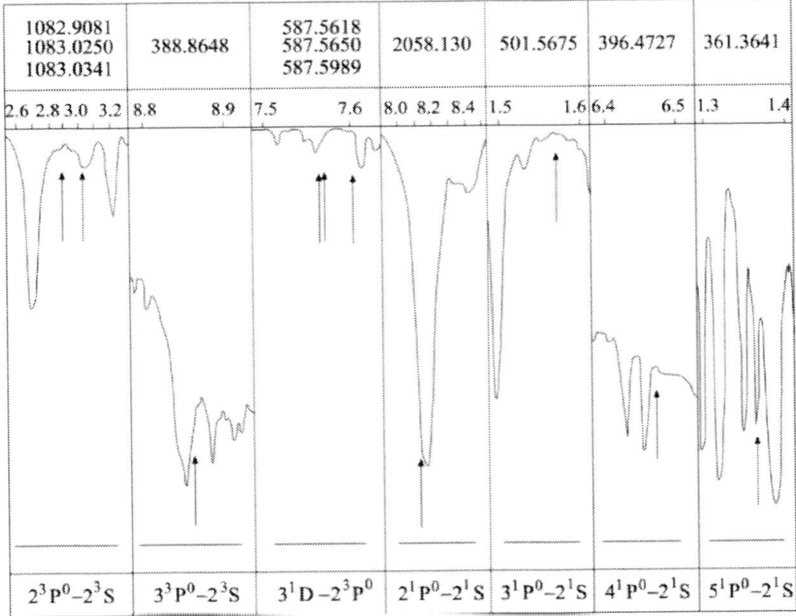

Figure 5. Emission intensity distribution in the solar spectrum [Minnaert et al. 1940; Molher et al. 1950] near the wavelengths (nm) corresponding to the orthohelium and parahelium lines [Khomich et al. 2008]

The processes of deactivation during collisions with molecules of the upper atmosphere are [Ferguson et al. 1964; Moussa et al. 1968; Lindinger et al. 1974; Coor et al. 1974]:

$$He(2^3S) + N_2 \rightarrow He(1^1S) + N_2^+ + e \qquad \beta(N_2) = 6.72 \cdot 10^{-10} \cdot \exp\left[-\frac{685}{T}\right]\ cm^3 \cdot s^{-1},$$

$$He(2^3S) + O_2 \rightarrow He(1^1S) + O_2^+ + e \qquad \beta(O_2) = 1.0 \cdot 10^{-9} \cdot \exp\left[-\frac{418}{T}\right]\ cm^3 \cdot s^{-1}$$

and atomic oxygen [Rundel and Stebbings 1974]

$$He(2^3S) + O \rightarrow He(1^1S) + O^+ + e \qquad \beta(O) = 1.03 \cdot 10^{-9} \cdot \left(\frac{T}{750}\right)^{\frac{1}{6}}\ cm^3 \cdot s^{-1}$$

The reaction of deactivation by thermal electrons

$$He(2^3S) + e \rightarrow He(2^1S) + e \qquad \beta_{2^3S,2^1S} = 5.4 \cdot 10^{-7} \cdot \left(\frac{T_e}{1000}\right)^{0.5} \cdot \exp\left(-\frac{9100}{T_e}\right)\ cm^3 \cdot s^{-1}$$

proceeds at a considerably lower rate.

Thus, according to the above mentioned processes, the emission rate, i.e., the rate of formation of the metastable He(2^3S) helium atom emission ($\mathrm{photon} \cdot \mathrm{cm}^{-3} \cdot \mathrm{s}^{-1}$) is determined by the relationship

$$Q(Z) = A_{1083} \cdot [He(2^3P)] = \frac{g \cdot \{q \cdot [He] + \alpha \cdot n_e \cdot [He^+]\}}{j + \beta(N_2) \cdot [N_2] + \beta(O_2) \cdot [O_2] + \beta(O) \cdot [O]} \cdot f(D) \cdot$$

Concentrations of the corresponding atmospheric components are shown in square brackets, and f(Z)) is the factor taking into account the influence of diffusion of metastable helium atoms.

In should be noted that the relation and the role of various processes depends on height. For example, the lifetime of excited metastable helium atoms due to photoionization (about 10 min) is implemented mainly above 700—800 km as follows from rate photoionization coefficient. Due to the processes of deactivation, this time is about 100 s, 10 s, and 1 s at altitudes below 600 km, near 400 km, and near 300 km, respectively [Bishop and Link 1993, 1999;Waldrop et al2005].

4. The Altitudinal Distribution of the Metastable Orthohelium Atoms

Bishop and Link [1993, 1999] and Waldrop et al. [2005] have anew considered various processes, which can in principle be responsible for the vertical distribution of the concentration of excited metastable orthohelium atoms and the observed emission 1083 nm. Patterson [1967] considered the effect of diffusion of metastable orthohelium atoms, first of all, in order to estimate their possible dissipation. Later the effect of this process on the content of He(2^3S) was considered by Bishop and Link [1993, 1999] and Waldrop et al. [2005]. However, these researchers did not indicate the quantitative contribution of this process at various heights. In this case we do not consider the influence of diffusion on the concentration of neutral components in the upper atmosphere [Pavlov, 1979]. The antisunward flux of He(2^3S) due to acceleration during scattering of the solar radiation [Waldrop et al. 2005] can be responsible only for horizontal transfer along the limb.

According to the cited above publications, the He(2^3S) content is defined by the following equations

$$\frac{d[He(2^3S)]}{dz} = -\frac{[He(2^3S)]}{H} - \frac{\Phi}{D},$$

$$\frac{d\Phi}{dz} = q \cdot [He(1^1S)] + \alpha \cdot [He^+] - [He(2^3S)] \cdot \{j + \beta(N_2) \cdot [N_2] + \beta(O_2) \cdot [O_2] + \beta(O) \cdot [O]\}$$

$$D = \frac{b_{He}}{[He(1^1S)]} - \frac{b_O}{[O]}.$$

Here, Φ is the vertical flux of orthohelium $He(2^3S)$, D is the diffusion coefficient ($cm^2 \cdot s^{-1}$), H is the scale height for $He(2^3S)$ atoms, $b_{He} = 9 \cdot 10^{17} \cdot T^{0.5}$ $cm^{-1} \cdot s^{-1}$ and $b_O = 3 \cdot 10^{18} \cdot T^{0.5}$ $cm^{-1} \cdot s^{-1}$ are the coefficients of binary collision of orthohelium atoms with parahelium and atomic oxygen, respectively. Bishop and Link [1999] emphasized that the contribution of $He(2^3S)$ collisions with atomic oxygen in the diffusion process is small.

According to the used approximations of the vertical distributions of helium and oxygen neutral atom concentrations [Patterson 1967; Rundel and Stebbings 1974],

$$\frac{1}{D} = 0.74 \cdot \left(\frac{T}{1000}\right)^{2.5} \cdot \frac{\exp\left[-\frac{16.8 \cdot (Z-400)}{T}\right]}{1 + 10.5 \cdot \left(\frac{T}{1000}\right)^3 \cdot \exp\left[-\frac{12.6 \cdot (Z-400)}{T}\right]} \ km^{-2} \cdot s \cdot$$

Therefore, the contribution of diffusion can numerically determined using the relationship:

$$f(D) = \frac{[He(2^3S), z]_D}{[He(2^3S), z]} = 1 - \frac{1}{[He(2^3S), z_{max}]} \cdot \int_{Z_0}^{Z} \exp\left(\int_{Z_0}^{Z} \frac{dZ}{H}\right) \cdot \frac{\Phi(Z)}{D(Z)} \cdot dZ \cdot$$

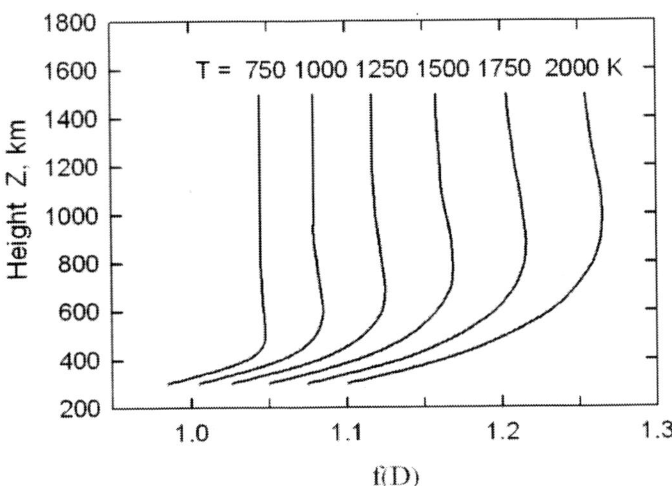

Figure 6. Vertical distribution of the $f(D) = \frac{[n(2^3S)]_D}{[n(2^3S)]}$ parameter for various values of the thermospheric temperature [Shefov et al. 2009 b].

The calculation showed that the influence of the diffusion process depends on the thermospheric temperature. The f(D) factor slightly varies with height. The values of this factor at an altitude of 100 km are: 1.05, 1.08, 1.17, 1.21, and 1.26 for T = 750, 1250, 1500, 1750, and 2000 K, respectively (Figure 6) [Shefov et al. 2009 a,b]. According to the calculations, performed by Bishop and Link [1993, 1999] and Waldrop et al. [2005] for various observational conditions and solar activity levels, the vertical profile of the $[He(2^3S)]$ concentration substantially decreases above heights of its maximal values. The measurements of an emission intensity of 1083 nm for greater heights of a shadow of the Earth of ($\chi_\odot > 120^o$, ~1500 km) indicate on their small values.

5. Processes of Excitation of the Parahelium Atoms

The excitation of helium atoms by photoelectrons is described by the reactions

$$He(1^1S) + e(E \geq 22\ \ eV) \rightarrow He(2^1P) + e \qquad q_{58.4} = 4.6 \cdot 10^{-9}\ \ s^{-1},$$

$$He(1^1S) + e(E \geq 23\ \ eV) \rightarrow He(3^1P) + e \qquad q_{53.7} = 1.2 \cdot 10^{-9}\ \ s^{-1},$$

$$He(1^1S) + e(E \geq 24\ \ eV) \rightarrow He(4^1P) + e \qquad q_{52.2} = 4.4 \cdot 10^{-10}\ \ s^{-1}.$$

It is necessary to note, that functions of cross section dependence from electron energy (for example, $He(2^1S)$ $\sigma_{2^1S} = 6 \cdot 10^{-18}\ \ cm^2$) have wider maxima than cross sections for processes of excitation of orthohelium [Massey and Burhop 1952].

For a solar ultraviolet radiation of the helium ions causing photoelectrons with energy about 30 eV, the optical thickness is practically identical.

The coefficients of absorption are

$$\sigma(He) = 3.0 \cdot 10^{-18}\ \ cm^2,\ \sigma(N_2) = 1.0 \cdot 10^{-17}\ \ cm^2,$$
$$\sigma(O_2) = 1.7 \cdot 10^{-17}\ \ cm^2,\ \sigma(O) = 1.0 \cdot 10^{-17}\ \ cm^2.$$

Excited states of parahelium atoms can arise due to recombination [Allen 1973]:

$$He^+(2^2S) + e \rightarrow He(2^1S) + h\nu(\lambda < 321\ \ nm) \quad \alpha_{He^+,2^1S} = 1.3 \cdot 10^{-12} \cdot \sqrt{\frac{1000}{T_e}}\ \ cm^3 \cdot s^{-1}$$

However, the contribution of this process is insignificant because of the small helium ion density [Shefov 1961 b; Ferguson and Schlüter 1962].

Rocket measurements have established that the intensity of the 58.4-nm He emission is about 1000 Rayleighs in the daytime (measured at ~180 km) and about 4–5 Rayleighs at night, and it is closely related to the illumination of the atmosphere by the Sun. The intensity of the interplanetary component of the 58.4-nm helium emission is 7–10 Rayleighs [Ajello and Witt 1979].

The parahelium emissions arise due to the transitions

$$He(2^1P) \rightarrow He(1^1S) + h\nu(\lambda\, 58.4 \;\; nm) \qquad A_{58.4} = 2.33 \cdot 10^9 \;\; s^{-1},$$

$$He(3^1P) \rightarrow He(1^1S) + h\nu(\lambda\, 53.7 \;\; nm) \qquad A_{53.7} = 5.66 \cdot 10^8 \;\; s^{-1},$$

$$He(4^1P) \rightarrow He(1^1S) + h\nu(\lambda\, 52.2 \;\; nm) \qquad A_{52.2} = 2.43 \cdot 10^8 \;\; s^{-1}.$$

The resonant absorption of the 58.4-nm, 53.7-nm and 52.2-nm solar emissions gives rise to $He(2^1S)$, $He(3^1S)$ and $He(4^1S)$ atoms

$$He(1^1S) + h\nu(\lambda\, 58.4 \;\; nm) \rightarrow He(2^1P) \qquad g_{58.4} = 1.8 \cdot 10^{-5} \;\; s^{-1},$$

$$He(1^1S) + h\nu(\lambda\, 53.7 \;\; \text{нм}) \rightarrow He(3^1P) \qquad g_{53.7} = 5.4 \cdot 10^{-7} \;\; c^{-1},$$

$$He(1^1S) + h\nu(\lambda\, 52.2 \;\; nm) \rightarrow He(4^1P) \qquad g_{52.2} = 1.0 \cdot 10^{-7} \;\; s^{-1}.$$

The coefficients of absorption of the neutral helium emissions by atmospheric components are presented in Tables 5 and 6.

The scattering factors g [Khomich et al, 2008] were calculated by the data on the intensity of UV solar radiation (Ivanov-Kholodny and Mikhailov 1980).

The differences in Doppler profiles of the solar and atmospheric emissions that are presented in Figure 7 have been taken into account.

Table 5. Coefficients of absorption of the neutral helium emissions by atmospheric species [Banks and Kockarts 1973 a,b; Khomich et al. 2008]

Component	σ, cm^2			
	58.4 nm	53.7 nm	52.2 nm	Factor
He	1.5(−13)	2.9(−14)	1.2(−14)	$\sqrt{1000/T}$
N_2	2.5(−17)	2.5(−17)	2.5(−17)	
O_2	2.7(−17)	2.7(−17)	2.7(−17)	
O	1.1(−17)	1.1(−17)	1.1(−17)	

Table 6. Coefficients of absorption of helium ion emissions by atmospheric components [Banks and Kockarts 1973 a, b; Khomich et al. 2008]

Component	σ, cm^2		
	30.4 nm	25.6 nm	24.3 nm
He	1.0(–18)	8.0(–19)	7.0(–19)
N_2	4.0(–18)	4.0(–18)	4.0(–18)
O_2	6.8(–18)	5.2(–18)	5.2(–18)
O	3.4(–18)	2.6(–19)	2.6(–18)

The above presented excited states are responsible both for the emissions that result in scattered resonant radiation in the Earth atmosphere, including the nightglow at high altitudes over the Earth's shadow, and for the 2058.1-nm and 501.6-nm emissions, respectively

$$\mathrm{He}(2^1\mathrm{P}) \rightarrow \mathrm{He}(1^1\mathrm{S}) + h\nu(\lambda\ 58.4\ \ \mathrm{nm}) \quad A_{58.4} = 2.33 \cdot 10^9\ \ \mathrm{s}^{-1},$$

$$\mathrm{He}(2^1\mathrm{P}) \rightarrow \mathrm{He}(2^1\mathrm{S}) + h\nu(\lambda\ 2058.1\ \ \mathrm{nm}) \qquad A_{2058.1} = 2.0 \cdot 10^6\ \ \mathrm{s}^{-1} \quad g_{2058.1} = 13.5\ \ \mathrm{s}^{-1}$$

$$\mathrm{He}(3^1\mathrm{P}) \rightarrow \mathrm{He}(1^1\mathrm{S}) + h\nu(\lambda\ 53.7\ \ \mathrm{nm}) \quad A_{53.7} = 5.6 \cdot 10^8\ \ \mathrm{s}^{-1},$$

$$\mathrm{He}(3^1\mathrm{P}) \rightarrow \mathrm{He}(2^1\mathrm{S}) + h\nu(\lambda\ 501.6\,\mathrm{nm})\ \ A_{501.6} = 1.4 \cdot 10^7\,\mathrm{s}^{-1} \quad g_{501.6} = 1.73\ \ \mathrm{s}^{-1}.$$

The photon scattering factors for the 2058.1-nm emission are smaller than for the 1083-nm emission [Shefov 1961 b, 1962 b], since the solar spectrum contains a strong Fraunhofer line, and they are greater for the 501.6-nm emission than for the 388.9-nm $3^3\mathrm{P} - 2^3\mathrm{S}$ emission, since the latter occurs in the range of the strong Fraunhofer lines of ionized calcium (Figure 5). However, in contrast to orthohelium atoms, absorption of 2058.1-nm and 501.6-nm photons according to fluorescence of parahelium atoms results in the main in 58.4-nm and 53.7-nm emissions because their transition probabilities are much higher. Therefore, the intensities of the 2058.1-nm and 501.6-nm parahelium emissions are considerably lower compared to those of the corresponding orthohelium emissions.

Besides the neutral helium emission, the airglow spectrum of the Earth atmosphere shows the presence of the 30.4-nm emission from He^+ helium ions. Satellites measurements [Meier 1974] have shown that the intensity of the atmospheric nightglow is ~12 Rayleigh and it is basically determined by the conditions of solar irradiation of the upper atmosphere, reflecting the structure of its illuminated part. The penetration of this emission into the atmosphere is restricted to altitudes of 130–140 km.

The emission from helium ions causes the highest excitation energy among all atmospheric constituents. Since the helium ion is a hydrogen-like atom, it can make a series of transitions similar to the Lyman and Balmer series of hydrogen. The wavelengths, transition probabilities, and absorption cross sections for helium emissions are presented in Table 4.

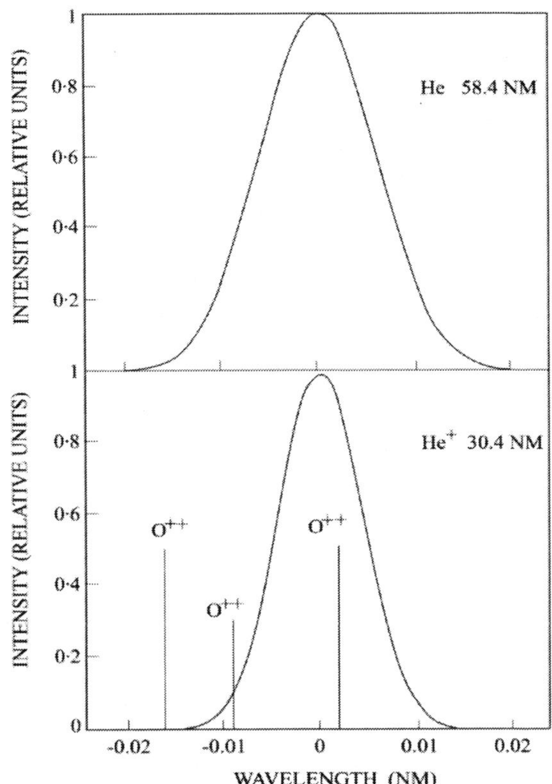

Figure 7. Profiles of the λ 58.4 nm (He) and of the λ 30.4 nm (He^+) emissions in solar radiation [Doschek et al. 1974]. For the 30.4 nm emission the positions of the multiplet components of the ionized oxygen atom are given [Khomich et al. 2008]

The emissions from helium ions are provided mainly due to fluorescence of solar radiation

$$He^+(2^2P^o) \rightarrow He^+(1^2S) + h\nu(\lambda\, 30.4 \text{ nm}) \qquad A_{30.4} = 1.00 \cdot 10^{10} \text{ s}^{-1},$$

$$He^+(3^2P^o) \rightarrow He^+(1^2S) + h\nu(\lambda\, 25.6 \text{ nm}) \qquad A_{30.4} = 2.68 \cdot 10^{9} \text{ s}^{-1},$$

$$He^+(4^2P^o) \rightarrow He^+(1^2S) + h\nu(\lambda\, 24.3 \text{ nm}) \qquad A_{30.4} = 1.09 \cdot 10^{9} \text{ s}^{-1}.$$

The solar emission intensities at these wavelengths [Ivanov-Kholodny and Mikhailov 1980; Ivanov-Kholodny and Nusinov 1987]:

$$S(\lambda\, 30.4 \text{ nm}) = 1.3 \cdot 10^{10} \text{ photon} \cdot \text{cm}^{-2} \cdot \text{s}^{-1},$$

$$S(\lambda\, 25.6 \text{ nm}) = 6.6 \cdot 10^{8} \text{ photon} \cdot \text{cm}^{-2} \cdot \text{s}^{-1},$$

$$S(\lambda\, 24.3\ \ \text{nm}) = 3.0 \cdot 10^{7}\ \ \text{photon} \cdot \text{cm}^{-2} \cdot \text{s}^{-1},$$

correspond to the 1 a. u. distance from the Sun and to the solar activity $F_{10.7} = 144$.

Following Ivanov-Kholodny and Mikhailov [1980], the dependence of the emission intensity on solar activity can be represented as

$$S_{30.4}(F_{10.7}) = S_{30.4}(144) \cdot 0.56 \cdot \ln\left(\frac{F_{10.7} - 17}{23}\right),$$

which is somewhat different from the average dependence for the entire ultraviolet radiation flux (in [Ivanov-Kholodny and Nusinov 1987] some more cumbersome relations are given).

For the excitation of helium due to resonant fluorescence, we have

$$He^{+}(1^{2}S) + h\nu(\lambda\, 30.4\ \ \text{nm}) \rightarrow He^{+}(2^{2}P^{o}) \qquad g_{30.4} = 4.1 \cdot 10^{-5}\ \ s^{-1},$$

$$He^{+}(1^{2}S) + h\nu(\lambda\, 25.6\ \ \text{nm}) \rightarrow He^{+}(3^{2}P^{o}) \qquad g_{25.6} = 3.4 \cdot 10^{-7}\ \ s^{-1},$$

$$He^{+}(1^{2}S) + h\nu(\lambda\, 24.3\ \ \text{нм}) \rightarrow He^{+}(4^{2}P^{o}) \qquad g_{24.3} = 5.5 \cdot 10^{-9}\ \ c^{-1}.$$

However, in this case, when performing calculations, one has to take into account the difference between the Doppler profiles of solar emission lines and the profiles of their absorption in the Earth atmosphere (Figure 7). According to the available measurements [Doschek et al. 1974; Dere 1977; Ivanov-Kholodny and Mikhailov 1980], the profiles of solar emission lines are very well described by the Doppler distribution [Khomich et al. 2008]

$$I = I_{0} \cdot \exp\left[-\frac{4 \cdot \ln 2 \cdot (\lambda - \lambda_{0})^{2}}{(\Delta\lambda)^{2}}\right] = 2\sqrt{\frac{\ln 2}{\pi}} \cdot \frac{S}{\Delta\lambda} \cdot \exp\left[-\frac{4 \cdot \ln 2 \cdot (\lambda - \lambda_{0})^{2}}{(\Delta\lambda)^{2}}\right],$$

where the profile's halfwidth (nm) is given by the formula

$$\Delta\lambda = 2 \cdot \frac{\lambda_{0}}{c} \cdot \sqrt{\frac{2 \ln 2 \cdot k \cdot N \cdot T}{M}} = 7.18 \cdot 10^{-7} \cdot \lambda_{0} \cdot \sqrt{\frac{T}{M}},$$

where T is the temperature and M is the atomic mass of helium.

For the helium lines in the solar spectrum, we have $\Delta\lambda$ (λ 30.4 nm) = 0.010 nm, $\Delta\lambda$ (λ 25.6 nm) = 0.0084 nm, and $\Delta\lambda$ (λ 24.3 nm) = 0.0080 nm [Doschek et al. 1974; Dere 1977; Ivanov-Kholodny and Mikhailov 1980]. These values correspond to a temperature of 845 000 K.

Figure 7 shows, for the 30.4-nm emission profile, the positions of the line triplet of the doubly ionized oxygen atom O^{++} whose wavelength, 30.3799 nm, practically coincides with the wavelength of the helium ion. Other emission line of 30.3693 nm is located inside

Doppler profile of helium emission. Other lines of 30.3625 nm, 30.3515 nm, 30.3460 nm and 30.3411 nm are outside of the profile. Their widths for T = 1000 K are, naturally, less than the width of the helium line.

As a result, a fluorescence occurs [Bowen 1934, 1947] that results in additional scattering of the solar radiation by oxygen ions.

At the same time, for the helium lines (M = 4) and atmospheric altitudes above 500 km (T ~ 1000 K), we have $\Delta\lambda$ (λ 30.4 nm) = 0.00034 nm, $\Delta\lambda$ (λ 25.6 nm) = 0.00029 nm, and $\Delta\lambda$ (λ 24.3 nm) = 0.00028 nm. The effective solar fluxes for these emissions (near the peaks of the profiles) are

$$S_m(\lambda\, 30.4\ \ \text{nm}) = 1.2\cdot 10^{12}\ \ \text{photon}\cdot\text{cm}^{-2}\cdot\text{s}^{-1}\cdot\text{nm}^{-1},$$

$$S_m(\lambda\, 25.6\ \ \text{nm}) = 7.5\cdot 10^{10}\ \ \text{photon}\cdot\text{cm}^{-2}\cdot\text{s}^{-1}\cdot\text{nm}^{-1},$$

$$S_m(\lambda\, 24.3\ \ \text{nm}) = 3.6\cdot 10^{9}\ \ \text{photon}\cdot\text{cm}^{-2}\cdot\text{s}^{-1}\cdot\text{nm}^{-1}.$$

For these values of solar fluxes, the scattering factors for the mentioned emissions were presented above. It should be noted that these values are somewhat different from those obtained by Chamberlain and Hunten [1987] based on theoretical profiles for the solar emission lines.

Excitation of He^+ states can occur due to photoionization of neutral helium atoms (ionization potential is 24.587 eV [Allen, 1973]):

$$He + h\nu(\lambda \le 19\ \ \text{nm}) \rightarrow He^+(2^2P^o) + e \qquad j_{30.4} = 1.1\cdot 10^{-8}\ \ s^{-1},$$

$$He + h\nu(\lambda \le 17\ \ \text{nm}) \rightarrow He^+(3^2P^o) + e \qquad j_{25.6} = 2.3\cdot 10^{-9}\ s^{-1},$$

$$He + h\nu(\lambda \le 16.5\ \ \text{nm}) \rightarrow He^+(4^2P^o) + e \qquad j_{24.3} = 1.1\cdot 10^{-9}\ \ s^{-1}.$$

As a result of collisions processes of neutral helium atoms with photoelectrons, the excited helium ions can be formed

$$He + e(E \ge 65\ \ \text{eV}) \rightarrow He^+(2^2P^o) + e + e \qquad q_{30.4} = 3.2\cdot 10^{-11}\ \ s^{-1},$$

$$He + e(E \ge 73\ \ \text{eV}) \rightarrow He^+(3^2P^o) + e + e \qquad q_{25.6} = 3.0\cdot 10^{-11}\ \ s^{-1},$$

$$He + e(E \ge 75\ \ \text{eV}) \rightarrow He^+(4^2P^o) + e + e \qquad q_{24.3} = 2.8\cdot 10^{-11}\ \ s^{-1}.$$

The processes of the spatial helium variability resulting from diurnal variations of the exospheric temperature and escape of helium are important for the helium density in the upper atmosphere. All this provides intraatmospheric migration of helium [Shefov 1970 a,b]. To account for the escape of helium, various chemical reactions were considered, such as the reaction of He^+ with O_2 and N_2 as well as the reaction of $He(2^3S)$ with O, O_2, and N_2 [Bates and Patterson 1962]. However, Patterson [1967] showed that the reactions related to metastable helium atoms can provide no more than 1% of the escape of helium. Moreover, Krasovsky [1969] noted that although these reactions results in a release of energy that exceeds the eluding energy level, these processes fail to provide the required escape of helium since the necessary quantities of oxygen and nitrogen molecules are far below the boundary of the exosphere from which helium atoms can escape. Therefore, the helium atoms that possess excessive energy will not be able to conserve it in their motion to the base of the exosphere because of their collisions with components of the dense atmosphere.

At the same time, the transport of ionospheric plasma from one hemisphere into another can occur due to a process which was proposed by Krassovsky [1959] and later considered by Rishbeth [1967]. Therefore, the transport of helium ions from the summer to the winter hemisphere followed by their neutralization can produce an excess of neutral helium.

6. Dependence of the Solar Uv Attenuation Degree on the Solar Zenith Angle $\chi_\odot$

The calculation of the vertical profile of the helium emission rate $(\text{photon}\cdot\text{cm}^{-3}\cdot\text{s}^{-1})$ was performed based on the relations presented in [Shefov 1963; Khomich et al. 2008; Shefov et al. 2009 b].

Since the formation of photoelectrons in the upper atmosphere is caused by the solar UV, under twilight conditions, one has to take into account the optical thickness of the atmospheric layer τ along the ray of its penetration to the intersection with the line of sight during measurements in the given direction at the zenith angle ζ_M. The thickness is determined by the expression

$$\tau = \sum_{k=1}^{n}\left\{\sigma_k \cdot \int_{-\infty}^{S_M}[n_k(Z)]\cdot dS\right\},$$

where n_k is the concentration of the absorbing components of the atmosphere, and σ_k is the absorption effective cross section. In this case the length of the ray from the limb to the line of sight is

$$S = \sqrt{(Z+R_E)^2-(Z_0+R_E)^2}\,.$$

where Z_0 is the distance of the solar radiation ray from the Earth's surface on the limb, and R_E is the Earth's radius.

From this, it follows that

$$\tau = \sum_{k=1}^{n} \left\{ \sigma_k \cdot [n_k(Z_0)] \cdot \int_{-\infty}^{Z_M} \frac{\exp\left(-\frac{Z - Z_0}{H_k(Z)}\right) \cdot dZ}{\sqrt{(Z + R_E)^2 - (Z_0 + R_E)^2}} \right\}.$$

After simple transformations,

$$\tau = \sqrt{\frac{\pi R_E}{2}} \cdot \sum_{k=1}^{n} \left\{ \sigma_k \cdot [n_k(200)] \cdot \exp\left(-\frac{Z_0 - 200}{H_k}\right) \cdot \sqrt{H_k(Z_M)} \cdot \left[1 + \mathrm{erf}\left(\frac{Z_M - Z_0}{H_k(Z_M)}\right)\right]\right\},$$

where H_k is the scale height for the k component.

For the given conditions of sighting at the solar azimuth,

$$\frac{Z_M - Z_0}{H(Z_M)} = \frac{m \cdot R_E}{H(Z_M)} \cdot \left[\frac{\sqrt{1 + \frac{\sin^2 \zeta_M}{m^2} - \frac{2 \cdot \sin \zeta_M \cdot \cos(\chi_{\odot M} - \zeta_M)}{m}}}{\sin(\chi_{\odot M} - \zeta_M)} - 1 \right],$$

where $m = 1 + \frac{Z_0}{R_E}$, $X_{\odot M}$ is the solar zenith at the observational site, and ζ_M is the zenith angle of the sighting.

During observations at an arbitrary azimuth, only the numerical solution is possible.

Figure 8 shows the height variations in the attenuation degree of the solar UV, which reaches point Z at the vertical line of sight for various solar zenith angles $X_\odot$ [Shefov et al. 2009 b].

The results of measurements of the helium 1083 nm emission intensity indicate that the intensity variations during the twilight period are most substantial when the solar zenith angle $X_\odot$ changes from 100° to 120°. In the latter case, this $X_\odot$ corresponds to altitudes higher than 1000 km, and the intensity of the 1083 nm emission becomes lower than 150—200 R. Even in the recent publications [Bishop and Link 1999; Waldrop et al. 2005], it is noted that the contribution of the recombination process of the helium ion recombination becomes comparable to the process of excitation by photoelectrons, but the registered intensity becomes low in this case (~10%). This is especially important when measurements at such solar zenith angles are performed under the conditions of helium emission (150—200 R) blended by a much more intense (1500—2000 R) hydroxyl emission $Q_1(1)$.

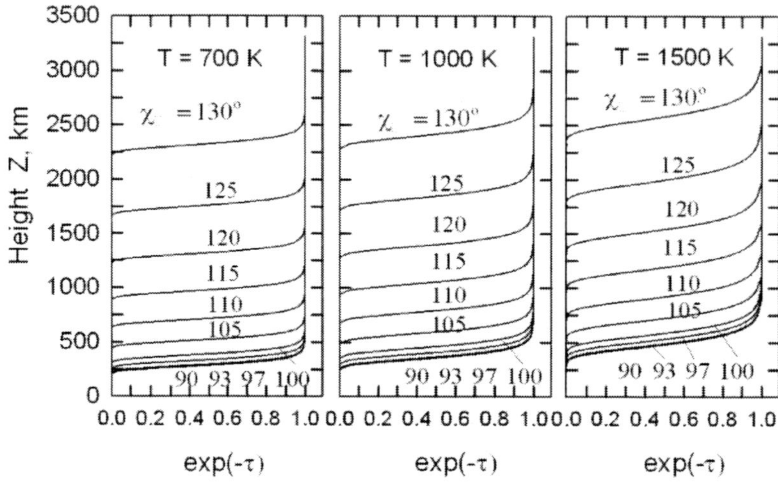

Figure 8. The vertical distribution of the dependence of the solar UV attenuation degree, causing the production of photoelectrons with an energy of ~ 25 eV, on the line of sight height and the solar zenith angle $X_{\odot}$ for various thermospheric temperatures [Shefov et al. 2009 b]

7. METHOD OF THE INTENSITY MEASUREMENTS

The rate constants of various processes g, q, and j depend on the optical thickness of the atmosphere for the solar radiation spectral regions responsible for these constants. Correspondingly, the emission intensity $I(\zeta, X_{\odot})$ for the given zenith angle of sight ζ and solar zenith angle $X_{\odot}$ is

$$I(\zeta, X_{\odot}) = \int_0^{\infty} Q(Z, \zeta, X_{\odot}) dZ$$

Consequently, emission intensity strongly depends on a solar activity level. Therefore, to obtain statistical regularities in the intensity variations, we should reduce the measurement data to homogeneous solar and geophysical conditions, e.g., F10.7 = 130. This value corresponds to the mean value of solar activity for a number of cycles in the last decades of the 20th century. Thus, this reduction is performed by multiplying measured intensities by the factor

$$\exp\left(-\frac{F10.7 - 130}{72}\right).$$

The measurements were performed in various geographic regions (see above). An important feature is that the emitting layer of the helium emission covers the height range above 300 km and has a thickness of several hundred kilometers at a level of $0.5 \cdot Q_{max}(Z)$. Therefore, during the processing of the obtained data, it is necessary to take into account information about the direction of the measuring device line of sight in order to determine the real geographic coordinates of the emitting region of the upper atmosphere and local time.

The scheme of the measurement conditions on the Earth's surface with geographic coordinates (φ_M , λ_M) in the direction having the zenith angle ζ_M and azimuth A_M, counted off from 0° to 360° from south to west, is shown in Figure 9 [Shefov et al. 2009 a]. In this case, the point at the line of sight, located at height Z_M over the Earth's surface, will be located on the Earth's surface at the distance corresponding to the angular distance ψ_M and will have the geographic coordinates (φ_M , λ_M) that are calculated using the formulas:

$$k_M = 1 + \frac{Z_M}{R_E},$$

$$\sin\psi_M = \frac{1}{k_M}\cdot\sin\zeta_M\cdot(\sqrt{k_M^2 - \sin^2\zeta_M} - \cos\zeta_M),$$

$$\cos\psi_M = \frac{1}{k_M}\cdot\cos\zeta_M\cdot(\sqrt{k_M^2 - \sin^2\zeta_M} + \sin\zeta_M\cdot tg\zeta_M),$$

$$tg\psi_M = tg\zeta_M\cdot\frac{k_M\cdot\sqrt{1+\left(1-\frac{1}{k_M^2}\right)\cdot tg^2\zeta_M} - 1}{k_M\cdot\sqrt{1+\left(1-\frac{1}{k_M^2}\right)\cdot tg^2\zeta_M} + tg^2\zeta_M},$$

$$\sin\varphi_M = \sin\varphi_O\cdot\cos\psi_M - \cos\varphi_O\cdot\sin\psi_M\cdot\cos A_M,$$

$$\sin(\lambda_O - \lambda_M) = \frac{\sin\psi_M\cdot\sin A_M}{\cos\varphi_M},$$

$$\cos(\lambda_O - \lambda_M) = \frac{\cos\psi_M\cdot\cos\varphi_O + \sin\psi_M\cdot\sin\varphi_O\cdot\cos A_M}{\cos\varphi_M}.$$

Here, R_E is the Earth's radius.

Therefore, during the observations with a certain zenith angle ζ_0, the intensity is defined by the formula

$$I(\zeta, X_\odot) = \int_0^\infty Q(Z, \zeta, X_\odot)dZ = \int_0^\infty Q(Z, \chi_{\odot M})\cdot\sec\zeta_M(Z)dZ,$$

where ζ_M is the zenith angle for the conditions of point M, depending on height Z

$$\sin\zeta_M = \frac{1}{k_M}\cdot\sin\zeta_0\,.$$

As it is known, photoelectrons coming from the geomagnetically conjugate atmospheric region affect both the helium emission and the atomic oxygen 630 nm emission [Cole 1965; Taranova 1967; Tinsley 1968; Toroshelidze 1971, 1984; Semenov 1975b; Krassovsky et al. 1976; Shefov et al. 2009 a].

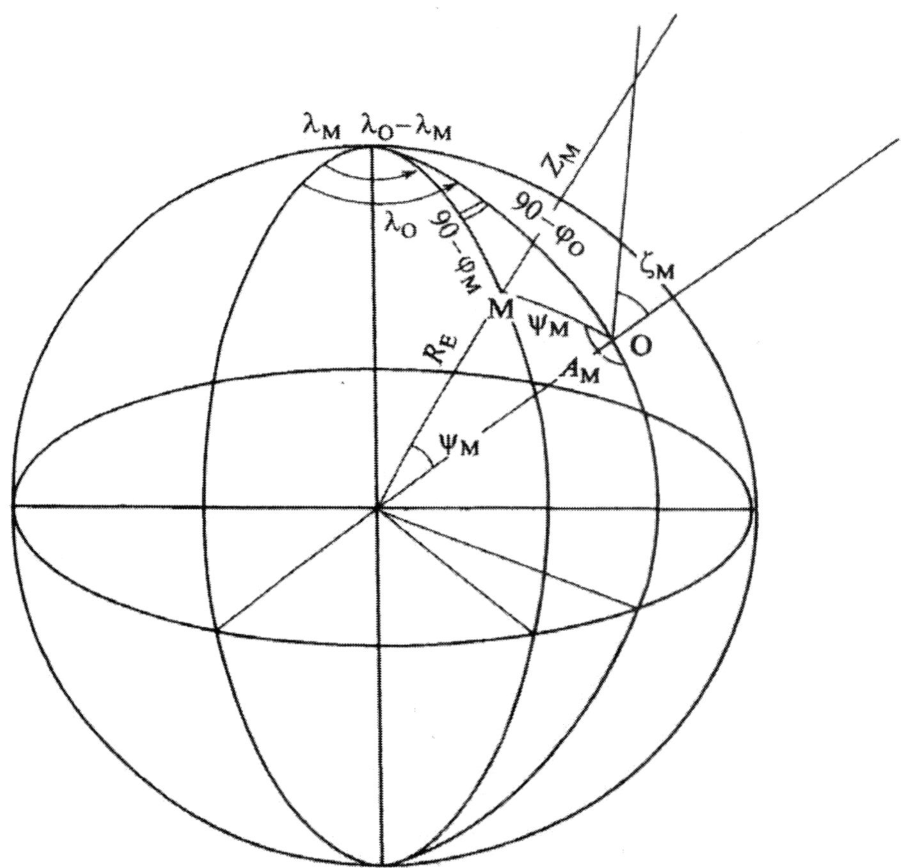

Figure 9. Geometric conditions of the emitting layer sighting at a height Z_M in the specified direction (ζ_M, A_M) from the observation site (φ_O, λ_O) [Shefov et al. 2009 a]

In the midlatitude region, where the central dipole approximation can be used to describe the magnetic field, the calculation of geomagnetic coordinates (Φ_M,Λ_M) and of the corresponding geographic coordinates of the conjugated atmospheric point $(\varphi_{Mcon}, \lambda_{Mcon})$ can be performed based on the spherical trigonometry formulas [Khomich et al. 2008].

The emission registered at the Z_M height on the line of sight has geomagnetic latitude Φ_M and corresponds to the geomagnetic field line (in the dipole approximation) crossing the Earth's surface at latitude Φ defined by the formula

$$\cos\Phi = \sqrt{k_M} \cdot \cos\Phi_M .$$

The hourly solar angle counted off from noon is connected to the local mean solar time τ and to the equation of time η by the relationship:

$$t_{\odot} = \tau - 12^h - \eta .$$

The equation of time η is a difference between mean and true solar times and reads

$$\eta = \tau_{\odot mean} - \tau_{\odot true} .$$

It is

$$\eta = 0^h.12833 \cdot \sin(\lambda_{\odot} + 78^{\circ}) - 0^h.15833 \cdot \sin 2\lambda_{\odot} ,$$

where $\lambda_{\odot}$ is the ecliptic solar longitude.

For the point on the line of sight, determined by coordinates (φ_M , λ_M) on tne basis of the observational point coordinates (φ_M , λ_M , τ_M) , the solar hourly angle is

$$t_{\odot M} = t_{\odot O} + \frac{\lambda_M - \lambda_O}{15} ,$$

and the local time (h) is

$$\tau_M = \tau_O + \frac{\lambda_M - \lambda_O}{15} .$$

For the conjugate point,

$$t_{\odot Mcon} = t_{\odot M} + \frac{\lambda_{Mcon} - \lambda_M}{15} .$$

The geographic coordinates of the conjugate point and the height Z_{Mcon} on at this site will be responsible for the conditions of illumination of the atmosphere, from which photoelectrons will flow. In this, it is necessary to take into account the screening height Z_{scr} for solar ultraviolet with λ = 30.4 nm, which is ~300 km.

Since real observations of the helium emission are conducted under twilight conditions in a certain specified direction of sighting (having zenith angle ζ_M and azimuth A_M), the determined intensity actually depends on the height of the Earth's shadow along the line of sight Z_M. For the given site, the solar zenith angle $\chi_{\odot}$ is defined by relationship:

$$\cos\chi_{\odot M} = -\frac{1+\dfrac{Z_{scr}}{R_E}}{1+\dfrac{Z_M}{R_E}} = -\frac{k_{scr}}{k_M}.$$

In the given case, Z_{scr} corresponds to the screening height of the solar radiation at a wavelength of 1083nm. However, $Z_{scr} \ll Z_M$ and therefore can be neglected, that is $k_{scr} \approx 1$.

According to the position of the Sun,

$$\cos\chi_{\odot M} = \sin\varphi_M \cdot \sin\delta_\odot + \cos\varphi_M \cdot \cos\delta_\odot \cdot \cos t_{\odot M}.$$

Here, $\delta_\odot$ and $t_{\odot M}$ are the declination and hourly angle of the Sun, respectively.

The geographic coordinates of point M are defined by the formulas presented above. Since $\frac{1}{k_M} = -\cos\chi_{\odot M}$, we can to a certain degree simplify the formulas presented above with an error not higher than 10%

$$\sin\psi_M \approx \frac{1}{2.3}\cdot\left(k_M - \frac{1}{k_M}\right)\cdot tg\zeta_M = -\frac{1}{2.3}\cdot\sin\chi_{\odot M}\cdot tg\chi_{\odot M}\cdot tg\zeta_M,$$

$$\cos\psi_M \approx \frac{2k_M}{k_M^2+1} = -\frac{2\cos\chi_{\odot M}}{1+\cos^2\chi_{\odot M}},$$

$$tg\psi_M \approx \frac{1}{4.6}\cdot\left(k_M^2 - \frac{1}{k_M^2}\right)\cdot tg\zeta_M = \frac{1}{4.6}\cdot(\sin^2\chi_{\odot M} + tg^2\chi_{\odot M})\cdot tg\zeta_M.$$

Thus, during observations toward the east and west, the registered region of the upper atmosphere will, on the average, be located slightly closer to the equator than during sighting toward the zenith. During observations toward the north, the region will shift poleward and equatorward in the evening and in the morning, respectively.

The specific feature of the recording of the helium 1083 nm emission consists in that it is necessary to select the emission lines of the triplet 1082.908, 1083.025, and 1083.034 nm, which have the relative intensities 1:3:5 among the Q branch lines of hydroxyl molecule (Figure 2). Spectrographs with a dispersion of -10 nm·mm^{-1} and Fabry–Perot interferometers were used for this purpose. In the first case, the helium emission intensity was determined by calculating the sum of the intensities of the $Q_1(1)$ and $Q_2(1)$) lines based on determining the rotational temperature of the OH (5–2) band. The helium emission intensity was calculated as a difference between the intensities measured in the region of the $Q_1(1)$ and $Q_2(1)$) lines and calculated in correspondence with the rotational temperature of OH (5-2).

The usage of an interferometer allowed one to select the helium line components among the hydroxyl lines. This made it possible to obtain the values of the intensity and Doppler temperature. However, Noto et al. [1998] indicated that the temperature profile of the helium emission is substantially affected by the recombination processes of He^+ ions reflecting the ion temperature of the upper atmosphere. Figure 2 indicates that the profiles of the two most intense components are overlapped, whereas the third (weak) component is blended by the duplet line of OH $Q_2(1)$). Table 1 presents the wavelengths of the indicated emissions. The annual mean intensities of the $Q_1(1)$ and $Q_2(1)$) lines, which are equal to 1000 and 120 R, respectively, and correspond to the rotational temperature of OH (5–2) equal to 200 K, are given for the hydroxyl emission of the OH (5–2) band. The total intensity of three lines of the helium emission is taken equal to 1000 R.

8. The Results of the Intensity Measurements

Regular measurements of helium emission intensity have begun during the International Geophysical Year (1957-1959). The numerous materials of the ground-based spectrophotometric measurements of the helium emission intensity have been obtained at low, middle and high latitudes. Based on this, the different types of regular variations have been revealed. The behavior caused by various sporadic disturbances was studied much less intensively. In this case, for creating the empirical dependences describing changes of intensity, the following expression has been used

$$I = I_0 \cdot \prod_i (I_i).$$

Here, the value of I_0 is taken equal to 1000 R for the following solar and geophysical conditions: geographic latitude $\varphi = 45^\circ N$, geographic longitude $\lambda = 40^\circ E$; time of the day corresponds to the emission intensity for the solar zenith angle $\chi_\odot = 105^\circ$; season of the year is equinox, day of the year t_d = 80; solar activity index F10.7 = 130; geomagnetic activity index Kp = 0; and the year is 1972.5. The systematic measurements of the helium emission intensity began during IGY (1957-1959).

Thus, we have the following regular variations I in the considered parameters: diurnal nighttime variations $I(\chi_\odot, \varphi)$ (here $\chi_\odot$ is the solar zenith angle, and (φ is geographic latitude); lunar variations during the synodic month (29.53 days) $I_{L\Phi}(t_{L\Phi})$ (here $t_{L\Phi}$ is the phase age of the Moon); seasonal variations $I_S(t_d)$ (here t_d is the day of the year); variations during the solar activity cycle $I_F(F10.7)$ (here F10.7 is the solar radioemission flux), and the 22-year variations taken into account by the F10.7 index.

The disturbed variations $I_{gm}(t_{gm}, K_p, \Phi)$ result from geomagnetic disturbances, where t_{gm} is the date of the geomagnetic storm beginning, Kp is the planetary disturbance index, and Φ is geomagnetic latitude.

1. Nighttime Variations

The diurnal variations were studied most thoroughly. The dependence of the emission intensity on the solar zenith angle $\chi_{\odot}$ in logarithmic scale is shown in Figure 10. In linear scale, this dependence is presented in Figure 11a. The observed change during the evening and morning twilight has a distinctly exponential character, corresponding to a decrease in the concentration of emitting atoms above the Earth's shadow boundary. Based on the systemization of available data, we present all intensity values relative to $I_0 = 1000\ R$, corresponding to a solar zenith angle of $\chi_{\odot} = 105^{\circ}$.

For $\chi_{\odot} \geq 100^{\circ}$, the regression equations have the form:

$$I_{dusk} = \exp\left(-\frac{\chi_{\odot} - 105}{13.6 \pm 1.1}\right),$$

for the evening period (the correlation coefficient r = – 0.912 ±0.031) and

$$I_{dawn} = 0.85 \cdot \exp\left(-\frac{\chi_{\odot} - 105}{15.7 \pm 1.4}\right) + 0.5 \cdot \exp\left[-\frac{(\chi_{\odot} - 110)^2}{(4 \pm 0.5)^2}\right].$$

for the morning period (the correlation coefficient r = – 0.925 ±0.032).

The second term in the formula is related to the influence of the photoelectrons from the conjugate atmospheric region.

The difference between the intensity variations in the evening and morning twilight has the following dependence (the correlation coefficient r = – 0.990 ± 0.007)

$$\Delta I_{duda} = 0.26 \cdot \exp\left(-\frac{\chi_{\odot} - 105}{11.3 \pm 0.4}\right),$$

which is caused by the influence of the helium bulge located in the region of the 1083 nm emission sighting in the period of evening measurements in the direction of the western part of the upper atmosphere.

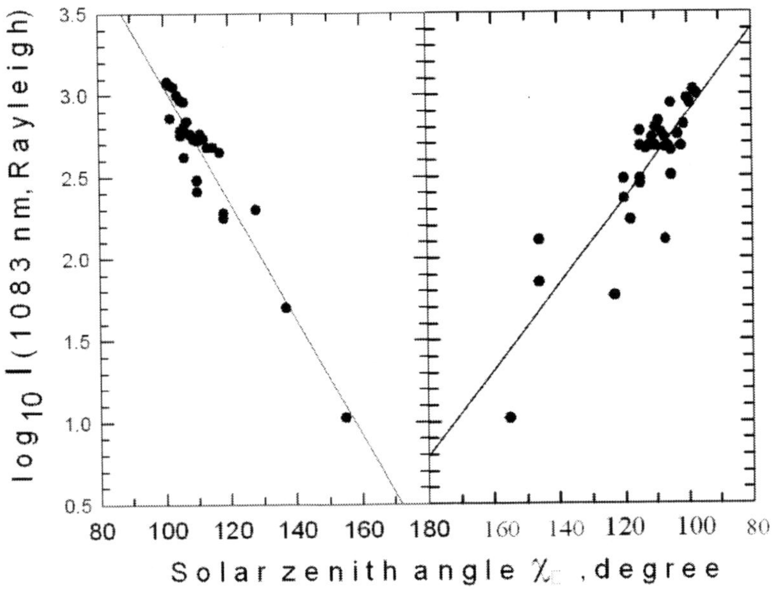

Figure 10. Dependences of the logarithms of the helium 1083 nm emission intensity (in Rayleighs) on the solar zenith angle $\chi_{\odot}$ for the solar activity level F10.7 = 130 for the evening and morning conditions at the winter solstice at midlatitudes, based on the measurements at various stations [Shefov et al. 2009 b]

2. Variations with the Moon Phase Age Period

After eliminating the effect of seasonal variations from the measuring data for a solar zenith angle of $\chi_{\odot} \geq 100^{\circ}$, we obtained the dependence on the phase age of the Moon with a period of 29.53 days (Figure 11 b). Because of the chosen solar zenith angle, the used measuring data mostly corresponded to the winter periods of the year. The Moon phase age $t_{L\Phi}$ (days) was approximately estimated to an accuracy of several tenths of a day, using the Meeus [1982] formula:

$$t_{L\Phi} = [(t_d / 365.25 + YYYY - 1900) \cdot 12.3685] \cdot 29.53 - 1.$$

Here, square brackets mean the fractional part of a number, YYYY is year value.

The approximation of the obtained data can be presented by the expression:

$$I_{L\Phi} = 1 + 0.10 \cdot \cos\left[\frac{2\pi}{29.53} \cdot (t_{L\Phi} - 23.4)\right] + 0.26 \cdot \cos\left[\frac{4\pi}{29.53} \cdot (t_{L\Phi} - 10)\right]$$

The value of the first harmonic with a phase of 23.4 days is in fact related to the influence of the asymmetry in the global distribution of the helium content in the Earth's atmosphere.

The data used to determine the lunar variations correspond to 1971–1972. The lunar declination was positive for almost all cases.

Thus, the sighted region of the helium bulge, which is located (for the available data) in the Northern Hemisphere if the Sun is located in the Southern Hemisphere, can be presented in the following form in the coordinates of the local solar time (t_{LT})

$$\Delta I_{bulge} = 0.10 \cdot \cos\left[\frac{2\pi}{24} \cdot (t_{LT} - 7)\right].$$

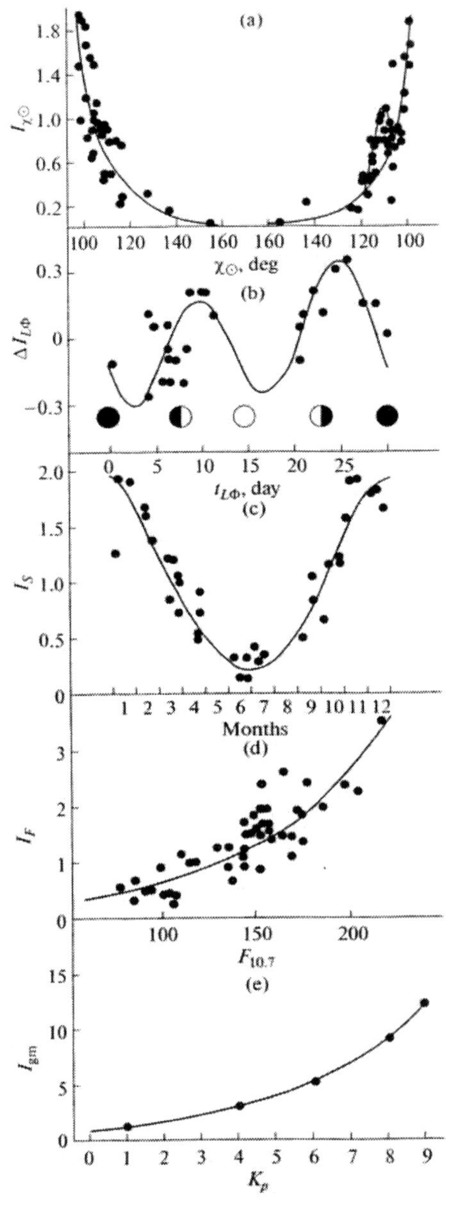

Figure 11. Regular variations in the helium emission intensity [Shefov et al. 2009 a]

The available data on the satellite measurements of the helium content in the upper atmosphere show that the maximum of helium concentration corresponds to approximately $t_{LT} \sim 9$ h of the local solar time [Cageao and Kerr 1984]. Since the observations of the helium emission were conducted for solar zenith angles of $\chi_{\odot} > 105^{\circ}$, this apparently means that the emission related to the helium concentration in the vicinity of the bulge maximum was registered.

3. Seasonal Variations

To determine the intensity variations during the annual cycle, we selected the data for a solar zenith angle of $\chi_{\odot} = 105^{\circ}$. Based on the results of midlatitude measurements, the seasonal changes are adequately approximated by the formula (Figure 11 c) depending on the day's serial number of the year t_d:

$$I_s = 0.9 \cdot \cos\left(\frac{2\pi}{365} \cdot (t_d + 10)\right).$$

The correlation coefficient is r = 0.912 ±0.031.

4. Dependence on Solar Activity

The dependence on the solar activity level was obtained based on the data for a solar zenith angle $\chi_{\odot} = 105^{\circ}$ after removing the effect of seasonal variations (see Figure 11 d). The mean variation, depending on the F10.7 index, is in a good agreement with the exponential dependence indicated above. The regression formula has the form (the correlation coefficient is r = 0.777 ± 0.059):

$$I_F = \exp\left(\frac{F10.7 - 130}{72 \pm 8}\right).$$

5. Dependence on Geomagnetic Disturbance

The influence of geomagnetic disturbance has been revealed earlier by Mironov et al. [1959], Shefov [1968, 1973]. This influence is shown in Figure 11e and can be described by the relationship (the correlation coefficient r = 0.998 ± 0.002):

$$I_{gm} = (0.80 \pm 0.01) \cdot 10^{\frac{K_p}{7.5 \pm 0.4}}.$$

If the presented relationships are used, the estimate of the helium emission intensity for the case of registration at Zvenigorod station during extreme sunlit aurora in the morning on February 11th in 1958, is determined by the following conditions: $\zeta_{vis} = 60^{o}$, $A_{vis} = 180^{o}$ from the southward direction, $\chi_{\odot} \sim 104^{o}$, $t_d = 42$, $t_{L\Phi} = 22.5$, F10.7 = 252, Kp = 8. In this case, we should bear in mind that since the line of sight was directed northward, the observed region of the helium content distribution around the Earth, caused by disturbances of the Moon and Sun, was in the opposite part of the upper atmosphere and corresponded to the minimum (Figure 11 b). Therefore, the expected intensity had to be 60÷70 kR. This value is in satisfactory agreement with the data of measurements [Shefov and Yurchenko 1970].

9. Interferometric Studies of the Temperature of Thermosphere

The intensity variations of helium emission as well as many other emissions of the upper atmosphere display not only of the rate of numerous photochemical processes at various heights but also a temperature of the atmosphere. One of measurement methods of the temperature of environment, in which occure various emission, is the interferometric measurement of the Doppler profiles of studied emissions. This method has been used for investigation of a thermosphere temperature basing on the helium emission [Kerr et al. 1996; Noto et al. 1998].

The spectral structure of a triplet of helium emission of 1083 nm has already been presented on Figure 2. As one can see, the triplet of the helium emission is blended by the rotational lines of the branch of the OH(5-2) band. The relation between the line intensities within the triplet is determined by their statistical weights that is 5:3:1. However, the most intense components of the triplet mutually overlap at temperatures higher than 600 K, corresponding to the upper atmosphere conditions forming a single emission line. Even the weakest line is not free from distortions due to the hydroxyl emission lines. This fact creates difficulties of interferometric measurements of the temperature, although the ratio of the Doppler width of the helium and hydroxyl lines is ~ 4.5. We should also bear in mind that the rotational hydroxyl lines have a doublet structure, with the distance between the lines being 0.0155 nm. Therefore, during spectrographic measurements of the helium emission intensity, one should take into account the variations of the rotational temperature of the hydroxyl emission during twilights [Shefov et al. 2009a].

In the common practice of interferometric determination of the Doppler profile temperature, the method proposed by [Semenov 1975a; Khomich et al. 2008] (Figure 12) is successfully applied. The advantage of the method consists in measuring the width of the Doppler profile $\delta\lambda_i$, at several given levels of its intensity I_i, relative to the intensity at its maximum I_0, naturally, under the condition of uniform wavelength scale. It is important to bear in mind this condition when using spectral registrograms obtained based on still interferometric images, resulted from measurements performed using such an emission receiver as a photographic film or CCD matrices. In this case, one should perform a preliminary transformation of the wavelength scale [Khomich et al. 2008].

The above indicated method makes it possible to substantially improve the accuracy in determining the calculated temperature, because the data on the profile width $\delta\lambda_i$ measurements in several parts of the Doppler line profile $\frac{I_i}{I_0}$ are used. In this case, after performing a linear regression, and considering that the regression line passes through the origin,

$$\delta\lambda_i = \rho \cdot \sqrt{\ln\frac{I_0}{I_i}},$$

which simultaneously controls the correctness of formation of the background intensity level in the interferogram, the regression coefficient ρ is calculated. This, in turn, is related to the temperature according to the emission line Doppler profile, that is halfwidth $\Delta\lambda_{HW}$ of Doppler profile

$$\rho = \Delta\lambda_{HW} = 7.18 \cdot 10^{-7} \cdot \lambda_0 \cdot \sqrt{\frac{T}{M}} = 1.665 \cdot \Delta\lambda_D.$$

Here λ_0 is the emission line wavelength (nm), M is the atomic mass, and T is the temperature. For T=1000 K and $\lambda_0 = 1083.03$ nm the halfwidth is $\Delta\lambda_{HW} = 0.0123$ nm.

From this, it follows that

$$T = \rho^2 \cdot M \cdot \frac{1.9398 \cdot 10^{12}}{\lambda_0^2}.$$

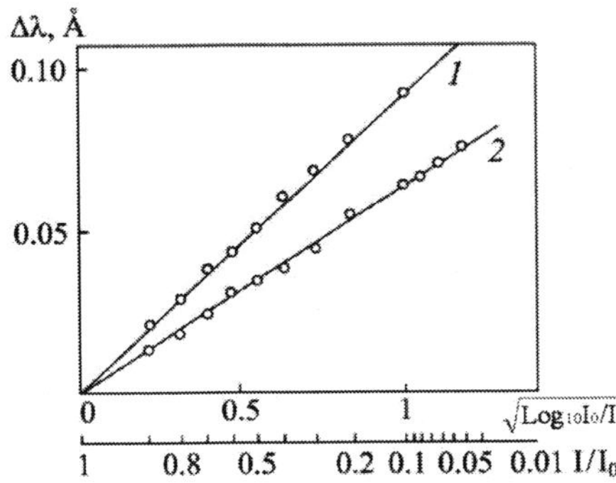

Figure 12. Relation between the measured profile width $\Delta\lambda$ (at the levels of the normalized intensity $\frac{I}{I_0}$) and $\sqrt{\log_{10}\left(\frac{I_0}{I}\right)}$. 1 – is emission of the upper atmosphere; 2 – is emission of the calibration source [Semenov 1975a; Khomich et al. 2008]

For a single component of the triplet of the helium 1083 nm emission, this relationship will have the form

$$T = 6.615 \cdot 10^6 \cdot \rho^2 .$$

The same method, slightly changed, can also be applied to overlapping doublet, the normalized profile of which is determined by the relationship

$$I(\lambda) = \frac{3}{8} \cdot \exp\left[-\frac{(\lambda - \lambda_1)^2}{(\Delta\lambda_D)^2} \right] + \frac{5}{8} \cdot \exp\left[-\frac{(\lambda - \lambda_2)^2}{(\Delta\lambda_D)^2} \right],$$

where λ_1 = 1083.0250 nm, λ_2 = 1083.0341 nm, and $\Delta\lambda_D$ is the Doppler width.

During measurements of the emission line Doppler profiles, one actually has to deal with small changes in wavelengths relative to the profile center. Therefore, for helium lines it is reasonable to take the average wavelength into account

$$\lambda_0 = \frac{\lambda_1 + \lambda_2}{2} = 1083.02955 \text{ nm}.$$

In this case $\lambda = \lambda_0 + \delta\lambda$, $\lambda - \lambda_1 = \delta\lambda + \frac{\Delta\lambda}{2}$, $\lambda - \lambda_2 = \delta\lambda - \frac{\Delta\lambda}{2}$, where $\Delta\lambda = \lambda_2 - \lambda_1 = 0.0091$ nm, $(\Delta\lambda_D)^2 = 5.47 \cdot 10^{-7} \cdot T$. The intensity distribution in the line profile depends only on $\delta\lambda$, which varies within ±0.0025 nm

$$I(\delta\lambda) = \frac{1}{8} \cdot \exp\left[-\frac{(\delta\lambda)^2 + \left(\frac{\Delta\lambda}{2}\right)^2}{(\Delta\lambda_D)^2} \right] \cdot \left\{ 3 \cdot \exp\left[-\frac{\Delta\lambda \cdot \delta\lambda}{(\Delta\lambda_D)^2} \right] + 5 \cdot \exp\left[\frac{\Delta\lambda \cdot \delta\lambda}{(\Delta\lambda_D)^2} \right] \right\}.$$

or on the scale of $\Delta\lambda$

$$I\left(\frac{\delta\lambda}{\Delta\lambda}\right) = \frac{1}{8} \cdot \exp\left[-\frac{\left(\frac{\delta\lambda}{\Delta\lambda}\right)^2 + 0.25}{\left(\frac{\Delta\lambda_D}{\Delta\lambda}\right)^2} \right] \cdot \left\{ 3 \cdot \exp\left[-\frac{\left(\frac{\delta\lambda}{\Delta\lambda}\right)}{\left(\frac{\Delta\lambda_D}{\Delta\lambda}\right)^2} \right] + 5 \cdot \exp\left[\frac{\left(\frac{\delta\lambda}{\Delta\lambda}\right)}{\left(\frac{\Delta\lambda_D}{\Delta\lambda}\right)^2} \right] \right\}$$

The numerical solution was used to apply the described method for practical determination of the temperature. For this purpose, the summarized profiles for various temperature values from 600 to 3000 K were calculated. It should be noted that the wavelength of the summarized profile maximum λ_{max} is shifted to the central wavelength λ_0 with increasing temperature. Thus, the temperature T, approximately with an error of ~ 100 K, can be estimated using the formula

$$T = \frac{3030}{1160 \cdot D\lambda - 1} \text{ K},$$

where $D\lambda = \lambda_{max} - \lambda_0$ in nm.

Based on the calculated profiles, the dependences of the width values $\delta_i = \frac{\delta\lambda_i}{\Delta\lambda}$ on $\sqrt{\ln\frac{I_0}{I_i}}$ were derived. Since the shape of the summarized profile has a complicated structure, it was found that it is reasonable to use the values $\frac{\delta\lambda_i}{\Delta\lambda}$ for the profile levels $\frac{I_i}{I_0}$ = 0.2, 0.3, 0.4, and 0.5 in order to determine the temperature. In this case, creating the correlation graph for the temperatures from 600 to 3000 K

$$\frac{\delta\lambda_i}{\Delta\lambda} = \left(\frac{\delta\lambda}{\Delta\lambda}\right)_0 + \rho \cdot \sqrt{\ln\frac{I_0}{I_i}},$$

the average correlation coefficient is equal to 0.9998 ±0.0002. The parameters of this regression equation make it possible to calculate the temperature (Figure 13), using the empirical relationships

$$\frac{T}{1000} = \frac{7.333 - 83.126 \cdot \delta_0 + 315.4516 \cdot \delta_0^2 - 407.574 \cdot \delta_0^3 + 42.954 \cdot \delta_0^4}{1 + 2.2542 \cdot \delta_0 - 113.787 \cdot \delta_0^2 + 543.649 \cdot \delta_0^3 - 739.617 \cdot \delta_0^4},$$

where $\delta_0 = \left(\frac{\delta\lambda}{\Delta\lambda}\right)_0$, the correlation coefficient is r = 0.9991, the approximating curve completely coincides with the initial data, and the mean error is 15 K at the error $\sigma_{\delta_0} = 0.022$ corresponding to approximation $\frac{\delta\lambda_i}{\Delta\lambda}$. Therefore, the temperature error for the experimental regression equation for $\frac{\delta\lambda_i}{\Delta\lambda}$ can be estimated as $\sigma_T = 680 \cdot \sigma_{\delta_0}$ K.

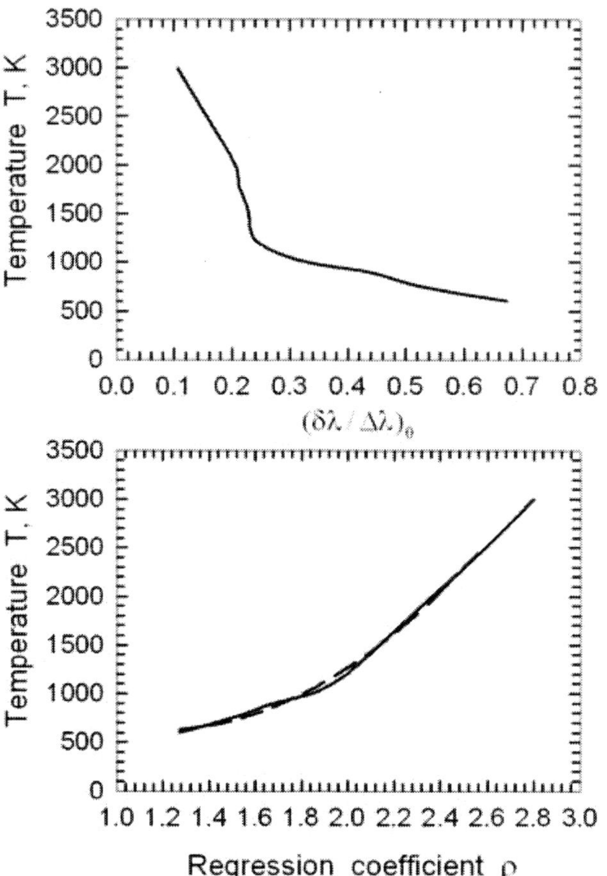

Figure 13. Dependences of temperature T on regression equation parameters: (a) the profile width $\left(\frac{\delta\lambda}{\Delta\lambda}\right)_0$ and (b) the regression coefficient ρ of the helium emission. Solid curves show the data obtained based on the regression lines for various temperatures. A dashed curve shows the approximation. In panel (a), the approximation line coincides with the initial curve [Shefov et al. 2009 b]

The second equation is

$$\frac{T}{1000} = 1.94275 - 2.24294 \cdot \rho + 0.951381 \cdot \rho^2,$$

where the correlation coefficient is r = 0.9982, and the approximating curve is shown by a dashed curve in Figure 13 b. In this case, the error is $\sigma_T = 410 \cdot \sigma_\rho$ K, i.e., being, of the order of 10 K.

The error of the temperature determination naturally depends on the error of the regression equation parameters δ_0 and ρ [Taylor 1982], the reliability of which depends on the quality of the experimental data on the helium emission profile.

10. EXPERIMENTAL INTERFEROMETRIC MEASUREMENTS OF TEMPERATURE

The interferometric measurements of temperature, using the helium emission, are quite scarce [Kerr al 1996; Noto et al. 1998]. In the latter paper, it was indicated that the conducted observations made it possible to find that the Doppler temperature of the 1083 nm emission increased with increasing Earth's shadow height. In the same paper a conclusion is made , what it is caused by substantial contribution of the recombination process of He^+ ions, the ion temperature of which is substantially higher than the neutral helium temperature.

If we separately consider the production of meta-stable helium atoms $He(2^3S)$ [Shefov 1961b, 1962a,b, 1963; Khomich et al. 2008; Shefov et al. 2009a] due to collisions of He atoms with photoelectrons and recombination of He^+ ions, we can obtain two relations:

$$Q_{phot}(Z) = \frac{g \cdot q \cdot \exp\left(\frac{F10.7 - 144}{72}\right) \cdot [He] \cdot f(D)}{j + \beta(N_2) \cdot [N_2] + \beta(O_2) \cdot [O_2] + \beta(O) \cdot [O]} \quad photon \cdot cm^{-3} \cdot s^{-1},$$

and

$$Q_{rec}(Z) = \frac{g \cdot \alpha_e \cdot [He^+] \cdot n_e \cdot f(D)}{j + \beta(N_2) \cdot [N_2] + \beta(O_2) \cdot [O_2] + \beta(O) \cdot [O]} \quad photon \cdot cm^{-3} \cdot s^{-1}.$$

Here f(D) is the factor taking into account the influence of the diffusion of helium metastable atoms (Figure 6), the characteristics of which and the rate constants of the excitation and deactivation processes g, q, j, α, and β were considered in several publications [Egorov et al. 1994; Massey and Burhop 1952; Rundel and Stebbings 1974; Shefov et al. 2009a]. The concentrations of the atmospheric components are shown in brackets.

According to the data of the publications on the helium ion content in the thermosphere [Bates and Patterson 1962; Heelis et al. 1990; Bishop and Link 1993; Erickson and Swartz 1994; Gonzalez and Sulzer 1996; Sulzer and Gonzalez 1996], which led to the conclusion of substantial contribution of helium ions to the emission intensity [Kerr et al. 1996; Noto et al. 1998], the portion of helium ions relative to the electron concentration at heights of 700–2000 km in the near-twilight time varies from 20% to 1% in the indicated height range. At the those heights, the electron, ion, and neutral temperatures are ~3500, ~ 3000, and ~1000 K, respectively.

In this case, the lifetime of helium atoms excited due to the helium ion recombination is only ~30 % as low as that of $He(2^3S)$ atoms produced due to collisions with photoelectrons at heights from 300 to 700 km because the deactivation coefficients β depend on the temperature, whereas above this level, the lifetimes are almost identical due to a sharp decrease in the atomic oxygen concentration.

According to all available models of the upper atmosphere, the concentration of helium at heights above 1000 km is 10^6 cm^{-3} and reaches ~10^4 cm^{-3} only at heights of ~ 2000 km.

Above 400 km, the He^+ ion concentration is ~1.5 x 10^4 cm^{-3} and gradually decreases to 3 x 10^3 cm^{-3} at heights of 2200 km [Erickson and Swartz 1994]. At heights above 600 km, H^+ is the predominant ion. The portion of He^+ ions at night has a maximum near 700 km, which amounts to 20% of the total ion content [Heelis et al. 1990]. In the twilight and at night, the He^+ concentration is usually not higher than $1 \cdot 10^4$ cm^{-3}.

Therefore, despite the growth of He^+ with height above 1000 km, a contribution of recombination processes (for the years of solar activity minimum (Figure 4 b)) in total helium emission intensity is approximately tens of Rayleighs. In this case, the interferometric measurements of temperature could not be reliable.

11. About Relation between Doppler and Kinetic Temperatures of the Upper Atmosphere

The problem of the influence of nonthermal components on the Doppler temperature of the emitting neutral atoms at heights of more than 700-800 km has been considered for the 630 nm emission of atomic oxygen in several studies [Schmitt et al. 1982; Yee and Dalgarno 1987; Yee 1988; Shematovich et al. 1999; Hubert et al. 2001; Kharchenko et al. 2005; Shefov et al. 2007]. Some portion of nonthermal metastable helium atoms will certainly exist. For metastable helium, this problem was considered by Kerr et al. [1996], Noto et al. [1998], and Waldrop et al. [2005]. At a small portion (~0.1-0.2) of the nonthermal component, this effect will be observed only in the wings of the Doppler profile for the intensities $< 0.1 \cdot I_{max}$.

In this case it is necessary to have in mind that the radiative life time of the excited oxygen $O(^1D)$ atoms is 110 s. As it was already considered above, for metastable orthohelium atoms, for which the two-photon transition into the ground state is determined by transition probability of $2.2 \cdot 10^{-5}$ s [Mathis, 1957] or $1.1 \cdot 10^{-4}$ s [Woodworth and Moos 1975], real life time is determined by the photoionization rate that equals to about 10 minutes above 700 km and also depends on the deactivation processes due to collisions with atoms and molecules of the atmosphere.

In recent theoretical publications, the problem of the absence of complete thermalization of $O(^1D)$ atoms formed in the upper atmosphere as a result of the dissociative recombination reaction was considered repeatedly [Schmitt et al. 1982; Yee and Dalgarno 1987; Yee 1988; Shematovich et al. 1999; Hubert et al. 2001; Kharchenko et al. 2005]. In these publications, it is stated that the calculated temperature of $O(^1D)$ exceeds the model values (based on interferometric measurements) and that (though it might also be due to other causes) a possibility that the interferometer Doppler temperature does not correspond to the temperature of the medium should be investigated. The theoretical calculations based on a numerical solution of the Boltzmann equation led the above authors to suggestions that there should be hot $O(^1D)$ atoms in the upper atmosphere at heights above 200 km, with temperatures considerably exceeding the kinetic temperature. For example, in the recent paper by Kharchenko et al. [2005], the results are presented indicating that there exist $O(^1D)$ atoms with the temperature equal to 137 % of the kinetic temperature in the daytime at an altitude of 200 km. However, as it follows from this paper, the amount of such atoms is only 5.5 %.

Schmitt et al. [1982] and Yee [1988] presented the results of calculations of the 630 nm metastable emission contours. From their calculations, it follows that a significant deviation from the thermalized profile begins to appear only when the relative intensity of the wings of a contour becomes less than 3-5 %. However, these parts of the emission line profile are knowingly at the level of the measurements accuracy. The same authors note that it is very difficult to detect such anomalous deviations in the spectral distribution of the line contour intensity.

The results of calculation of the vertical concentration distribution of $He(2^3S)$ atoms, which are produced during the excitation of neutral helium by photoelectrons and the recombination of helium ions, are presented by Waldrop et al. [2005] and distinctly show that the winter contribution of the excitation by photoelectrons is by several orders of magnitude higher than the contribution of ion recombination for solar zenith angles of 90°–110°. However, only at zenith angles of ~120° and more, the roles of these processes become comparable in the spring-summer-fall periods. The height of the Earth's shadow (~1000 km) for the 1083 nm solar radiation, providing the fluorescence of metastable helium atoms, corresponds to these conditions; therefore, the emission intensity is only several tens of Rayleighs.

Kerr et al. [1996] and Noto et al. [1998] present examples (apparently, the among the best ones) of interferograms of the helium line doublet 1083.025 and 1083.034 nm, obtained using stepped scanning using changes in the pressure in the interferometer during 2.5 min and then averaged over 10 min under the conditions of the highest emission intensity, when the intensity changes most rapidly. Unfortunately, these interferograms had insufficient quality, especially for determining the level of the background intensity, which varies from one wing of the profile to the other by approximately 20% relative to the maximum of the helium line itself in the presented examples. Naturally, the situation should be even less favorable when the intensity decreases with increasing Earth's shadow height. This can apparently cause a substantial increase in the calculated temperature during the evening twilights and a temperature decrease during the morning measurements.

Therefore, the temperature was estimated based on the data presented in [Kerr et al. 1996; Noto et al. 1998] and on the method of temperature determination for the doublet of helium lines described above. However, an analysis of (Figure 2 in [Kerr et al. 1996]) and (Figure 4 in [Noto et al1998]) revealed inconsistency between the shown profile of the helium emission doublet and the scale of wavelengths, which is shifted toward the LF part by 0.04 A, which is very substantial. Moreover, the shape of the profile, which has a distinct maximum in the right-hand part, is typical for the temperatures not higher than ~1000 K, when the asymmetry of the doublet structure is still observed, disappearing at higher temperatures (2000-3000 K) indicated by Kerr et al. [1996] and Noto et al. [1998]. Therefore, no data on the solar zenith angle for the indicated profile at that moment are presented. These papers had no quantitative information about the characteristics of the real instrumental profile. The presented value of 0.117 A, based on the calculated parameters of the interferometer, is evidently the width of an ideal instrumental profile according to the Airy formula [Tolansky 1947] for the 86% reflective property of the plates. It determines the effective width of the ideal instrumental profile, the temperature of which corresponds to 900 K. However, no possibility that the profile of the measured emission is widened due to purely instrumental causes and the time variations in the intensity during the time of scanning is taken into account and nothing is said about a compensation of these effects. This information is completely absent. The direct estimate of the temperature from the above indicated profile gives a value of ~3000 K. However, it is impossible to introduce the true correction in the measured temperature due to the instrumental profile because necessary information is not given.

The results of calculation of the vertical distributions of the helium emission intensity depending on the height of the Earth's shadow during the observation to the zenith for the conditions of solar activity maximum and minimum are shown in Figure 4 a. Figure 4 b shows the ratios of the component intensity for the same conditions. For various mechanisms by which metastable $He(2^3S)$ atoms are produced, the vertical distributions of helium ions according to Noto et al. [1998] (1994 is the year of a minimum when $F10.7 = 85$) and Erickson and Swartz [1994] (1991 is the year of a maximum when $F10.7 = 208$) were taken. One can see in Figure 4, where the dashed curves show the intensities due to the recombination of He^+ ions, that the contribution of this process is ~10%. For the conditions of solar activity minimum, the role of the recombination processes should apparently predominate to a certain degree. However, the helium emission intensity is several tens of Rayleighs for these conditions, and the interferometric measurements can most probably not provide the needed accuracy of temperature determination.

In this connection, the problem of temperature estimation using summation of the Doppler profiles with various temperatures has been considered. A natural understanding of the determined value of the temperature is that this value corresponds to the weighted mean of the temperatures within the emitting layer. As far as the vertical distribution of the emission rate serves as a weight function, the weighted mean values of the temperature almost correspond to the height of the maximum emission rate. As was mentioned above, Semenov [1975a] and Khomich et al. [2008] proposed a method of the Doppler temperature determination in which the correctness of withdrawing the background intensity (a small portion of the maximum intensity of the emission line) was simultaneously controlled during processing of registrograms of the interference image (Figure 12).

In papers [Kerr et al. 1996; Noto et al. 1998], the method of determination of temperature with a use the profile of helium emission was not presented. Abnormal values of intensity observed by them in wings of the measured profile could be less than 10 % from value of intensity in a maximum of a profile. It could be caused, in particular, by wrong determination of a background level of registrogram. A method presented in sudies [Semenov 1975a; Khomich et al. 2008], allows one to solve this problem. It allows to carry out the control of correctness of a choice of a background level of registrogram (see Section 9).

It is easy to show that, if the ratio of the content of the nonthermal component n_{nth} to the kinetic one n_{kin} $\frac{n_{nth}}{n_{kin}}$ is 0.1, then the ratio of the amplitudes of the maximums of the emission Doppler profile is equal to

$$k = \frac{I_{nth}}{I_{kin}} = \frac{n_{nth}}{n_{kin}} \cdot \sqrt{\frac{T_{kin}}{T_{nth}}} = \frac{0.1}{\sqrt{2}} = 0.07 .$$

It follows from that $\int_{-\infty}^{\infty} \exp\left(-\frac{x^2}{a^2}\right)dx = \sqrt{\pi}\cdot a$, and in this case

$$a = \Delta\lambda_D = 7.40\cdot 10^{-3}\cdot\sqrt{\frac{T}{1000}}\ \text{nm}.$$

In such a case, one can present the summation of two Doppler contours with different amplitudes and different temperatures and their sum in the form of a Doppler profile with some effective temperature

$$\exp\left[-\left(\frac{\Delta\lambda}{\Delta\lambda_{Dkin}}\right)^2\right] + k\cdot\exp\left[-\left(\frac{\Delta\lambda}{\Delta\lambda_{Dnth}}\right)^2\right] = (1+k)\cdot\exp\left[-\left(\frac{\Delta\lambda}{\Delta\lambda_{Deff}}\right)^2\right].$$

After simple transformations, we obtain from this expression

$$\frac{T_{kin}}{T_{eff}} = 1 - \left(\frac{\Delta\lambda_{Dkin}}{\Delta\lambda}\right)^2\cdot\ln\left\{\frac{1+\frac{n_{nth}}{n_{kin}}\cdot\sqrt{\frac{T_{kin}}{T_{nth}}}\cdot\exp\left[\left(\frac{\Delta\lambda}{\Delta\lambda_{Dkin}}\right)^2\cdot\left(1-\frac{T_{kin}}{T_{nth}}\right)\right]}{1+\frac{n_{nth}}{n_{kin}}\cdot\sqrt{\frac{T_{kin}}{T_{nth}}}}\right\}.$$

As far as the temperature is determined not by one part of the Doppler profile (though it is formally possible) but by a series of points of the entire contour (actually from the maximum to the intensity level not less than 5÷10%), the average relation between the above indicated temperatures can be calculated introducing the relation $\Delta\lambda = x\cdot\Delta\lambda_{Dkin}$ and using the following formula

$$\left\langle\frac{T_{kin}}{T_{eff}}\right\rangle = \frac{1}{m}\cdot\int_0^m \frac{T_{kin}}{T_{eff}}dx = 1 - \frac{1}{m}\cdot\int_0^m \frac{1}{x^2}\cdot\ln\left\{\frac{1+\frac{n_{nth}}{n_{kin}}\cdot\sqrt{\frac{T_{kin}}{T_{nth}}}\cdot\exp\left[x^2\cdot\left(1-\frac{T_{kin}}{T_{nth}}\right)\right]}{1+\frac{n_{nth}}{n_{kin}}\cdot\sqrt{\frac{T_{kin}}{T_{nth}}}}\right\}dx,$$

where m actually determines the number of $\Delta\lambda_{Dkin}$ in the integration interval. In reality m ~ 2. The application of the L'Hospital's rule confirms that the integrand has no singularities at x = 0. Further simplification of the analytical expression is not reasonable because it becomes less convenient for numerical calculations. The result of the simplest way of calculation of the effective Doppler temperature using the line profile measured by a Fabry-Perot interferometer and evaluations of its difference from the kinetic temperature T by a numerical summation of two Doppler profiles are made for T_{kin} = 1000 K. The resulting temperature (without any signs of a considerable deviation of the line profile in the wings for the values $\Delta\lambda = 2\cdot\Delta\lambda_D$)

was found to be less than 1020 K. Such differences (~20 K) are certainly less than the measurement errors.

The calculation using the above presented analytical formula shows that for the profile levels from the maximum to 3% for the above indicated conditions $\left\langle \frac{T_{eff}}{T_{kin}} \right\rangle \sim 1.015$. Even for the above indicated nighttime conditions (for which $\frac{T_{nth}}{T_{kin}} \sim 2$) $\left\langle \frac{T_{eff}}{T_{kin}} \right\rangle \sim 1.021$.

This makes it possible to conclude that the interpretation of the helium emission temperature measurements with increasing Earth's shadow height during twilights [Noto et al., 1998] as a consequence of the increasing Doppler profile width, the cause of which is unclear for the authors themselves [Noto et al. 1998] but is interpreted by them as a result of influence of high helium ion temperature, looks overestimated. We should also note that the later publication with participation of the same group of authors [Waldrop et al. 2005], discussing the role of helium ion recombination in formation of metastable helium atoms, indicated that their interpretation of the interferometric temperature measurements was doubtful.

When solving the problem connected with measurement of temperature with use of metastable helium emission considered in works [Kerr et al. 1996; Noto et al. 1998; Waldrop et al. 2005], it becomes clear that this effect will be exhibited only in the wings of Doppler profile for intensities of $\leq 0.1 \cdot I_{max}$. It will take place at a small portion (0.1÷0.2) of not thermal orthohelium atoms.

Thus, one can conclude, that the existence of the non-thermal component formally creates an increase of the measured Doppler temperature as compared to the kinetic temperature within the maximum of the helium line emitting layer. However, this increase is only a few percents (or tens of K) and this value is within the measurements accuracy and, moreover, sinks in the variations caused by various dynamical processes. Therefore, there is almost no reasons for statements on the discrepancy between the measured Doppler and kinetic temperatures of the metastable helium atoms for altitudes of the emitting layer maximum [Shefov et al. 2009 b].

12. Variations of the Helium Emission Intenisy as a Function of the Solar Zenith Angle for Various Thermospheric Temperatures

Since the interferometric measurements of the helium emission temperature are accompanied with some difficulties, measurements of variations in the intensity during twilight period are quite accessible method of temperature determination. This method was described earlier [Shefov 1963]. In this case, when the measurements are performed in the zenith, the usage of formulas $Q_{phot}(Z)$ and $Q_{rec}(Z)$, described the yelds of photoelectron and recombination processes, respectively, makes it possible to calculate the intensity variations according to solar zenith angle

$$I(\chi_\odot, F10.7) = \int_{Z(\chi_\odot)}^{\infty} [Q_{phot}(Z)\cdot e^{-\tau(\chi_\odot)} + Q_{rec}(Z)]\cdot dZ$$

where the height $Z(\chi_\odot)$ is determined from the relation

$$1+\frac{Z_0}{R_E} = \left(1+\frac{Z(\chi_\odot)}{R_E}\right)\cdot \sin\chi_\odot .$$

From this

$$\frac{Z(\chi_\odot)}{R_E} = \frac{1+\frac{Z_0}{R_E}}{\sin\chi_\odot} - 1 .$$

The obtained data on the variations in the helium intensity (R) during the dusk and dawn for the conditions of midlatitudes and moderate solar activity (F10.7 =130), based on the systematization of the data from various stations (including Zvenigorod [Shefov et al. 2009a]) presented on the logarithmic scale, demonstrate an almost linear dependence on the solar angle within 100-120° (Figure 10). These variations are described by the regression equations

$$\lg I_{1083} = (6.586 \pm 0.257) - (0.0354 \pm 0.0023)\cdot \chi_\odot ,$$

with a correlation coefficient of r = – 0.952 ± 0.018 for the evening conditions, and by

$$\lg I_{1083} = (6.487 \pm 0.351) - (0.0354 \pm 0.0032)\cdot \chi_\odot ,$$

with a correlation coefficient of r = – 0.872 ± 0.044 for the morning conditions, where $\chi_\odot$ is given in degrees.

The results of the calculations for the dusk at midlatitudes agree with the above behavior of the intensity logarithm (Figure 14). The calculations performed by Waldrop et al. [2005] for low latitudes also indicate that the intensity logarithm changes during the dawn.

The approximation of the calculated data has the form

$$\lg I_{1083} = (5.4311 \pm 0.0180) + (0.5379 \pm 0.0255)\cdot \cos\left(\frac{2\pi}{365}\cdot t_d\right) -$$

$$-(0.0344 \pm 0.0002)\cdot \chi_\odot + (0.8405 \pm 0.0359)\cdot \frac{T}{1000} .$$

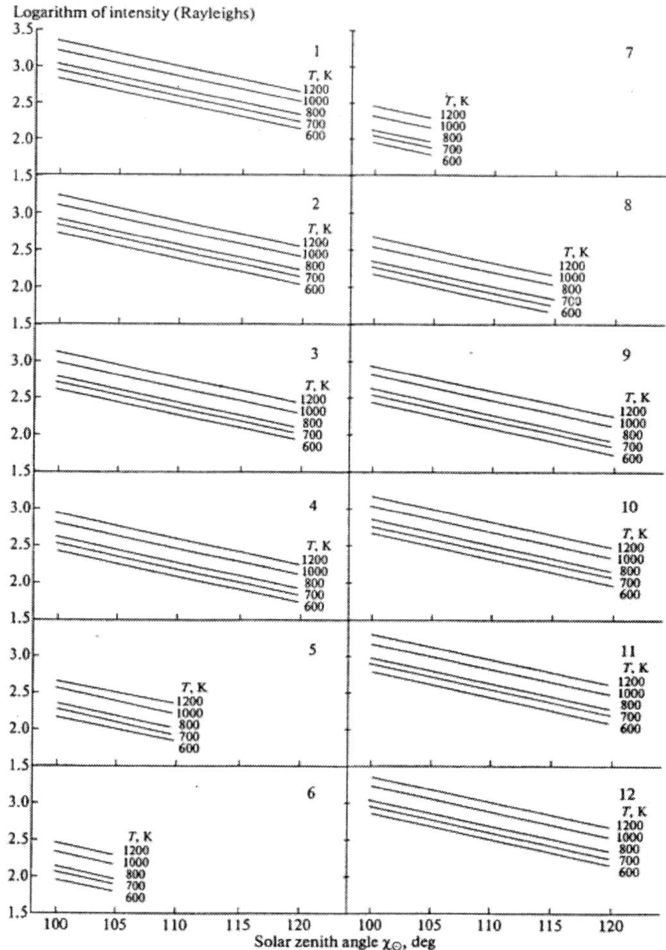

Figure 14. Dependences of the logarithm of the monthly mean intensities (in Rayleighs) of the 1083 nm emission on the solar zenith angle $\chi_\odot$ for various thermospheric temperatures and months of the year at midlatitudes [Shefov et al. 2009 b].

From this, it follows that the temperature of the atmosphere T, based on measurements of the variations in the helium emission intensity I_{1083} (Rayleighs) during twilights (as a function of solar zenith angle $\chi_\odot$) for a given day of year t_d , can be determined using the equation

$$\frac{T}{1000} = (1.1898 \pm 0.0254) \cdot \lg I_{1083} - (6.4617 \pm 0.0214) +$$

$$+(0.6400 \pm 0.0303) \cdot \cos\left(\frac{2\pi}{365} \cdot t_d\right) + (0.0409 \pm 0.0002) \cdot \chi_\odot \, .$$

The temperature determination accuracy depends on the error in the regression equations similar to relationships of $\log I_{1083}$ but based on the data of measurements at a particular geographical location. When error of the measurements of the helium emission intensity can be estimated as ~ 10–15 %, the error of the logarithm estimate should be ~ 0.04. Thus, the possible error in the temperature estimation is ~ 50–100 K.

CONCLUSION

The presented work describes the basic results of studies of helium emission of the upper atmosphere of the Earth. The reviewed results reveal considerable difficulties and problems, both theoretical and experimental in studying metastable atom of helium in a terrestrial atmosphere. In this overview we did not aim to resolve all these problems. The task was rather to to show how the accumulated data of ground-based measurements of characteristics of helium atom emission in the upper atmosphere during many years can be used for retrieving information about characteristics of the atmosphere at emission heights. On the basis of the analysis of the numerous available measurements obtained at low, middle and high latitudes, the information about parameters of the upper atmosphere at heights of exosphere above 300 km, and also about intensity of ultraviolet radiation of the Sun for various levels of solar activity are presented. The obtained data, naturally, still leaves a broad field for further research. It includes studying of long-term changes of structure of this range of heights, propagation of the disturbances from a polar zone to equator, arising after geomagnetic storms etc. The obtained conclusions and models of variations of characteristics of helium atom emission for various helio and geophysical conditions can be of interest for geophysicists and the astronomers who study the problems of airglow and its applied aspects, connected with a use of this emission for studying structural and dynamic characteristics of the upper atmosphere and a near-Earth space environment.

REFERENCES

Ajello, J. M. & Witt, N. (1979). *Simultaneous H(1216 A) and He(584 A) observations of the interstellar wind by Mariner 10*. Space Research / ed. MJ Rycroft. Pergamon Press. Oxford *V. 19,* P. 417–420.

Allen, C. W. (1973). Astrophysical Quantities. *The Athlone*, London, *448*.

Banks, P. M. & Kockarts, G. (1973a). *Aeronomy*. Pt A. Academic Press, New York, 430.

Banks, P. M. & Kockarts, G. (1973b). *Aeronomy*. Pt B. Academic Press, New York, 356.

Bates, D. R. & Patterson, T. N. L. (1962). Helium atoms in the upper atmosphere. *Planet Space Sci*, *V. 9*. N 10, 599-605.

Bely, O. (1968). The two–photon emission from the triplet metastable state of HeI. *J Phys B ser*. 2. *V. 1*. N 4. P. 718-723.

Bishop, J. & Link, R. (1993). Metastable He 1083 nm intensities in the twilight: a reconsideration. *Geophys Res Lett, V. 20,* N 11. P. 1027-1030.

Bishop, J. & Link, R. (1999). He(2^3S) densities in the upper thermosphere: Updates in modeling capabilities and comparisons with midlatitude observations. *J Geophys Res*, *V. 104*, N A8. P. 17157-17172.

Bowen, I. S. (1934). The excitation of the permitted OIII nebular lines. Publ *Astron Soc Pacific*, *V. 46*, N 271. P. 146-148.

Bowen, I. S. (1947). Excitation by line coincidence. *Publ Astron Soc Pacific*, *V. 59*, N 2. P. 196–198.

Cageao, R. P. & Kerr, R. B. (1984). Global distribution of helium in the upper atmosphere during solar minimum. *Planet Space Sci*, *V. 32*. N 12. P. 1523-1529.

Chamberlain, J. W. & Hunten, D. M. (1987). *Theory of planetary atmospheres.* Academic Press, New York, *481*.

Christensen, A. B., Patterson, T. N. L. & Tinsley, B. A. (1971). Observation and computation of twilight helium 10830 A emission. *J Geophys Res, V. 76*, N 7. P. 1764-1777.

Christensen, A. B., Tinsley, B. A., Teixeira, N. R. & Angreji, P. D. (1972). Tropical He I 10830-A observations. *J Geophys Res*, *V. 77*. N 4. P. 784-787.

Cole, R. D. (1965). The pre-dawn enhancement of 6300 A airglow. *Ann Géophys*, *V. 21*. N 1. P. 156-158.

Cook, T. B., West, W. P., Dunning, F. B., Rundel, R. D. & Stebbings, R. F. (1974). Absolute cross sections for Penning ionization of atomic oxygen by helium metastable atoms. *J Geophys Res*, *V. 79*, N 4. P. 678–680.

Dere, K. P. (1977). Extreme ultraviolet spectra of solar active region and their analysis. *Solar Phys*, *V. 82*, N 1-2. P. 77–93.

Doschek, G. A., Behring, W. E. & Feldman, U. (1974). The profiles of the solar HeI and HeII lines at 584, 537 and 304 *Å*. *Astrophys J* , *V. 190*, N 3. P. L141–L142.

Egorov, V. S., Tolmachev, Y. U. A. & Klyucharev, A. N. (1994). *A handbook of the elementary process constants with participation of atoms, ions, electrons, photons* / ed. AG Zhiglinsky. S. Peterburg Univ Press, S.Peterburg, 336.

Erickson, P. J. & Swartz, W. E. (1994). Mid-latitude incoherent scatter observations of helium and hydrogen ions. *Geophys Res Lett*, *V. 21*. N 24. P. 2745-2748.

Fedorova, N. I. (1961a). On emission λ 10830 A in aurorae. *Planet Space Sci V. 5*. N 1. P. 75.

Fedorova, N. I. (1961b). On the enhancement of λ 10830 A emission in auroral spectra. *Spectral electrophotometrical and radar researches of aurorae and airglow* / ed VI Krassovsky. USSR Acad Sci Publ House, Moscow N 5. P. 44-48.

Fedorova, N. I. (1962). Twilight fluorescence of the 10830 A helium emission. Izvestiya USSR *Acad Sci. Geophys serie* , N 4. P. 538-547.

Fedorova, N. I. (1967). *Twilight emission of helium at high latitudes. Aurorae and Airglow* / ed. VI Krassovsky. USSR Acad Sci Publ House, *Moscow N,* 13. P. 53-63.

Ferguson, E. E. & Schlüter, H. S. (1962). Metastable helium atoms concentration in the Earth's atmosphere. *Planet Space Sci*, *V. 9,* N 10. P. 701-712.

Ferguson, E. E., Fehsenfeld, F. C., Dunkin, D. B., Schmeltekopf, A. L. & Schiff, H. I. (1964). Laboratory studies of helium ion loss processes of interest in the ionosphere. *Planet Space Sci, V. 12*, N 12. P. 1169–1171.

González, S. A. & Sulzer, M. P. (1996). Detection of He^+ layering in the topside ionosphere over Arecibo during equinox solar minimum conditions. *Geophys Res Lett, V. 23,* N 18. P. 2509-2512.

Harrison, A. W. (1969). Some observations of twilight He (1.083) emission. *Can J Phys, V. 47,* N 13. P. 1377-1380.

Harrison, A. W. & Cairns, C. D. (1969). Helium emission (1.083 μ) in sunlit aurora. *Planet Space Sci, V. 17.* N 11. P. 1213-1219.

Hays, P. B. & Roble, R. G. (1971). A technique for recovering Doppler line profiles from Fabry-Perot interferometer fringes of very low intensity. *Appl Opt, V. 10,* N 1. P. 192–200.

Hedin, A. E. Extension of the MSIS thermospheric model into the middle and lower atmosphere. *J Geophys Res, V. 96,* N A2. P. 1159-1172.

Heelis, R. A., Hanson, W. B. & Bailey, G. J. (1990). Distribution of He^+ at middle and equatorial latitudes during solar maximum. *J Geophys Res, V. 95,* N A7. P. 10313-10320.

Henriksen, K. & Sukhoivanenko, P. Y. A. (1982). The detection and interpretation of the orthohelium emission at 5876 A in aurora. *Planet Space Sci, V. 30,* N 7. P. 695-699.

Hubert, B., Gérard, J. C., Killeen, T. L., Wu, Q., Bisikalo, D. V., Shematovich, V. (2001). Observation of anomalous temperatures in the daytime $O(^1D)$ 6300 A thermospheric emission: A possible signature of nonthermal atoms. *J Geophys Res, V. 106.* N A7. P. 12753-12764.

Ivanov-Kholodny, G. S. & Mikhailov, A. V. (1980). Ionospheric state forecast. *Hydrometeoizdat,* Leningrad, 192.

Ivanov-Kholodny, G. S. & Nusinov, A. A. (1987). Ultraviolet radiation of the Sun and its influence on the upper atmosphere and ionosphere. Total results of the Science. Investigations of the cosmic processes. VINITI, Moscow, *V. 26,* P. 80-154.

Kerr, R. B., Noto, J., Lancaster, R. S., Franco, M., Rudy, R. J., Williams, R. & Hecht, J. H. (1996). Fabry Perot observations of helium 10830 A emission at Millstone Hill. *Geophys Res Lett, V. 23.* N 22. P. 3239-3242.

Kharchenko, V., Dalgarno, A. & Fox, J. L. (2005). Thermospheric distribution of fast $O(^1D)$ atoms. *J Geophys Res, V. 110.* N A12. doi:10.1029/ 2005JA011232.

Khomich, V. Y. U., Semenov, A. I. & Shefov, N. N. (2008). Airglow as an indicator of upper atmospheric structure and dynamics. Springer–Verlag, *Berlin Heidelberg,* 739.

Krassovsky, V. I. (1959). Energy sources of the upper atmosphere. *Planet Space Sci, V. 1,* N 1. P. 14-19.

Krassovsky, V. I. & Galperin, G. I. (1960). Review of observational results on the airglow and aurorae. *Trans Intern Astron Union* (Moscow 1958). Cambridge Univ Press, Cambridge, *V. 10A.* P. 327-328.

Krassovsky, V. I. (1969). Intra-atmospheric migration and escape of H and He under the action of MHD waves. *Geomagnetism Aeronomy, V. 9,* N 4. P. 689-692.

Krassovsky, V. I., Semenov, A. I., Shefov, N. N. & Yurchenko, O. T. (1976). Predawn emission at 6300 A and super–thermal ions from conjugate points. *J Atmos Terr Phys, V. 38.* N 9-10. P. 999-1001.

Krinberg, I. A. (1978). *Kinetics of electrons in the Earth's ionosphere and plasmosphere.* Nauka, Moscow, 215.

Lindinger, W., Schmeltekopf, A. L. & Fehsenfeld, F. C. (1974). Temperature dependence of de-excitation rate coefficients of $He(2^3S)$ by Ne, Ar, Xe, N_2, O_2, NH_3, and CO_2 . *J Chem Phys, V. 61,* N 7. P. 2890-2895.

Massey, H. S. W. & Burhop, E. H. S. (1952). *Electronic and ionic impact phenomena.* Clarendon Press. Oxford, 604.

Mathis, J. S. (1957). Statistical equilibrium of triplet levels of neutral helium. *Astrophys J, V. 125*, N 2. P. 318–327.

McElroy, M. B. (1965). Excitation of atmospheric helium. *Planet Space Sci, V. 13*, N 5, P. 403–433.

Meeus, J. (1982). Astronomical formulae for calculators. Willmann-Bell, *Richmond*, 168.

Mironov, A. V., Prokudina, V. S. & Shefov, N. N. (1959). Auroral observations on 10-11 February, 1958, Moscow. *Spectral electrophotometrical and radar researches of aurorae and airglow* / ed VI Krassovsky. USSR Acad Sci Publ House, *Moscow*, N 1. P. 20-24.

Minnaert, M., Mulders, G. F. W. & Houtgast, J. (1940). *Photometric atlas of the solar spectrum from λ 3612 to λ 8771 Å*. D Schnabel. Amsterdam Kampert Helm, Amsterdam, 161.

Molher, O. C., Pierce, A. M., McMath, R. R. & Goldberg, L. (1950). Photometric atlas of the infra-red solar spectrum λ 8465 to λ 25242 Å. Michigan Univ Press, *Ann Arbor, 196.*

Moussa, H. R. M., de Heer, F. J. & Schutten, J. (1968). Excitation of helium by 0.05-6 keV electrons and polarization of the resulting radiation. *Physica, V. 40*, N 3. P. 517–549.

Neo, Y. P. & Rundle, H. N. (1969). An identification of twilight helium 10,830 A emission with increased resolution and preliminary photometric measurements. *Planet Space Sci, V. 17*, N 4. P. 715-724.

Noto, J., Kerr, R. B., Shea, E. M., Waldrop, L. S., Fisher, G., Rudy, R. J., Hecht, J. H., González, S. A., Sulzer, M. P. & Garcia, R. (1998). Evidence for recombination as a significant source of metastable helium. *J Geophys Res, V. 103,* N A6. P. 11595-11603.

Patterson, T. N. L. (1967). Metastable helium in the upper atmosphere. *Planet Space Sci, V. 15*, N 7. P. 1219-1222.

Pavlov, A. V. (1979). On thermal diffusion in the Earth's upper atmosphere. *Geomagnetism Aeronomy, V. 19*, N 6. P. 1050-1057.

Rishbeth, H. (1967). Transequatorial diffusion in the topside ionosphere. *Planet Space Sci, V. 15*, N 8. P. 1261-1265.

Rundel, R. D. & Stebbings, R. F. (1974). Metastable helium in the Earth's upper atmosphere. *J Geophys Res, V. 79*, N 4. P. 681-684.

Rundle, H. N. (1960). Ionization of a static interplanetary gas and expected emission lines from this gas. *Planet Space Sci, V. 2*, N 2-3. P. 86–98.

Schmitt, G. A., Abreu, V. L. & Hays, P. B. (1982). Line shape of the non-thermal 6300 A $O(^1D)$ emission. *Planet Space Sci, V. 30*, N 5. P. 457-461.

Semenov, A. I. (1975a). Interferometric measurements of the upper atmosphere temperature.I. *Application of the cooled image converters.* Aurorae Airglow / ed VI Krassovsky. Nauka, Moscow, N 23, P. 64-65.

Semenov, A. I. (1975b). Pre-dawn variations of the temperature and intensity of the 6300 *A emission. Astron Circ USSR Acad Sci,* N 882. P. 6-7.

Shcheglov, P. V. (1962a). Twilight enhancement of the infrared helium line 10830 *A. Astron Rep, V. 39*, N 1. P. 158-159.

Shcheglov, P. V. (1962b). Observation of the twilight helium emission λ 10830 A with Fabry-Perot interferometer. Aurorae Airglow / ed VI Krassovsky. USSR Acad Sci Publ House, *Moscow*, N 9. P. 59-60.

Shefov, N. N. (1961a). On the nature of helium emission λ 10830 A in aurorae. *Planet Space*

Sci, V. 5, N 1. P. 75-76.

Shefov, N. N. (1961b). Émission de l'helium dans la haute atmosphere. *Ann Géophys*, *V. 17*, N 4. P. 395-402.

Shefov, N. N. (1961c). On the nature of helium emission λ 10830 A in aurorae. *Spectral electrophotometrical and radar researches of aurorae and airglow* / ed VI Krassovsky. USSR Acad Sci Publ House, Moscow N 5. P. 47-48.

Shefov, N. N. (1961d). Twilight enhancement of the λ 10830 A helium emission. Astron Circ USSR *Acad Sci N*, 222, P. 11-12.

Shefov, N. N. (1962a). Sur l'émission de l'helium dans la haute atmosphere. *Ann Géophys*, *V. 18*, N 1. P. 125.

Shefov, N. N. (1962b). *The helium emission in the upper atmosphere. Aurorae Airglow* / ed VI Krassovsky. USSR Acad Sci Publ House, Moscow, N 8. P. 50-65.

Shefov, N. N. (1963a). The behaviour of the helium λ 10830 A emission in twilight. Aurorae Airglow / ed VI Krassovsky. USSR Acad Sci Publ House, Moscow N 10. P. 56-64.

Shefov, N. N. (1963b). Helium in the upper atmosphere. *Planet Space Sci* V. 10, P. 73-77.

Shefov, N. N. (1967). *Statistical properties of the helium emission.* Aurorae Airglow / ed VI Krassovsky. USSR Acad Sci Publ House, Moscow N 13. P. 64-68.

Shefov, N. N. (1968). Twilight helium emission during low and high geomagnetic activity. *Planet Space Sci, V. 16*, N 9. P. 1103-1107.

Shefov, N. N. (1970). Migration of the H and He inside the atmosphere and their escape. *Space Research* / eds TM Donahue, PA Smith and L Thomas. North-Holland Publ Co, Amsterdam, *V. 10*, P. 623-632.

Shefov, N. N. & Yurchenko, O. T. (1970). Absolute intensities of the auroral emissions in Zvenigorod. *Aurorae Airglow* / ed VI *Krassovsky. Nauka, Moscow*, N 18. P. 50-96.

Shefov, N. N. (1973). Hydrogen and helium emissions and concentrations in the upper atmosphere. *Aurorae Airglow* / ed VI Krassovsky. Nauka, Moscow, N 20. P. 40-56.

Shefov, N. N., Semenov, A. I., Yurchenko, O. T. & Sushkov, A. V. (2007). Empirical model of variations in the 630 nm atomic oxygen emission. 2. Temperature. *Geomagnetism Aeronomy, V. 47*, N 5. P. 654-663.

Shefov, N. N., Semenov, A. I. & Yurchenko, O. T. (2009a). An empirical model of the helium emission variations 1083 nm. 1. *Intensity. Geomagnetism Aeronomy*, *V. 49*, N 1. P. 93-104.

Shefov, N. N., Semenov, A. I. & Yurchenko, O. T. (2009b). An empirical model of the helium emission variations 1083 nm.2.Temperature. *Geomagnetism Aeronomy.*, *V. 49*, N 5 . P.670-678.

Shematovich, V., Gérard, J. C., Bisikalo, D. V. & Hubert, B. (1999). Thermalization of $O(^1D)$ atoms in the thermosphere. *J Geophys Res*, *V. 104*, N A3. P. 4287-4295.

Shklovsky, I. S. (1951). *The solar corona.* Gostekhizdat, Moscow, 338.

Shouyskaya, F. K. (1963). An attempt to detect the proper glow of atmosphere during the solar eclipse on February 15, 1961. *Aurorae Airglow* / ed VI Krassovsky. USSR Acad Sci Publ House, Moscow N 10. P. 44-53.

Stebbings, R. F., Dunning, F. B., Tittel, F. K. & Rundel, R. D. (1973). Photoionization of helium metastable atoms near threshold. *Phys Rev Lett*, *V. 30*, N. P. 815–817.

Stoffregen, W. (1969). Transient emissions on the wavelength of helium I, 5876 A recorded during auroral break – up. *Planet Space Sci*, *V. 17*, N 12. P. 1927-1935.

Striganov, A. R. & Odintsova, G. A. (1982). Tables of the spectral lines of the atoms and ions. *Handbook. Energoizdat*, Moscow 312.

Sukhoivanenko, P. Y. a. & Fedorova, N. I. (1976). Fast registration of the λ 10830 Å helium emission. Studies of the upper atmospheric emission / eds NN Shefov, AP Savrukhin. Ylym, *Ashkhabad P.* 12-16.

Sulzer, M. P. & González, S. A. (1996). Simultaneous measurements of O^+ and H^+ temperatures in the topside ionosphere over Arecibo. *Geophys Res Lett, V. 23*, N 18. P. 3235-3238.

Suzuki, K (1983). Observation of helium 10830 A airglow emission in midlatitude. *J Geomagn Geoelectr, V. 35*, N 9. P. 321-330.

Taranova, O. G. (1967). *On diurnal variations of helium emission. Aurorae and Airglow* / ed VI Krassovsky. USSR Acad Sci Publ House, Moscow N, 13, P. 50-52.

Taylor, J. R. (1982). *An introduction to error analysis*. Univ Science Book, Mill Valley, California, 272.

Teixeira, N. R., Angreji, P. D., Sahai, Y., Tinsley, B. A. & Christensen, A. B. (1975). Tropical twilight HeI 10830 emission. *Planet Space Sci, V. 23*, N 10. P. 1425-1430.

Telegin, G. G. & Yatsenko, A. S. (2000). *The optical spectra of the atmospheric gases* / ed SG Rautian. Nauka, Novosibirsk, 241.

Tinsley, B. A. (1968). Measurements of twilight helium 10830 A emission. *Planet Space Sci, V. 16*, N 1. P. 91-99.

Tinsley, B. A. & Christensen, A. B. (1976). Twilight helium 10,830 A calculations and observations. *J Geophys Res, V. 81*, N 7. P. 1253-1263.

Tolansky, S. (1947). *High resolution spectroscopy*. Methuen and Co, London 292.

Toroshelidze, T. I. (1970). Twilight emission of the helium by the observations of Abastumani. *Geomagnetism Aeronomy, V. 10*, N 6. P. 1037-1042.

Toroshelidze, T. I. (1971). The emission of atmospheric helium 10830 A at the predawn period. Astron Circ *USSR Acad Sci N*, 652, P. 1-3.

Toroshelidze, T. I. (1976). On the certain particularities of the 10830 A helium emission in twilight. Studies of the upper atmospheric emission / eds NN Shefov, AP Savrukhin. Ylym, *Ashkhabad*, P. 22-32.

Toroshelidze, T. I. (1984). On structural features of the thermosphere according to the 10830 A helium emission observations. *Bull Acad Sci Georgian SSR, V. 113*, N 3. P. 531-534.

Toroshelidze, T. I. (1991). The analysis of the aeronomy problems on the upper atmosphere glow / ed NN Shefov. Metsniereba, Tbilisi 217.

Waldrop, L. S., Kerr, R. B., González, S. A., Sulzer, M. P., Noto, J. & Kamalabadi, F. (2005). Generation of metastable helium and the 1083 nm emission in the upper thermosphere. *J Geophys Res, V. 110,* N A8. P. A08304, doi:1029/2004LA010855.

Wiese, W. L., Smith, M. W. & Glennon, B. M. (1966). *Atomic transition probabilities H through Ne.* NSRDS-NBS 4. Washington DC, *V. 1*, 154.

Woodworth, J. R. & Moos, H. W. (1975). Experimental determination of the single-photon transition rate between the 2^3S_1 and 1^1S_1 states of He. *Phys Rev A, V. 12*, N 12. P. 2455–2463.

Yee, J. H. & Dalgarno, A. (1987). Energy transfer of $O(^1D)$ atoms in collision with $O(^3P)$ atoms. *Planet Space Sci, V. 35*, N 4. P. 399-404.

Yee, J. H. (1988). Non-thermal distribution of $O(^1D)$ atoms in the night-time thermosphere. *Planet Space Sci, V. 36*, N 1. P. 89-97.

In: Helium: Characteristics, Compounds...
Editors: Lucas A. Becker

ISBN: 978-1-61761-213-8

Chapter 2

HELIUM IN METALS-DIFFUSION AND EQUATION OF STATE

Benny Glam*
Soreq Nuclear Research Center, Yavne, Israel

ABSTRACT

In the first part of this paper experimental and analytical investigation of helium diffusion and bubble formation and growth in aluminum is presented [1;2]. A theoretical model for equation of state of aluminum with helium bubbles and its validation in shock wave experiments are presented in the second part [3].

A pure aluminum with 0.15% wt of ^{10}B was neutron-irradiated in a nuclear reactor to get homogeneous helium atoms in the metal according to the reaction $^{10}B + n \rightarrow {}^{7}Li + {}^{4}He$. Formation and growth of helium bubbles was observed in situ by heating the post-irradiated metal to 470ºC in TEM with a hot stage holder. It was found that above 400ºC the time scale for bubble's shape change is seconds. In other experiments the Al-^{10}B was first heated in its bulk shape and then observed in TEM at room temperature. In this case the helium bubble formation takes hours.

Analytical evaluation of the diffusion processes in both cases was done to explain the experimental results. The number of helium atoms in a bubble was calculated from the electron energy loss spectrum (EELS) measurements. These measurements confirmed the hard sphere equation of state (EOS) for inert gases that was used in the analytical diffusion calculations.

It was also found that the helium-rich area expands due to helium migration. Electron beam diffraction revealed that the preferred orientation of the helium atoms migration is normal to plane $(0\bar{2}2)$. The results are consistent with models for helium atoms migration between interstitial sites for an fcc metal.

At the second part of this paper a theoretical model for equation of state of aluminum with helium bubbles is presented. Based on this equation of state, the influence of helium bubbles on shock loading is examined. The Hugoniot curve (temperature vs. pressure as

* Corresponding author: Tel: +972-8-943-4914, Mobile phone: +972-52-850-1508, Fax: +972-8-943-4346, e-mail: benny.glam@gmail.com

well as shock velocity vs. particle velocity) for aluminum containing bubbles is calculated for various bubbles mass, bubbles percentage and helium equation of state models. The bubble mass and concentration seem to affect measurably the Hugoniot curve. The equation of state model implied for the helium in the bubbles has minor significance, which means the model is not sensitive to the details of the helium EOS. These findings were confirmed in shock wave experiments.

1. Helium Diffusion and Bubbles Formation in Aluminum

1.1. Introduction

The creation of helium atoms in metals is very significant, since their precipitation into bubbles can substantially deteriorate the mechanical properties of materials, particularly in metals at high homologous temperatures ($T>0.5T_m$) where drastic embrittlement due to helium bubble formation at the grain boundary is found [4].

In most of the research on helium–metal interaction, the helium bubbles are induced by implantation [5-7] and tritium decay [8]. The disadvantages of those techniques are the very low near surface penetration of the helium with a non-homogenous bubble growth within a depth of a few hundred nanometers for the implantation technique, and the long-term preparation needed for the tritium decay technique. Introduction of helium in metals based on neutron irradiation of aluminum-boron samples [9] was used in this work. The advantage of this technique is the ability to get fairly uniform distribution of helium atoms in the metal. While heating the post-irradiated metal, the helium atoms combine, producing clusters and bubbles. Tiawari and Singh [10] investigated in this way the effect of temperature on the final helium bubble radius in aluminum and copper.

In order to see the formation process of bubbles, in situ observation during heating is needed. Work with in situ ion irradiation of metals in TEM showed a Brownian motion of helium bubbles in the metal [11;12]. But in this case the mechanism of bubble growth and motion is not only a result of the temperature, it is a combination of the irradiation effects and the temperature.

In this part an investigation of the influence of heating conditions on helium diffusion and bubble formation and growth in aluminum with ^{10}B after neutron irradiation is presented [1]. The helium bubble formation in Al-^{10}B metal was observed in TEM in two types of experiments: **A.** during in situ heating using a hot stage holder. **B.** The helium bubbles were formed by heating an Al-^{10}B bulk sample in a furnace and then prepared for TEM observation at room temperature.

1.2. Alloy Preparation

Pure aluminum (99.9999%) was melted with 0.15% wt ^{10}B powder in an arc furnace. This amount of ^{10}B is solute in the aluminum at the elevated temperatures in the arc furnace.

The prepared metal was then neutron-irradiated in the Soreq nuclear reactor with a flux of $\phi = 3\times10^{17}\left(n/m^2\,s\right)$ for 20 hours. The irradiation time was limited to 20 hours to avoid radioactivity of the sample. The concentration of helium atoms N_{He} that were created in the bulk of the sample from the reaction $^{10}B+n \rightarrow {}^{7}Li+{}^{4}He$ is given by

$$N_{He} = \phi\sigma N_{^{10}B}t = 1.2\times10^{24} \quad (m^{-3}) \tag{1}$$

where $\sigma = 4.0\times10^{-25}(m^2)$ is the cross-section of the ^{10}B atom, $N_{^{10}B} = 7.7\times10^{25}\ m^{-3}$ is the number of ^{10}B atoms per unit volume, and t = 20 (hr) is the irradiation time. The Li product at this small concentration is solute in the Al matrix and therefore can be neglected [13].

After irradiation the Al-^{10}B alloy was rolled to a 2.7 mm thickness plate. From this plate two groups of specimens were taken:

Group A samples were prepared for the in situ TEM observations and investigation of the helium bubble formation and growth. The Al-^{10}B plate was cut with a low speed saw to obtain 500 μm thickness foils. Disks of 3 mm diameter were punched from the foils and were thinned to 100 μm by grinding. Subsequently, the disks were thinned to produce a hole in their center by an electro-polishing jet in a Tenupol 50 device. The combined procedure of graded grinding and electro-polishing at low temperatures ensures a minimum influence of the preparation procedure on the specimen's microstructure. The estimated thickness of the observed region near the hole, where the electron beam can be transmitted, is about 50–150 nm. The TEM observations were made at different temperatures up to 470°C.

Group B samples were prepared for the bulk diffusion investigation of helium in aluminum. Rolled Al-^{10}B bulk samples (2.7 mm thick) were heated to different temperatures and times. TEM characterization was carried out at room temperature.

1.3. Experimental Results

1.3.1. In situ observation of bubble formation and growth

The observations were carried out in an FEI T20 model TEM with an acceleration of 200 keV. The group A specimens were heated in discrete steps to 470°C. At room temperature no helium bubbles were observed except for a restricted strip of 50 nm with a few nanometric bubbles. At 200°C bubbles started to appear. The bubbles grew, changed their shape, and merged. These processes increased with temperature, while above 400°C they became faster and the bubble formation front moves toward the Al bulk.

Pictures of TEM selected areas of a specimen heated to 470°C are shown in Figure 1. The pictures in Figure 1a to d were taken every second. In Figure 1a the bubbles have a faceted shape with typical length of 4~30 nm; ^{10}B inclusions that did not dissolve in the Al remain, as can be seen at the bottom of the figure. Below them the specimen is thicker and no bubbles were observed. The circle in Figure 1a marks a 50 nanometer bubble. In Figure 1b the bubble has blown up to 12 small bubbles (~6 nm diameter) that recombine to four bubbles of 15 nm in diameter (Figure 1c) and then to a single 40 nm diameter bubble shown in Figure 1d.

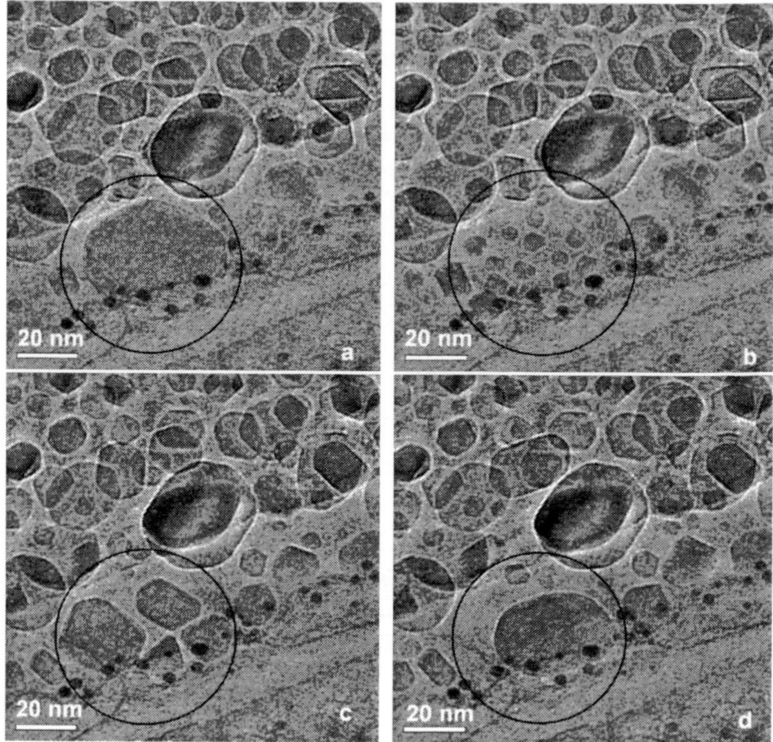

Figure 1.TEM pictures of helium bubble splitting (a, b) and recombination (c, d) after heating to 470°C in the hot stage

Another specimen that was heated to 400°C is shown in Figure 2a. Figure 2b was taken 5 seconds later at the same temperature. The diagonal line at the bottom right side of the picture is the edge of the sample. The thickness of this area is about 60 nm and increases in thickness as it moves into the bulk. The left area is darker because of this thickness, e.g., it is less transparent to the microscope electrons than the thinner area in the right. The average bubble diameter is ~20 nm. The bubbles at the boundary (bottom right) are larger, about ~50 nm diameter, and have a faceted shape. The two bubbles that are marked with a blue circle coalesce to an 80 nm long bubble as shown in Figure 2b.

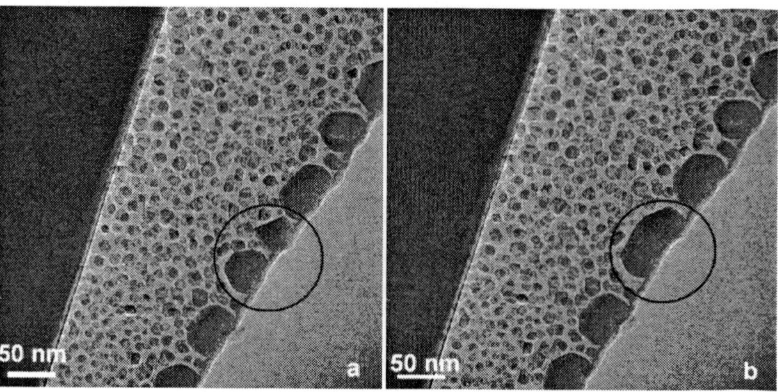

Figure 2. Helium bubble formation and coalescence after heating to 400°C in the hot stage

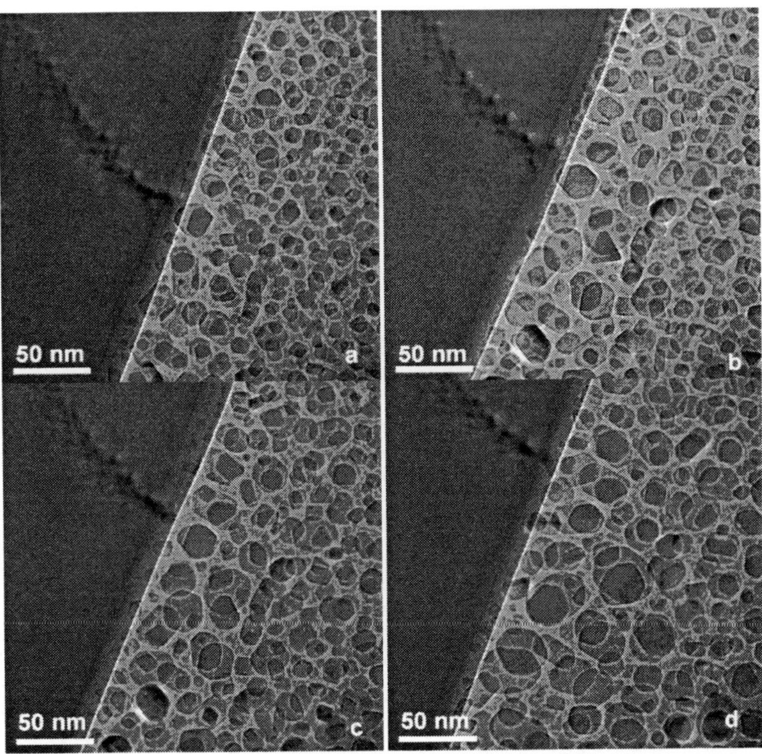

Figure 3. Helium bubble formation and growth after heating to 400°C in the hot stage. The bubble region expands (from a to d) due to diffusion along the bulk and swelling of the grain boundary

Figure 3a–d is from a different selected area of the same specimen. The pictures were taken every minute from 3a to 3d. Bubbles 3 to 30 nm in diameter and various polygonal shapes are shown in Figure 3a. During the heating the bubbles' diameter grows, they combine, and change their shape, while new bubbles are created. Bubbles can be observed at different depths, e.g., one above another with partial overlapping. The swelling in the dark at the left side of the pictures in Figure 3 was caused by accumulation of helium atoms in a grain boundary that were diffused to the bubbles area. During the heating the swelling structure varies and the grain boundary seems to be used as a channel for the helium atoms. New bubbles are formed in the sharp boundary between the bulk (dark region on the left) and the bubble region (brighter on the right). The formation of new bubbles improves the transparence of the material for the electron beam and the brighter area with bubbles becomes wider.

These observations reveal that there is a diffusion of helium from the bulk to the edge. An examination of the specimen with lower magnification as shown in Figure 4 reveals that the bubbles were created in the area that was exposed to the condensed TEM beam with greater magnifications. In the other areas there is no sign of bubble formation, leading to the conclusion that the electron beam strongly influenced the process.

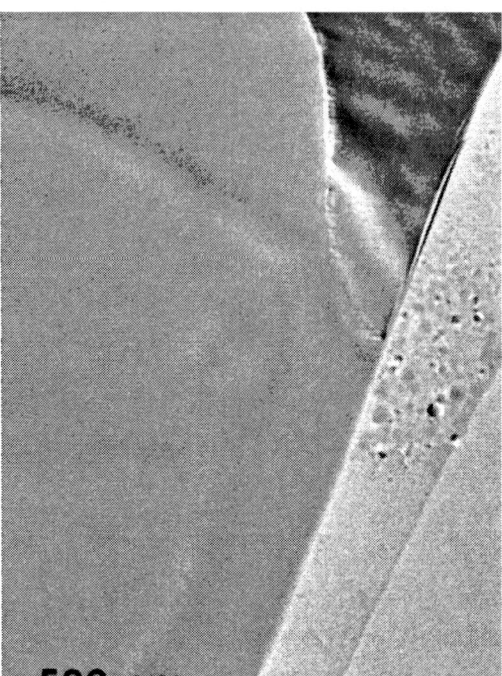

Figure 4. The diagnosed area at low TEM magnification. The bubbles were formed in the restricted area that was subjected to the high magnification electron beam

One of the possible explanations is that the electron beam creates local heating and a temperature gradient in the metal that causes the helium atoms to move to the hotter area by thermal diffusion. This possibility was rejected since calculation of the local heating by the formula suggested by Egerton et al. [14] shows that the temperature rises only by ~1°C. A radiation damage mechanism that is not accompanied by drastic temperature rise can be caused by displacement of the aluminum atoms due to the bombardment by the TEM energetic electrons. Measurements of the electrons bombard threshold energy made by Pelles and Phillips [15] give a value of 175±25keV for the aluminum ion displacement. These measurements were obtained with $\sim 5\times 10^4$ (A/m^2) current density. In the present experiments a 200keV electron beam was applied and the current density in the magnification to obtain Figures 1–4 was $\sim 5\times 10^4$ (A/m^2). Since it is above the threshold damage found by Pelles and Phillips, the following explanation for the dependence of helium bubble formation and growth under the electron beam is suggested. The bubbles are formed by accumulation of helium atoms in flaw sites. The electron beam in the TEM causes displacement of the aluminum ions and local flaws are used as trap sites for the helium atoms that are moving by diffusion. At room temperature the diffusion rate of the helium in the aluminum is very low and exponentially increases with temperature [16]. Since the diffusion coefficient of helium atoms in aluminum at 400°C is higher by two orders of magnitude than at room temperature, the helium atoms are moving faster in the metal until they are trapped in the flaws. Bubble formation and growth in these conditions continuously develop in minutes and can be observed in situ by the TEM.

Table 1. Measurements of helium bubble radii in bulk aluminum after heating to different temperatures

Specimen	Heating	Heating Bubble radius	
	Temperature (°C)	Duration (hr.)	(nm)
A	400	0.8	Undetectable
B	500	0.8	Undetectable
C	550	23	~5
D	600	23	~10
E	600	48	~30

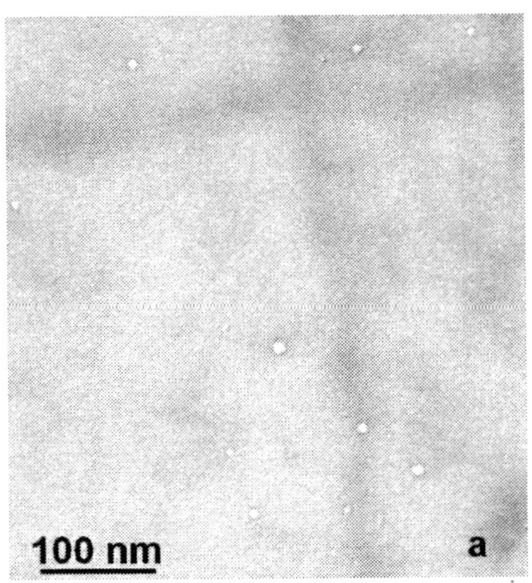

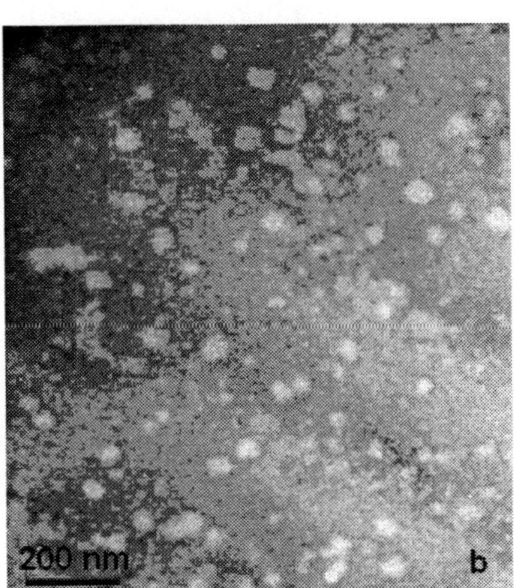

Figure 5. Helium bubble formation after bulk heating to 550°C for 23 hours (a) and to 600°C for 48 hours (b).

1.3.2. Bubble formation in bulk aluminum

Five samples of 2.7 mm thickness were cut from the irradiated Al-^{10}B plate. Each piece was heated to different temperatures for different times as shown in Table 1. After cooling to room temperature, TEM specimens were prepared from each slice and examined to detect helium bubbles. No bubbles were observed in specimens after heating to 400°C for 50 minutes, compared to the case of in situ heating where the bubbles formed in seconds at the same temperature. Heating to 500 °C for 50 minutes (specimen B) also did not reveal any bubbles. The helium bubbles were detected only in the specimens that were heated for longer durations. In specimen C that was heated to 550°C for 23 hours (Figure 5a), the average bubble radius is 5 nm and in specimen D (600°C for 23 hours, Figure 5b) the average radius is 10 nm. Heating for 48 hours at the same temperature (specimen E) caused formation of 30 nm radius helium bubbles.

In both cases of heating in the TEM with hot stage and bulk heating, temperature dependence of the helium formation and growth was observed, but the time scale is different: seconds or hours, respectively.

1.3.3. Number of helium atoms in a bubble

The number of helium atoms in a bubble was measured in TEM with an electron energy loss spectrum (EELS) device. According to EELS spectrum, the number of helium atoms in a unit volume is given by

$$N = \frac{I_k}{I_0} \cdot \frac{1}{\sigma_k} \cdot \frac{1}{d} \tag{2}$$

where I_k is the intensity of the electron loss spectrum at 21.5 eV (first helium ionization energy) and I_0 is the intensity of the un-scattered electrons, $\sigma_k = 3\times10^{-23}$ m^2 is the electron cross section of the appropriate scattered electrons and $d = 70$ nm is the foil thickness in the TEM experiment. An electron intensity spectrum from EELS measurement in a bubble with a radius r_b =5 nm is shown in Figure 6. Substituting the intensity ratio (I_k/ I_0), σ_k, and d into equation 2 we get the helium density $N=4.2\times10^{28}$ m^{-3}. Multiplying N by the bubble volume, $V_b = (4/3)\pi r_b^3$, one gets that in a 5 nm radius bubble there are $N_b = (2.2\pm0.2)\times10^4$ helium atoms.

It is interesting to compare the measurements with the theoretical calculation of helium atoms in a bubble assuming the hard sphere equation of state (EOS) suggested by Brearley and MacInnes [17]:

$$N_b = \frac{P_b V_b}{zkT} = \frac{(2\gamma / r_b)(4/3\pi r_b^3)}{zkT} = \frac{8\pi\gamma r_b^2}{3zkT} \tag{3}$$

where P_b is the pressure inside the bubble as given by $2\gamma/r_b$ for a bubble radius r_b with a volume V_b, and γ = 1 J/m^2 is the aluminum surface tension. T is the temperature, k is Boltzmann's constant, and z is a compressibility factor that is in general a function of temperature and pressure and can be calculated from the Lennard-Jones potential. For a 5 nm radius the helium pressure inside the bubble is 400 MPa. The appropriate compressibility factor for P_b=400 MPa and T=300K is given by Brearley and MacInnes [17] to be z= 2.2. Using r_b = 5 nm, T = 300K, and z = 2.2 we get a theoretical estimation of $N_b = 2.3\times10^4$ helium atoms in the bubble, in excellent agreement with EELS measurements. It is interesting to point out that the agreement between EELS measurement and the theoretical estimate is actually a proof for the correctness of the EOS used for the helium nanoparticle.

Under the assumption that all of the helium atoms in the specimen were accumulated into uniform bubbles, the concentration of the helium bubbles and the inter-bubbles distance s can be estimated by the relation:

$$N=1/s^3=N_{He}/N_b\approx1.2\times10^{18}\ \mathrm{m}^{-3} \tag{4}$$

where $N_{He}=6.7\times10^{23}$ (m^{-3}) is the concentration of helium atoms in the specimen and N_b is the number of helium atoms in one bubble. The number of helium atoms in 5 nm bubble is $N_b\approx2.3\times10^4$ and the average inter-bubble distance is 220 nm, in fairly good agreement with the TEM observation (figure 5a). For 30 nm bubble $N_b\approx6\times10^5$ and the average distance between the bubbles is ~1300 nm, different from the TEM observation (figure 5b). This difference reveals that the TEM image at figure 5b does not represents the bubbles distribution in the target that was heated to 600 °C for 48 hours.

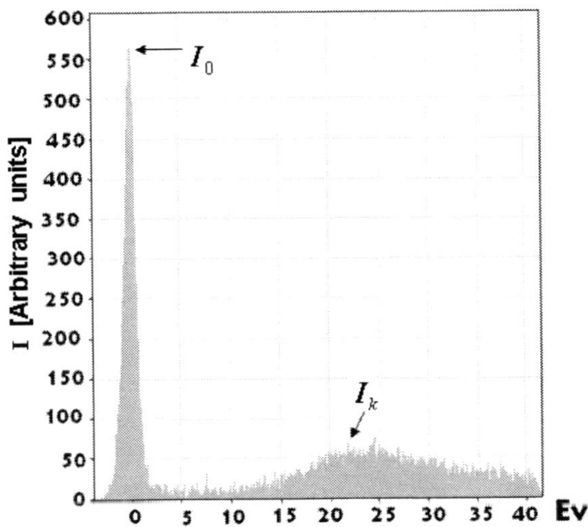

Figure 6. Electron energy loss spectrum (EELS) measurements in 5 nm helium bubble

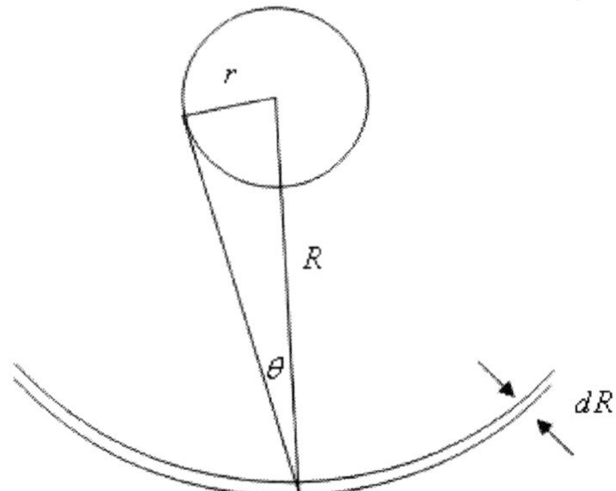

Figure 7. Schematic description of helium atom diffusion and trapping in a sink

1.4. Theoretical Estimation of Helium Bubble Formation by Diffusion in Aluminum: Bulk Heating and Hot Stage Heating

1.4.1. Bubble formation during bulk heating

In order to understand the experimental results, an analytic approximation of the solution to a diffusion equation with a sink was carried out as follows.

Let us consider a spherical volume with radius R. Assuming that the helium atoms are randomly moving in all directions in this volume, the probability Γ that helium atoms will pass through a spherical sink with a radius r (as shown in Figure 7) is given by the relation:

$$\Gamma = d\Omega / 4\pi \tag{5}$$

Under a further assumption that all the atoms that hit the spherical sink will also be trapped inside to create a bubble, the number of helium atoms in the bubble N_b can be evaluated by:

$$N_b = \int_R \int_\Omega \left(N_{He} 4\pi R^2 dR \right) \left(\frac{d\Omega}{4\pi} \right) \tag{6}$$

where N_{Ne} is the concentration of helium atoms per unit volume in the aluminum. It is important to emphasize that equations 3 and 5 are independent derivations of N_b. Equation 3 is based on equation of state knowledge of the helium state, while equation 5 expresses the diffusion effect with a sink.

For small values of θ in Figure 7, $d\Omega = 2\pi r dr / R^2$. Substituting it into equation 5 gives:

$$N_b = \pi r^2{}_b N_{He} R \tag{7}$$

For 3D diffusion, the dependence of the average diffusion distance R in time t is given by $R^2 = 6Dt$. Substituting it into equation 6 gives an expression for the bubble growth time:

$$t = \left(\frac{1}{6\pi^2} \right) \left(\frac{N_b}{r_b^2} \right)^2 \left(\frac{1}{N_{He}} \right)^2 \left(\frac{1}{D} \right) \tag{8}$$

The diffusion coefficient of helium in aluminum D was taken from Glyde [16] and the number of helium atoms in a bubble N_b can be estimated by the hard sphere EOS (equation 3). Now the diffusion time can be calculated by:

$$t = \left(\frac{32}{27} \right) \left(\frac{\gamma}{zkT} \right)^2 \left(\frac{1}{N_{He}} \right)^2 \left(\frac{1}{D} \right) \tag{9}$$

Note that since the compressibility factor z is a function of the pressure inside the bubble that is given by $2\gamma / r_b$, the diffusion time is also a function of the bubble radius r_b.

Figure 8 represents the bubble radius growth in time as calculated from equation 8 for 500°C, 550°C, 600°C, and experimental data from TEM measurements. The theory is in good agreement with the experimental results. The solution reveals a nonlinear dependence of the bubble's radius in time and that heating time of several hours is needed for the formation of the smallest helium bubbles that can be observed in the TEM (~3 nm), unlike the case of heating in the hot stage TEM holder.

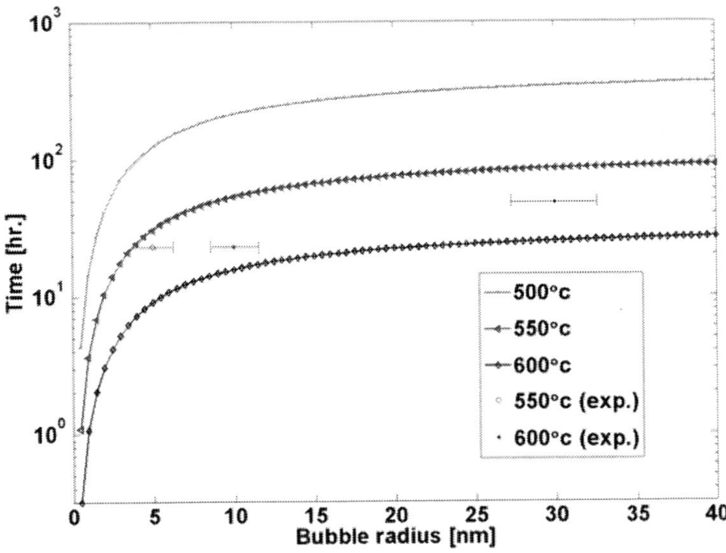

Figure 8. Time for helium bubble growth in bulk aluminum at different temperatures, calculated and experimental

1.4.2. Bubble formation during heating in the hot stage TEM holder

From the results of the experiments it is clear that helium bubble formation during heating of a TEM specimen in the hot stage holder is faster by orders of magnitude than heating a bulk specimen. The reason is probably the different thicknesses of the heated specimen. A TEM specimen is a round disc with a small hole in its center. The thickness of the diagnosed area changes from a few nanometers near the hole to 100 μm, as shown in Figure 9. The maximum penetration thickness of the electron beam is ~200 nm.

From geometric relations, the diffusion distance L of helium atoms from the aluminum to N bubbles can be calculated by:

$$\left(\frac{1}{2}\right)N_{He}\left(\frac{d}{2}\right)\left(\frac{\pi \quad L^2}{2}\right)=N_bN\rightarrow L=\sqrt{\frac{8N_bN}{\pi \quad N_{He}d}} \qquad (10)$$

Substituting equation 9 into the expression for the average diffusion distance $L=\sqrt{6Dt}$ will give an expression for the diffusion time of helium in aluminum while heating it in the hot stage holder:

$$t=\frac{4}{3\pi}\frac{N_bN}{N_{He}}\frac{1}{Dd} \qquad (11)$$

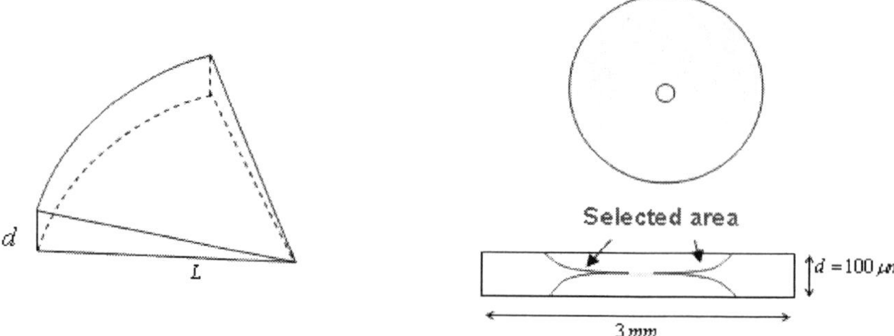

Figure 9. Scheme of TEM specimen.

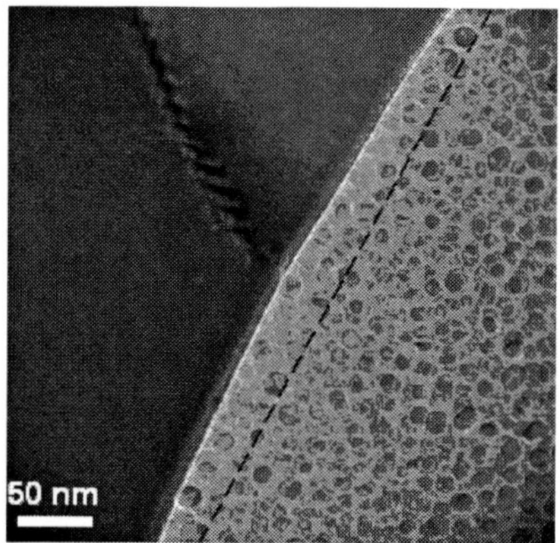

Figure 10. TEM picture of the helium bubble region after heating to 400ºC in the hot stage. The red line denotes the boundary of the helium bubble region as observed one minute before

In Figure 10 one can see helium bubbles that were formed during heating to 400ºC in the hot stage holder. The dashed line represents the boundary of the bubble region as observed one minute before. During this minute about $N \approx 130$ new bubbles with 8 nm average radius were formed. From the hard sphere EOS the calculated number of helium atoms in one bubble is $N_b \approx 6.5 \times 10^4$. At 400ºC the coefficient of helium diffusion in aluminum is $D \approx (8.9 \pm 8) \times 10^{-16} \left[m^2 / \sec \right]$. At these conditions the calculated diffusion time (from equation 10) is 60 seconds, in very good agreement with the experimental results.

1.4.5. Conclusions

Helium bubble formation in aluminum was investigated experimentally and analytically. The formation and growth process was observed in situ when heating the metal in TEM with a hot stage holder. It was found that the electron beam in the TEM influences the process and the time scale for bubble formation is seconds. Further TEM observations at room temperature of post-heated bulk aluminum reveal that the time for bubble formation in this case is hours. Analytical calculations of the diffusion time for bubble formation in both of the cases explain the experimental results. The number of helium atoms in a bubble was calculated from EELS measurements. These measurements confirm the hard sphere EOS that was used for the diffusion calculations.

2. Preferred Helium Migration Direction in Aluminum

Theoretical investigation of the atomistic behavior of helium-vacancy (He_nV_m, where He_n are n helium atoms and V_m are m voids) clusters was carried out analytically and by molecular dynamics codes during the last few decades [18-20]. Despite many years of research regarding helium's effects in metals there are still many unknowns, especially regarding the initial stages of voids and bubble nucleation.

The TEM observations that are presented in this part are consistent with the theoretical models that predict that the migration of helium atoms in an fcc aluminum lattice is done through interstitial sites.

2.1. Experimental Results

The observations were carried out in an FEI Titan model TEM with an acceleration of 300 keV. TEM images of selected areas of a specimen that had been heated in the TEM hot stage holder to 470°C are shown in Figure 11a–f. A polygonal area with sharp boundaries was observed (Figure 11a–e). A defocused image under magnification of this area revealed a formation of nanometric helium bubbles, as shown in Figure 11f. The edge contrast arises because the helium displaces the aluminum reflecting planes. Similar contrast effect of stacking fault has been observed by TEM in metals [21]. According to EELS measurement, the number of helium atoms in the helium area is $N_{He} = (3\pm1)\times10^{28}\ m^{-3}$ helium atoms, which are (30–65)% of the atoms in the observed area. For comparison, the number of helium atoms in the rest of the aluminum matrix is $(1.2\pm0.2)\times10^{24}\ m^{-3}$, which is 20 appm.

The formation of an area with a high concentration of helium is attributed due to the existence of 5nm diameter segregated boron particles that were found in the vicinity of the polygonal area. During the neutron irradiation a high helium concentration region was formed around the boron particles according to the nuclear reaction $^{10}B + n \rightarrow {}^{7}Li + {}^{4}He$. At room temperature the diffusion rate of helium in the aluminum is very low. It increases exponentially with temperature [16]. The diffusion coefficient of helium atoms in aluminum at 470 °C is higher by two orders of magnitude than at room temperature, therefore helium atoms are moving fast in the metal until they are trapped in the lattice defects. The helium migration and bubble formation in these conditions continuously develop in minutes and can be observed by in situ TEM.

The images in Figure 11a–e were taken at one second intervals from a to e. In Figure 11b the helium area started to expand and the horizontal boundary at the bottom moved downwards. Although the diagonal boundary at the bottom seems to be also moving, Figure 11c shows that this shape was preserved. The helium area continued to expand downward along a direction perpendicular to the lower boundary as shown in Figure 11a–e. The expansion of the other boundaries is more moderate.

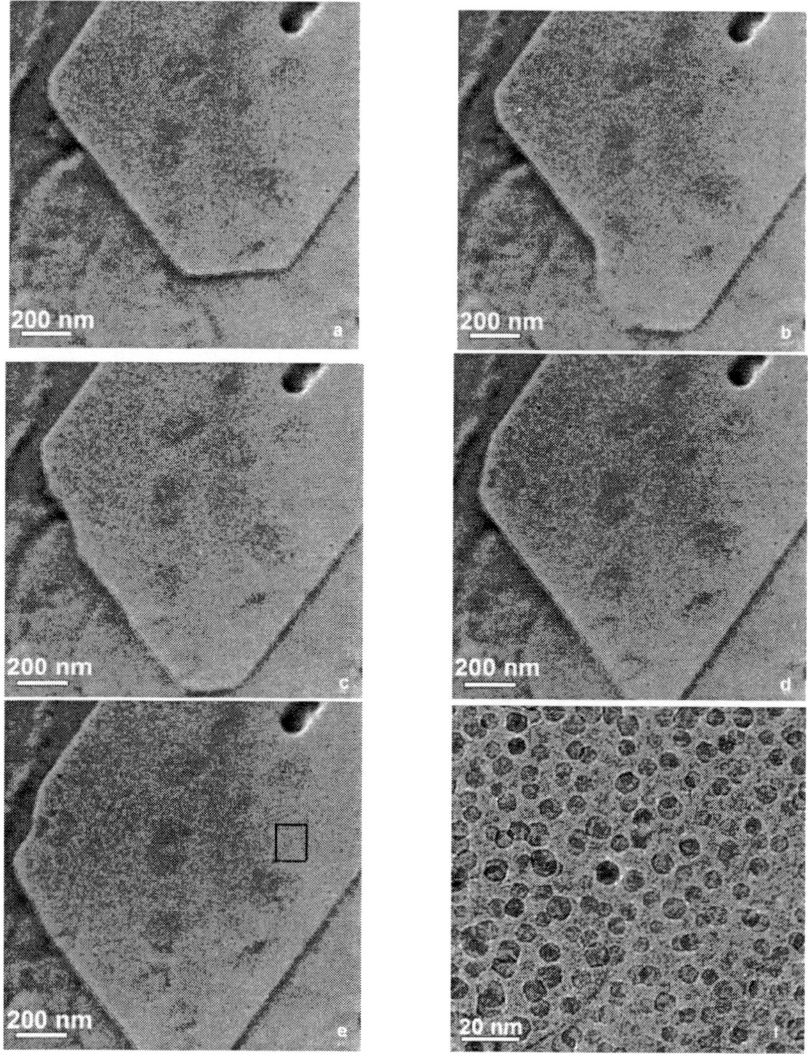

Figure 11. TEM images of migration of the helium-rich area in aluminum during heating to 470 °C in the hot stage holder (a–e). A defocused image under magnification of this area revealed a formation of nanometric helium bubbles (f)

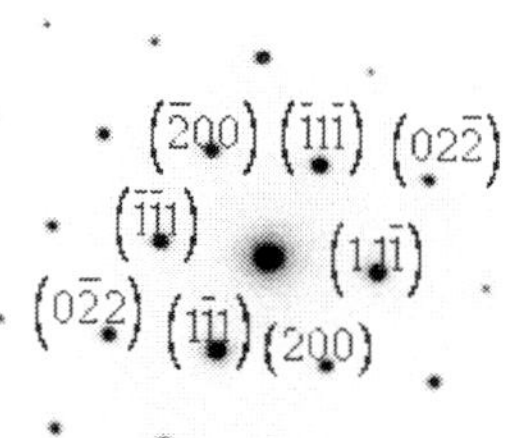

Figure 12. Diffraction pattern of the selected area in Figure 11

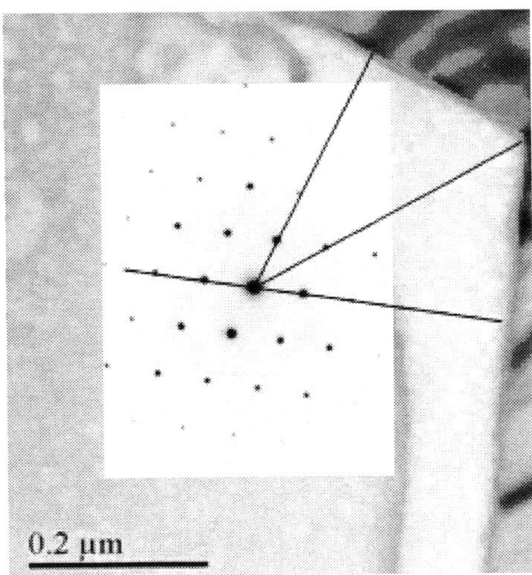

Figure 13. TEM image with the diffraction pattern of the selected area after cooling to room temperature. The vectors to the diffraction spots indicate that the boundaries of the helium-rich areas are parallel to the aluminum crystallographic planes ($\bar{1}\bar{1}1$), ($1\bar{1}1$) and ($0\bar{2}2$)

The sharp boundaries and the migration of the helium area in a preferred direction imply that this phenomenon is related to the aluminum crystallographic directions. A diffraction image was taken after cooling the specimen to room temperature, in order to stop the helium migration and to stabilize the diagnosed area. The electron beam diffraction pattern and the relevant crystallographic planes notations are shown in Figure 12. It was found that the zone axis lies along the [011] direction; hence the observed plane in the TEM is (011). The diffraction image combined with the TEM image of the observed area is shown in Figure 13. One can see that the vectors to the spots reflected from planes $(\bar{1}\bar{1}1)$ and $(1\bar{1}1)$ in the diffraction image are perpendicular to the boundaries of the helium rich area in the TEM image. The vector to plane $(0\bar{2}2)$ is pointing to the corner between the two boundaries. This is the preferred migration direction of the helium area. These results reveal that the boundaries of the helium area are parallel to the corresponding crystallographic planes of the aluminum bulk. Analysis of the angles between the diffraction spots supports these findings. A schematic description of the diffraction image as received in the experiment is shown in Figure 14 [22]. A comparison of the calculated angles between crystallographic planes in an fcc metal as shown in Figure 14, and the measured angles between the polygon sides in the experiment (figure 15) is given in table 2. The agreement between the experimental and calculated angles indicates that the boundaries are parallel to the crystallographic planes as shown in Figure 1 The arrow in figure 15 shows the migration direction of the lower boundary. It is parallel to the plane $(0\bar{2}2)$. The direction of the migration is normal to this plane.

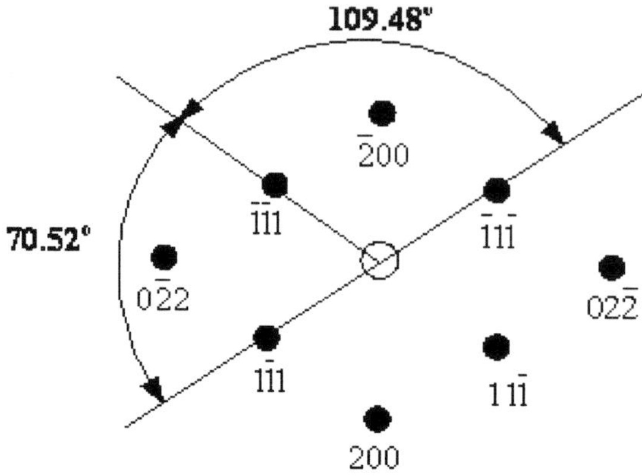

Figure 14. Theoretical diffraction pattern in fcc metal with zone axis $[011]$

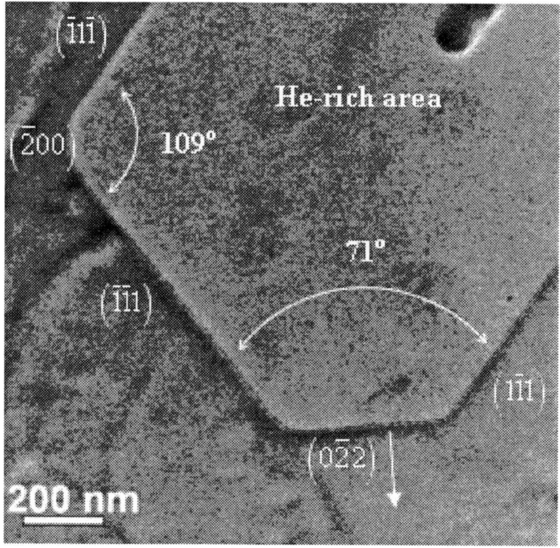

Figure 15. TEM image of the helium-rich area in aluminum during heating to 470°C. The angles between the boundaries reveal that the boundaries are parallel to the crystallographic planes

Table 2. A comparison of the calculated angles between crystallographic planes in an fcc metal (figure 14) and the measured angles between the polygon sides (figure 15)

Between Planes	Theoretical angle	Measured angle in the TEM image
$(1\bar{1}1)$ and $(\bar{1}\bar{1}1)$	70.52°	71±1 °
$(\bar{1}1\bar{1})$ and $(\bar{1}\bar{1}1)$	109.48 °	109±1 °

2.2. Analysis and Discussion

The experimental results show that helium migrates in the aluminum in a direction normal to the family of planes {111} and the plane $(0\bar{2}2)$. The expansion rate of the boundaries seems to be higher in the direction parallel to plane $(0\bar{2}2)$ than in the other directions. Helium bubbles are formed in the helium-rich area behind the migration front.

To understand this behavior it is essential to examine the helium–metal interaction. Past research in this field has found that helium atoms get into the interstitial sites and create clusters in the metal defects that become nucleation sites for the bubble formation [23]. It was also suggested that the helium atoms in the aluminum matrix increase the stability of vacancy clusters and act as a catalyst for the formation of He-void clusters. [24;25]

Aluminum has an fcc atomic structure with two types of interstitial voids: the larger voids are known as octahedral sites and the smaller voids are known as tetrahedral sites [26]. In the fcc metal the octahedral interstitial site is the favorable position for interstitial helium [24;25;27]. Ab initio calculations made by Yang et al. [25] showed that the migration energy of interstitial helium atoms between two octahedral sites without crossing the tetrahedral site is 0.16 eV, while passing via the tetrahedral site is 0.13 eV. Since the difference between tetrahedral and octahedral diffusion path is only 0.03 eV, the helium migration in the experiments at 470 °C (=0.064 eV) will be probably a combination of both types of transitions.

According to this model, and the behavior of the sample under the TEM, the following explanation is suggested. Due to neutron irradiation of a specimen containing nanometric boron segregates helium-rich regions are initially present in the sample. After irradiation the specimen was heated in the TEM and the diffusion rate of helium in the aluminum matrix is accelerated exponentially as a function of the temperature. The amount of helium that can be absorbed by a specific region of the material depends on the number of vacancies available. Therefore the helium migrates to areas that contain smaller proportion of helium atoms, as observed in the experiment.

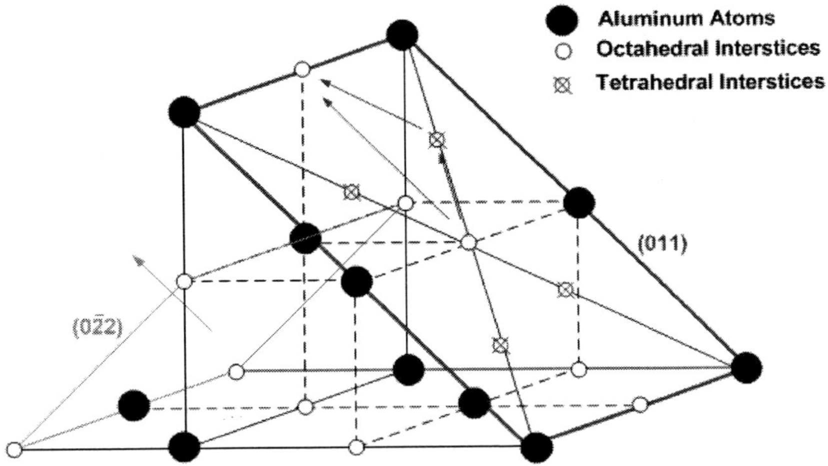

Figure 16. A schematic fcc lattice with jump paths between two octahedral sites

The initial step of the helium migration is a jump between octahedral sites either directly or via tetrahedral sites. In the TEM experiments the observation direction is normal to the plane (011). While looking on this plane, the jump of helium atoms from the octahedral site at the center [222] to the next octahedral site, as shown by the arrows in Figure 16, will appear as a migration of the helium front in a direction normal to plane ($0\bar{2}2$). The other possible vacant octahedral sites that the helium atoms could jump into are not lying in the plane (011) and therefore the migration of helium in directions normal to planes {111} looks less dominant by the TEM two-dimensional observations.

This interpretation of helium atom migration in aluminum, based on helium-vacancy models is consistent with the TEM experiment as shown in Figure 11a–e.

2.3. Conclusions

Migration of helium in aluminum was observed in situ by TEM with a hot stage holder while heating the metal. The TEM observation and calculation of the number of helium atoms from EELS revealed a helium-rich area in the vicinity of boron segregates. During the specimen heating, the helium starts to migrate to other areas and the helium-rich area expands. Electron beam diffraction reveals that the migration direction is normal to the planes ($0\bar{2}2$) and {111} while the observed direction is normal to plane (011). These findings can be explained by the migration of helium atoms in an fcc aluminum lattice through interstitial sites.

3. Equation of State for Aluminum Containing Helium Bubbles

3.1. Introduction

The change in the mechanical properties of metals may be a combined effect of several defects, such as bubbles, voids, and dislocation loops, and it is difficult to experimentally identify the pure effect of bubbles on the dynamic behavior of metals. Theoretical and numerical methods are appropriate to study the influence of a single factor determining the properties of metals containing helium. The dynamic strength of aluminum containing an initial distribution of microscopic defects was calculated by Kubota et al. [28], using molecular dynamics methods. The elastic constants of helium embedded aluminum were recently calculated by molecular dynamics simulations by Wang et al. [29]. A high density equation of state (EOS) for helium and its application to bubbles in solids was reported by Wolfer [30]. Based on this equation of state, the influence of helium bubbles on shock loading is examined. The Hugoniot curve (temperature vs. pressure as well as shock velocity vs. particle velocity) for aluminum containing bubbles is calculated for various bubbles mass, bubbles percentage and helium equation of state models [3].

3.2. Theoretical Model

3.2.1. General framework

The EOS of aluminum containing helium bubbles suggested here is based on realistic EOS for aluminum – SESAME tables [31], and two high density EOS for helium, Carnahan-Starling hard sphere EOS [32] and Trinkhaus semi-empiric EOS [33].

The model for aluminum with helium bubbles presented here includes two input parameters defining the bubbles, which can be measured experimentally:

1. α - the mass ratio between the helium bubbles and the total sample mass, which is connected to the helium atomic percentage in the metal.
2. m - the helium bubble mass, that can be calculated from Transmission Electron Microscopy (TEM) with Electron Energy Loss Spectroscopy (EELS) measurement. This mass is assumed to be constant during shock loading.

Both quantities are determined during the formation process of the bubbles in the metal. It is assumed that all the bubbles are identical and that all the helium in the sample is contained into bubbles. Moreover, the model is founded on the assumption that the bubbles do not ruin the metal crystalline structure. For this reason, a reasonable constrain would be that the radius of the bubbles should be smaller by at least an order of magnitude than the distance between bubbles d. This premise imposes an upper limit for the value of α for which the model is adequate. The relation between α and the distance between bubbles d is:

$$\alpha \sim \frac{1}{7} \cdot \frac{4\pi}{3} \left(\frac{r_b}{d} \right)^3 \tag{12}$$

Where r_b is the bubble radius and the factor $\frac{1}{7}$ comes from the mass ratio of helium and aluminum (supposing the number density of helium and aluminum is of the same order of magnitude). This implies that maximal value α for which the model is valid is $\alpha_{max} = (1/7)[4\pi/3](1/10)^3 \approx 5 \cdot 10^{-4}$.

Additionally, the mechanical stability limit [4] $P \leq 0.2G$ at which the matrix would yield by spontaneous plastic deformation, where G = 270 kbar is the shear modulus of aluminum [34] and $P = 2\gamma / r_b$ is the equilibrium pressure in the bubble before shock loading, is satisfied for radius of the order of 1 nm. γ is the surface tension and its value is typically 1 (J/m^2).

The EOS can be formulated as the dependence of the pressure and internal energy on temperature and average density. In the above two materials model the average density is given as a function of the densities of aluminum and helium ρ_{Al} and ρ_{He}:

$$\rho = \left[(1-\alpha)\frac{1}{\rho_{Al}} + \alpha\frac{1}{\rho_{He}} \right]^{-1} = \left[(1-\alpha)\frac{1}{\rho_{Al}} + \alpha\frac{4\pi r_b^{\,3}}{3m} \right]^{-1} \quad (13)$$

The internal energy and pressure of aluminum with helium bubbles are given by:

$$P = P_{Al}(\rho_{Al}, T),\; E = \alpha E_{He}(\rho_{He}, T) + (1-\alpha)E_{Al}(\rho_{Al}, T) \quad (14)$$

E_{He} and E_{Al} denote the internal energy per unit mass of the helium and aluminum, respectively, and P_{Al} is the pressure of the aluminum which is assumed to drive the shock (since the pressure on the bulk surface is that of the aluminum). From equation 14 it is clear that if one want to know E, P for a given average density, a supplementary relation (in addition to equation 13) between ρ_{Al}, ρ_{He} is required.

The bubble dynamics obeys the Rayleigh equation [35]:

$$\ddot{r}r + \frac{3}{2}\dot{r}^2 + \frac{4\eta(\rho_{Al}, T)\dot{r}}{\rho_{Al}r} + \frac{2\gamma(\rho_{Al}, T)}{\rho_{Al}r} + \frac{P_{Al}(\rho_{Al}, T) - P_{He}(\rho_{Al}, T)}{\rho_{Al}} = 0 \quad (15)$$

In equation 15 P_{He} is the pressure inside the bubble and η is the viscosity.

Neglecting time dependent terms the helium pressure in equilibrium bubble of radius r_b becomes:

$$P_{He}(\rho_{He}, T) = \frac{2\gamma(\rho_{Al}, T)}{r_b} + P_{Al}(\rho_{Al}, T) \quad (16)$$

The surface tension is weakly dependant on density and temperature and therefore a constant typical value $\gamma = 1\,J/m^2$ (for aluminum) was used. In order to display the importance of the surface tension, let us consider an example: for bubble radius r_b =1nm the surface tension contribution to the helium pressure is of the order of 10^4 times the atmospheric pressure. For this reason, supplying quantitative description of the bubble even without shock compression requires high pressure EOS for the helium.

Finally, we have a closure (equations (13), (14), (16)) by which we can evaluate the total energy and pressure for a given ρ, T and thus obtain the total EOS. The only remaining issue to address is the helium EOS, which is essential for the sake of obtaining P_{He} in equation 16.

3.2.2. Helium EOS

For discussing shock loading of the order of hundred of kbars, an EOS for helium in this region is necessary. Comprehensive theoretical work had been done concerning EOS of dense liquids [36], and particularly on EOS of helium [17;30;33;37]. Two options for the liquid helium EOS were considered: Carnahan-Starling EOS [32] with hard sphere which is determined using the inter-atomic potential [38] and Trinkaus semi empiric EOS [33]. Both are briefly described below.

Both EOS assume the pressure and the internal energy can be written as following:

$$P_{He} = nk_B Tz \,,\ E_{He} = \frac{3}{2} P_{He} V \tag{17}$$

T is the temperature, k_B is Boltzmann's constant , $n = \rho_{He} / m_{He}$ is the number density of the helium atoms in the bubbles, $m_{He} = 4m_p$ is the helium atom mass and z is the compressibility. The two EOS differ in the formula for z.

3.2.2.1. Carnahan-starling EOS

Carnahan-Starling hard sphere EOS is given by:

$$z(y) = \frac{\left(1 + y + y^2 - y^3\right)}{(1 - y)^3} \quad , \quad y = \frac{\pi \rho_{He}}{6m_{He}} d^3 \tag{18}$$

Where y is the packing ratio. The diameter of the hard sphere is d and it can be approximated [36] by

$$d = \int_0^{\sigma_e} \left\{ 1 - \exp\left[-\frac{V(r)}{k_B T} \right] \right\} dr \tag{19}$$

With $V(r)$ the interatomic Lennard-Jones potential

$$V(r) = 4\varepsilon_e \left[\left(\frac{\sigma_e}{r} \right)^{12} - \left(\frac{\sigma_e}{r} \right)^{6} \right] \tag{20}$$

$$\sigma_e = 0.2556nm \qquad \varepsilon_e / k_B = 10.22K$$

The values of σ_e, ε_e are taken from Hirschfelder [38].

The Carnahan-Starling EOS is adequate for $y \le 0.36$, which corresponds approximately to pressures up to 20kbar in the helium. Indeed, the implementation of this EOS to helium had been done by Brearley and MacInnes [17] and was found to reproduce experimental results up to pressures of 20kbar. Nevertheless, in section 3.3 we shall see that practically, using this EOS does not cause serious error also at high pressures because in this region the total EOS is mainly governed by the aluminum EOS.

3.2.2.2. Trinkaus EOS

Here the semi empiric EOS developed by Trinkaus [33] for liquid helium is presented.

According to Trinkaus the helium compressibility takes the form:

$$z = (1-x)(1+x-2x^2)+(1-x)^2 x\frac{B}{v_l}+(3-2x)x^2 z_l+(1-x)x^2 z_l' v_l \quad (21)$$

Where $x \equiv \rho v_l$ and z_l, v_l are the specific volume and compressibility of the liquid at the melting curve. It can be estimated by:

$$v_l = 56T^{-1/4}\exp(-0.145T^{1/4}) \ (in \ \text{A}^3) \qquad (22)$$

$$z_l = 0.1225 v_l T^{0.555} \qquad (23)$$

B(T) is the second virial coefficient. Evaluating the integral by which B is defined results in the following expression:

$$B(T) = 170T^{-1/3} - 1750/T \quad (\text{in A}^3) \qquad (24)$$

The derivative of the compressibility in the region $100K < T < 1000K$ is approximated by

$$z_l' v_l \approx -50 \qquad (25)$$

This EOS was compared with the experimental and theoretical data in several works [37;39;40].

3.2.3. Hugoniot calculation

After deriving the EOS, we can explore the influence of the bubbles on the dynamical behavior of the metal. For this purpose, we are interested in the pressure and internal energy along the Hugoniot curve of aluminum containing helium bubbles.

The Hugoniot relation reads:

$$\tfrac{1}{2}\left(P(\rho,T)+P_0\right)\left(1/\rho_0 - 1/\rho\right) = E(\rho,T) - E_0 \qquad (26)$$

Where P, E are the pressure and internal energy previously obtained and ρ_0, E_0, P_0 the initial density, internal energy and pressure correspondingly.

Solving numerically equations (13), (14), (16) and (26) for ρ_{Al}, T, r, we can obtain the pressure vs. temperature and average density behind the shock.

After P, ρ, T are known, the particle velocity u and the shock velocity D are obtained by [41]:

$$u = D(1 - \rho_0/\rho) \quad (27)$$

$$D = \sqrt{\frac{P - P_0}{\rho_0(1 - \rho_0/\rho)}} \quad (28)$$

Where D is the shock velocity and u is the particle velocity behind the shock.

3.3. Results and Discussion

3.3.1. EOS

In order to demonstrate the effect of the bubbles on the total EOS the T = 300K isotherm was calculated (figure 17). For high pressures (hundreds of kbar) the EOS of pure aluminum and aluminum containing helium bubbles are quite similar and therefore this region was omitted from the graph.

The behavior of the isotherm in figure 17 is accounted for by the following argument: for atmospheric pressure, the helium pressure is larger by orders of magnitude than that of the aluminum because of the surface tension (equation 16), so the average pressure is clearly larger than that of pure aluminum. As we compress the metal, the surface tension term importance decreases and the average pressure is mainly due to the aluminum (α is very small), meaning it is primarily determined by ρ_{Al}. For obvious reasons, $\rho_{av} < \rho_{Al}$ and therefore for the same average density the pure aluminum pressure will be smaller than that of aluminum with bubbles. For extremely high pressure $\rho_{av} \approx \rho_{Al}, P_{He} \approx P_{Al}$ and hence the total EOS is only weakly affected by the bubbles. To conclude this argument, pressure of aluminum with bubbles is always higher than that of pure aluminum, where the diversity is maximal around the atmospheric pressure and decreases for high pressures.

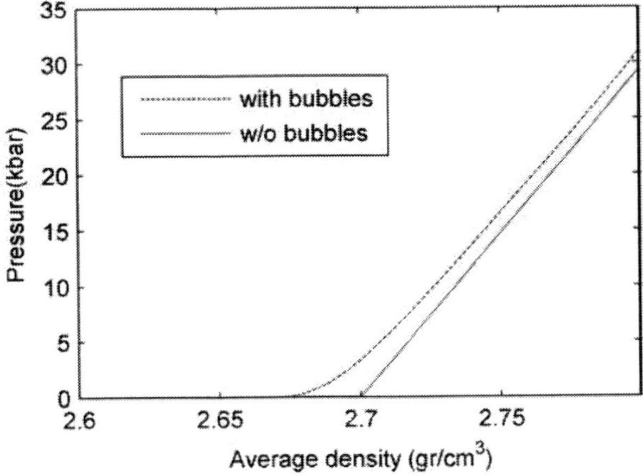

Figure 17. Isotherm of pure aluminum (dashed line) and aluminum containing $\alpha = 5\times10^{-4}$ helium bubbles (solid line). Bubble mass: $m = 2\times10^{-18}\,gr$. Helium EOS: Trinkaus

3.3.2. Hugoniot curves

The main objective is exploring the bubbles effect on dynamical behavior of the metal. Hence, the influence of α, m and the helium EOS on the Hugoniot curve will be examined. Figure 18 demonstrates the impact of the helium concentration, α, on the P-T Hugoniot. For small amounts of helium ($\alpha = 10^{-5}$) the Hugoniot is practically the same as the one of pure aluminum (about 0.1% difference), and consequently omitted from the graph. However, for $\alpha = 5 \times 10^{-4}$ (the maximal alpha for which the model is valid), the difference between the curves is observable, i.e. a diversity of about 5% both for a given temperature and for a given pressure.

It should be mentioned, that the distinction exists also at high pressures (hundreds of kbars) although at this region the EOS of aluminum with bubbles is the same as the one of pure aluminum. It is reasonable due to the fact that all the Hugoniot curves start with the same pressure, which means that ρ_0 depend on the EOS, and decreases with α. ρ_0 has major influence on the Hugoniot even at asymptotically high pressures since then equation (26) can be approximated by:

$$\tfrac{1}{2} P\left(1/\rho_0 - 1/\rho\right) = E \qquad (29)$$

The meaning is that metal containing helium bubbles behaves analogously to porous material. Consequently, in the P-T Hugoniot, higher temperatures are attained for a given pressure.

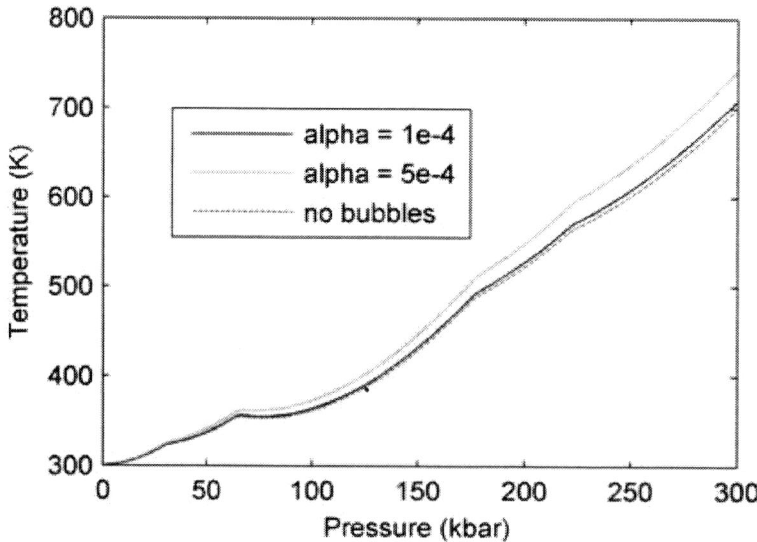

Figure 18. P-T Hugoniot curve for aluminum with several helium concentrations. Bubble mass: $m = 2 \times 10^{-18} gr$. Helium EOS: Trinkaus

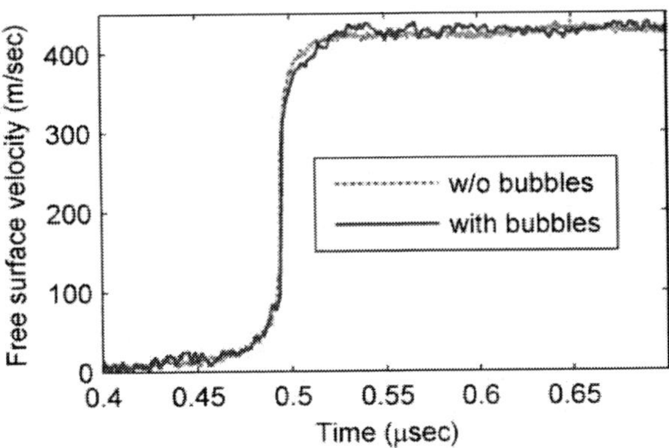

Figure 19. The free surface velocity versus the time in shock experiment [42]. There are no differences in the asymptotic velocity due to the low helium percentage $\alpha = 10^{-6}$. This result is consistent with the prediction

The theoretical calculations above, obtaining negligible differences for $\alpha = 10^{-5}$, is consistent with the experimental results described in details elsewhere [42]. This experiment was performed by accelerating aluminum impactor into Al-0.15 wt % 10B targets with/without helium bubbles (for the sake of the comparison, we suppose that the 10B atoms, located in the crystal matrix, do not alter the aluminum EOS crucially). The bubbles were produced by irradiating aluminum sample containing 0.15 wt % of 10B with neutrons and afterwards heating it in a furnace, a method which creates relatively uniform bubbles distribution and does not damage the crystalline structure of the metal. The experiment was conducted with small amounts of helium $\alpha = 10^{-6}$ to avoid radioactivity of the sample occurring for longer irradiation times required to reach high helium percentage.

In figure 19 one can see, that no differences in the asymptotic free surface velocity were detected (disregard what happens after secondary waves arrive at the surface). The surface velocity is connected to the EOS of both the impactor and the target [43], so we can deduce that at this helium percentage, the bubbles did not made measurable effect on the metal EOS.

The mass of the bubble has significant influence on the P-T Hugoniot - as much as 15% in the temperature for a given pressure (figure 20). In order to account for this fact, let us assume for simplicity an ideal gas EOS. Hence, without the surface tension term, 2 light bubbles of mass $m/2$ will occupy the same volume as one heavy bubble of mass m . Now we shall add the surface tension. It clearly compresses the light bubbles more affectively than the heavy one. For this reason, the heavy bubble will occupy larger volume; its helium density will be lower and the total density as well. Therefore, heavier bubbles cause lower ρ_0 and higher temperature on the Hugoniot curve (in the same manner as for increasing α).

Although the bubbles influence on the P-T Hugoniot is not negligible, it cannot be measured experimentally due to the considerable errors in temperature measurement by methods existing today. Hence, analogous plots were produced (using equations (26), (27)) of the shock velocity versus the particle velocity. Both quantities can be measured with satisfactory precision: the particle velocity with the aid of VISAR diagnostic (Velocity Interferometer System for Any Reflector), and the shock velocity with stepwise targets experiment. The calculation shows (figure 21) that for $\alpha = 5\times10^{-4}$ maximal difference of more than 10% is acquired, which is certainly measurable.

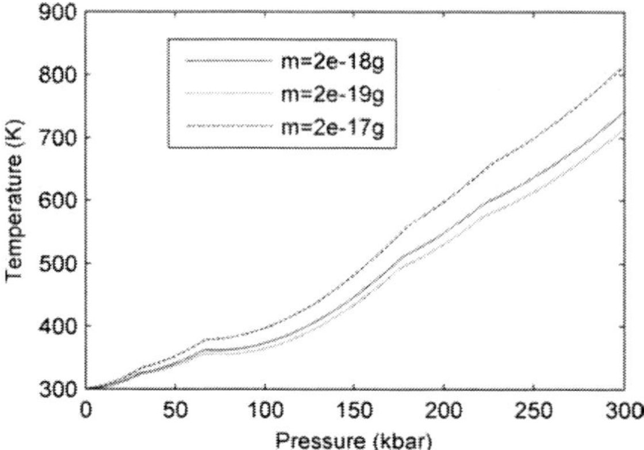

Figure 20. P-T Hugoniot curve for aluminum with $\alpha = 5\times10^{-4}$ helium bubbles for several bubble masses. Helium EOS: Trinkaus

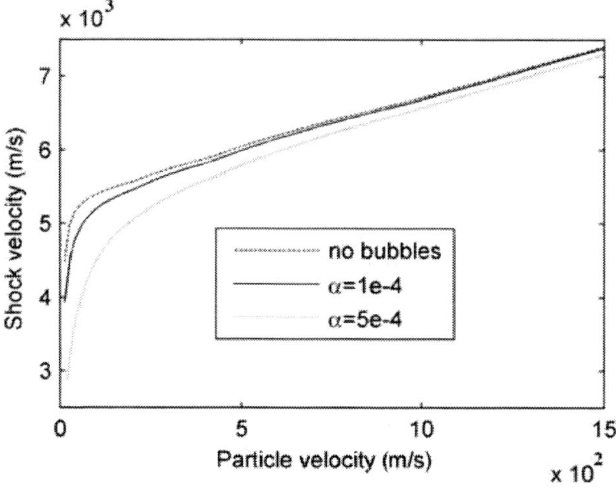

Figure 21. Shock velocity vs. particle velocity along the Hugoniot curve for aluminum with several helium concentrations. Bubble mass: $m = 2\times10^{-18} gr$. Helium EOS: Trinkaus

In order to figure out why the shock velocity is smaller for metal with bubble we shall address eq. (26). The EOS is approximately the same for high pressures, and therefore equal particle velocity corresponds to equal shocked material density ρ. Nevertheless, the initial density ρ_0 is smaller for metal with bubbles, which means that the shock velocity is smaller (for a given particle velocity). Moreover, a confirmation that for high particle velocity (corresponding to high pressure) the linear phenomenological relation [31] holds for pure aluminum was made.

$$D = c_0 + au, \quad c_0 = 5.35\, km/s, \quad a = 1.4 \tag{31}$$

Identical rationalization can be applied for the trends in figure 22, where heavier bubbles cause smaller shock velocity.

For the benefit of assessing the theoretical uncertainties, the calculations for the same bubble size and concentration but with different helium EOS models were repeated.

Figure 23 exhibits Hugoniot curves (temperature vs. pressure and shock velocity vs. particle velocity) for two EOS for helium (Carnahan-Starling EOS and Trinkaus EOS). Although for pressures above 20kbar the EOS's differ drastically, no significant difference in the Hugoniot curve is observed. The reason is that for low pressures, where the bubbles do affect the total EOS (as was mentioned before) the differences between the above EOS of helium are not crucial. In the shock velocity vs. particle velocity the maximal divergence is only 0.05% and thus there is no reason plotting it. This result is encouraging since it implies that the theoretical predictions presented above are not sensible to the EOS, which is the main uncertainties source in the model.

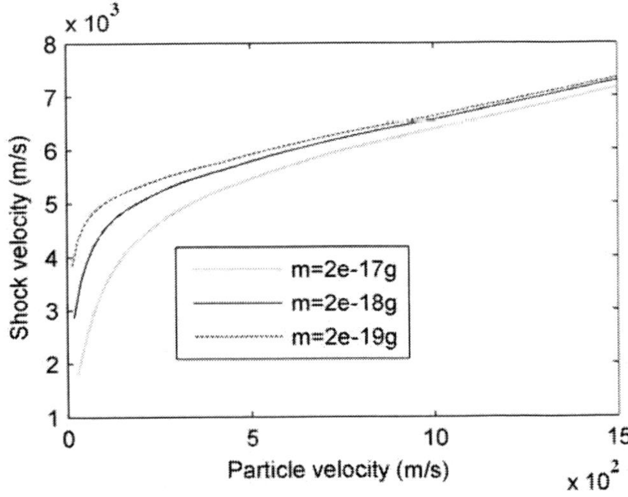

Figure 22. Shock velocity vs. particle velocity along the Hugoniot curve for aluminum with $\alpha = 5 \times 10^{-4}$ helium bubbles for several bubble masses. Helium EOS: Trinkaus

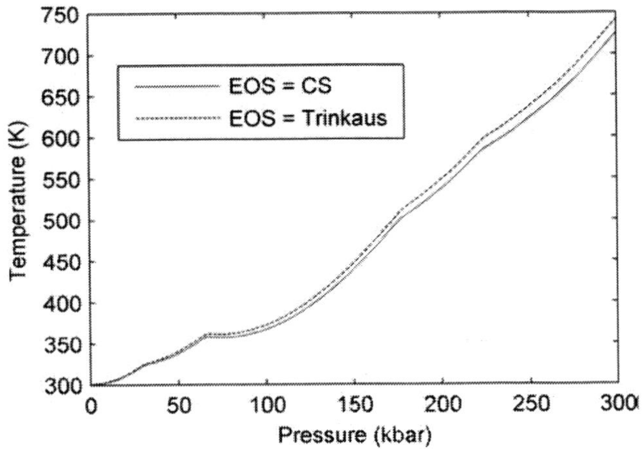

Figure 23. P-T Hugoniot curve for aluminum with $\alpha = 5\times10^{-4}$ helium bubbles for different helium EOS (Trinkaus and Carnahan Starling). Bubble mass: $m = 2\times10^{-18}gr$.

3.4. Conclusions

A new model for EOS of metal with helium bubbles was formulated. By applying this model, the Hugoniot curves at high pressures for various bubble mass, concentration and helium EOS models were investigated. The calculations predict that helium mass fraction of at least $\alpha = 5\times10^{-4}$ is required for measurable changes to occur on the Hugoniot. In addition, a significant sensitivity of the Hugoniot to the bubble mass was recognized: as the bubble size increases, its influence is more noticeable. The helium EOS, however, did not seem to have major influence on the Hugoniot calculated by the model: the maximal diversity is 2% in the P-T Hugoniot and 0.05% in the shock velocity vs. particle velocity Hugoniot.

REFERENCES

[1] Glam, B., Eliezer, S., Moreno, D. & Eliezer, D. (2009). "Helium bubbles formation in aluminum: Bulk diffusion and near-surface diffusion using TEM observations," *Journal of Nuclear Materials*, *vol. 392*, no. 3, 413-419, Aug.

[2] Glam, B., Moreno, D., Eliezer, S. & Eliezer, D. (2009). "Experimental investigation of helium migration in an fcc aluminum matrix," *Journal of Nuclear Materials*, *vol. 393*, no. 2, 230-234, Sept.

[3] Raicher, E., Glam, B., Henis, Z., Pecker, S., Eliezer, S. & Moreno, D. (2009). "Equation of state for aluminum containing helium bubbles," *Journal of Applied Physics*, *vol. 106*, no. 8, 083519-5, Oct.

[4] Trinkaus, H. & Singh, B. N. (2003). "Helium accumulation in metals during irradiation - where do we stand?," *Journal of Nuclear Materials*, *vol. 323*, no. 2-3, 229-242, Dec.

[5] Foreman, A. J. E. & Singh, B. N. (1985). "Bubble nucleation in grain interior and its influence on helium accumulation at grain boundaries," *Journal of Nuclear Materials*, *vol. 133-134*, 451-454, Aug.

[6] Marian, J., Wirth, B. D. & Perlado, J. M. (2002). "Mechanism of Formation and Growth of <100> Interstitial Loops in Ferritic Materials," *Phys. Rev. Lett.*, *vol. 88*, no. 25, 255507, June.

[7] Moreno, D. & Eliezer, D. (1996). "Structural changes in a copper alloy due to helium implantation," *Scripta Materialia*, *vol. 35*, no. 12, 1385-1389, Dec.

[8] Satou, M., Koide, H., Hasegawa, A., Abe, K., Kayano, H. & Matsui, H. (1996). "Tensile behavior of helium charged V---Ti---Cr---Si type alloys," *Journal of Nuclear Materials*, vol. 233-237, no. Part 1, pp. 447-451, Oct.

[9] Pati, S. R. & Barrand, P. (1969). "The influence of precipitates on the formation of helium bubbles in irradiated copper-boron alloys," *Journal of Nuclear Materials*, *vol. 31*, no. 1, 117-120, May.

[10] Tiwari, G. P. & Singh, J. (1990). "Effect of temperature on the growth of inert gas

bubbles in metals," *Journal of Nuclear Materials*, *vol. 172*, no. 1, 114-122, June1990.

[11] Ono, K., Furuno, S., Hojou, K., Kino, T., Izui, K., Takaoka, O., Kubo, N., Mizuno, K. & Ito, K. (1992). "In-situ observation of the migration and growth of helium bubbles in aluminum," *Journal of Nuclear Materials*, *vol. 191-194*, no. Part 2, 1269-1273, Sept.

[12] Ono, K., Arakawa, K., Hojou, K., Oohasi, M., Birtcher, R. C. & Donnelly, S. E. (2002). "Quantitative study of Brownian motion of helium bubbles in fcc metals," *J Electron Microsc (Tokyo)*, *vol. 51*, no. suppl_1, S245-S251, Mar.

[13] Nowak, S. K. (1956). *Journal of Metals*, 553-556,.

[14] Egerton, R. F., Li, P. & Malac, M. (2004). "Radiation damage in the TEM and SEM," *Micron*, *vol. 35*, no. 6, 399-409, Aug.

[15] Pells, G. P. & Phillips, D. C. (1979). "Radiation damage of [alpha]-Al2O3 in the HVEM: I. Temperature dependence of the displacement threshold," *Journal of Nuclear Materials*, *vol. 80*, no. 2, 207-214, May.

[16] Glyde, H. R. & Mayne, K. L. (1965). "Helium Diffusion in Aluminum," *Philosophy Magazine*, *vol. 12*, 997-1003.

[17] Brearley, I. R. & MacInnes, D. A. (1980). "An improved equation of state for inert gases at high pressures," *Journal of Nuclear Materials*, *vol. 95*, no. 3, 239-252, Dec.

[18] Adams, J. B. & Wolfer, W. G. (1989). "Formation energies of helium-void complexes in nickel," *Journal of Nuclear Materials*, *vol. 166*, no. 3, 235-242, Aug.

[19] Morishita, K., Sugano, R. & Wirth, B. D. (2003). "MD and KMC modeling of the growth and shrinkage mechanisms of helium-vacancy clusters in Fe," *Journal of Nuclear Materials*, *vol. 323*, no. 2-3, pp. 243-250, Dec.

[20] Morishita, K., Sugano, R., Wirth, B. D. & az de la Rubia, T. (2003). "Thermal stability of helium-vacancy clusters in iron," *Nuclear Instruments and Methods in Physics Research Section B: Beam Interactions with Materials and Atoms*, *vol. 202*, pp. 76-81, Apr.2003.

[21] Smallman, R. E. & Bishop, R. J. (1999). *Modern Physical Metallurgy and Materials Engineering*, sixth ed. Oxford: Butterworth Heinemann.

[22] Edington, J. W. (1991). *Practical Electron Microscopy in Materials Science*. New York: Van Nostrand Reinhold Co.

[23] Reed, D. J. (1977). "A review of recent theoretical developments in the understanding of the migration of helium in metals and its interaction with lattice defects," *Radiation Effects and Defects in Solids*, *vol. 31*, no. 3, 129-147.

[24] Ao, B. Y., Yang, J. Y., Wang, X. L. & Hu, W. Y. (2006). "Atomistic behavior of helium-vacancy clusters in aluminum," *Journal of Nuclear Materials*, *vol. 350*, no. 1, 83-88, Mar.

[25] Yang, L., Zu, X. T. & Gao, F. (2008). "Ab initio study of formation, migration and binding properties of helium-vacancy clusters in aluminum," *Physica B: Condensed Matter*, *vol. 403*, no. 17, 2719-2724, Aug.

[26] Barrett, C. & Massalski, T. B. (1982). *Structure of Metals*. Oxford: Pergamon.

[27] Nielsen, B. B. & Veen, A. V. (1985). "The lattice response to embedding of helium impurities in BCC metals," *Journal of Physics F: Metal Physics*, *vol. 15*, no. 12, 2409-2420.

[28] Kubota, A., Reisman, D. B. & Wolfer, W. G. (2006). "Dynamic strength of metals in shock deformation," *Applied Physics Letters*, *vol. 88*, no. 24, 241924-2, June.

[29] Wang, H. Y., Zhu, W. J., Liu, S. J., Song, Z. F., Deng, X. L., Chen, X. R. & He, H. L.

(2009). "Atomistic simulations of the elastic properties of helium bubble embedded aluminum," *Nuclear Instruments and Methods in Physics Research Section B: Beam Interactions with Materials and Atoms*, *vol. 267*, no. 5, 849-855, Mar.

[30] Wolfer, W. G. (1981). "High Density Equation of State for Helium and its Applications to bubbles in Solids," *ASTM International*. 201-202.

[31] Holian, K. S. (1984). "T-4 Handbook of Material Properties Databse: Vol. I Equation of State," LANL,LA-10160-MS.

[32] Carnahan, N. F. & Starling, K. E. (1969). "Equation of State for Nonattracting Rigid Spheres," *The Journal of Chemical Physics*, *vol. 51*, no. 2, 635-636, July.

[33] Trinkaus, H. (1983). "Energetics and formation kinetics of helium bubbles in metals," *Radiation Effects and Defects in Solids*, *vol. 78*, no. 1, 189-211.

[34] Steinberg, D. J. (1991). "Equation of State and Strength Properties of Selected Materials," Lawrence Livermore,UCRL-MA-106439.

[35] Curran, D. R., Seaman, L. & Shockey, D. A. (1987). "Dynamic failure of solids," *Physics Reports*, *vol. 147*, no. 5-6, 253-388, Mar.

[36] Barker, J. A. & Henderson, D. (1976). "What is "liquid"? Understanding the states of matter," *Rev. Mod. Phys.*, *vol. 48*, no. 4, 587, Oct.

[37] Young, D. A., McMahan, A. K. & Ross, M. (1981). "Equation of state and melting curve of helium to very high pressure," *Phys. Rev. B*, *vol. 24*, no. 9, 5119, Nov.

[38] Hirschfelder, J. O., Curtiss, C. F. & Bird, R. B. (1954). *Molecular Theory of Gases and Liquids*. New York: Wiley.

[39] Mills, R. L., Liebenberg, D. H. & Bronson, J. C. (1980). "Equation of state and melting properties of He4 from measurements to 20 kbar," *Phys. Rev. B*, *vol. 21*, no. 11, 5137, June.

[40] Zha, C. S., Mao, H. K. & Hemley, R. J. (2004). "Elasticity of dense helium," *Phys. Rev. B*, *vol. 70*, no. 17, 174107, Nov.

[41] Zeldovich, Y. B. & Raizer, Y. P. (2002). *Physics of Shock Waves and High Temperature Hydrodynamics Phenomena*. New York: Dover.

[42] Glam, B., Eliezer, S., Moreno, D., Perelmutter, L., Sudai, M. & Eliezer, D. (2010). "Dynamic fracture and spall in aluminum with helium bubbles," *International Journal of Fracture*,.

[43] Eliezer, S. & Mima, K. (2009). *Application of Laser Plasma Interaction*. New York: CRC.

In: Helium: Characteristics, Compounds…
Editors: Lucas A. Becker

ISBN: 978-1-61761-213-8

Chapter 3

RELEASE OF MANTLE HELIUM AND ITS TECTONIC IMPLICATIONS

Koji Umeda**, Atusi Ninomiya and Koich Asamori

Tono Geoscientific Research Unit, Geological Isolation Research and Development Directorate, Japan Atomic Energy Agency, Tokio, Japan

ABSTRACT

Helium is the lightest noble gas and both stable isotopes, ^{3}He and ^{4}He, are produced in the crust in a ratio of ~ 0.02 RA, with RA being the atmospheric ^{3}He/^{4}He ratio of 1.4×10^{-6}. Higher values are an indication of helium from the mantle where ^{3}He captured during planetary accretion has been stored. It has been suspected for some time that degassing of the planet does not occur homogeneously over the Earth's surface, but is rather concentrated along plate boundaries, where the dynamics of the lithosphere are more intense and mantle helium from the Earth's interior can be more easily transported to the surface. We indicate that the spatial distribution of ^{3}He/^{4}He ratios in gas samples from crustal fluids are considered to provide potentially useful information for determining not only latent magmatic activity but also potential pathways for mantle volatiles, such as in tectonically active zones.

1. INTRODUCTION

The Earth has been continually losing helium through the degassing of rocks from its interior over the past of 4.5 billion years. Any helium left on the Earth at all is largely due to its replenishment in the interior through radioactive decay. The decay of uranium and thorium produces ^{4}He, whereas ^{3}He in the Earth's mantle is not produced by radioactive decay and was only incorporated during accretion, that is, it is primordial (Porcelli and Ballentine, 2002). Since the discovery of mantle helium in oceanic waters over the East Pacific Rise in 1968 (Clarke et al., 1969), considerable efforts have been made to identify, quantify and

* Corresponding author: umeda.koji@jaea.go.jp

elucidate the variation of helium isotope ratios ($^3He/^4He_{air}$ = 1.4×10^{-6}; $^3He/^4He_{mantle}$ = 1-1.4×10^{-5}; $^3He/^4He_{radiogenic}$ = 1-5×10^{-8}: Lupton, 1983). In the most widely accepted mantle helium evolution model, the $^3He/^4He$ ratios observed in the mid-ocean ridge basalts (MORBs) (Craig et al., 1975) and ocean island basalts (OIBs) (Kurz et al., 1983) are interpreted in terms of mixing two reservoirs (e.g., Porcelli and Ballentine, 2002). OIBs are thought to sample undegassed, high $^3He/^4He$ mantle that has retained most of its ^3He since the origin of the Earth and are often associated with the lower mantle. In contrast, MORBs are thought to sample degassed, low $^3He/^4He$ mantle that has lost essentially all of its ^3He through melting (Kurz et al., 1982) and is associated with the upper mantle.

It is common knowledge that mantle degassing does not occur homogeneously over the Earth's surface. In continental settings, the flux of radiogenic ^4He from the crust has been investigated and the distribution of mantle helium and its manifestation in tectonically active areas has been mapped (e.g., Marty et al., 1989). The continental crust is a low permeability barrier to mantle degassing, and radiogenic helium production within the crust dilutes any trans-crustal mantle helium. This is reflected in the low $^3He/^4He$ ratios observed in crustal fluids from stable platforms (e.g., Torgersen and Clarke, 1987). In contrast, $^3He/^4He$ ratios higher than the typical crustal values found in volcanic regions and tectonically active areas, are interpreted to indicate that transfer of mantle volatiles into the crust by processes or mechanisms such as magmatic intrusion, continental underplating and lithospheric rifting (O'Nions and Oxburgh, 1988; Torgersen, 1993). Mantle melting is considered to be the most likely mechanism responsible for the transfer of mantle helium from the subcrustal lithosphere where the volatiles would be released directly from a magma body into the crustal fluids (e.g. Torgersen, 1993). Significantly higher $^3He/^4He$ ratios have been observed in seismically active zones compared to non-active zones (e.g., Güleç et al., 2002). Consequently, the presence or the absence of mantle helium in crustal fluids could be efficiently used to trace the inter-relationship of tectonics, magmatism and fluids circulating in active fault zones. Recently, Kennedy and van Soest (2007) proposed that helium isotope would be a new tool, without any drilling, for identifying potential geothermal energy resources arising not from volcanism but from the flow of surface fluids through deep fractures that penetrate the lower crust. In addition, temporal changes in the helium isotope ratios have been examined as a predictive indicator in studies monitoring earthquakes and/or volcanic eruptions (e.g., Sano et al., 1998; Shimoike and Notsu, 2000).

For the last decade, we have been acquiring helium isotope data from hot springs and water wells throughout the Japanese Islands, which are above active subduction zones in this island are system. These activities led to the elucidation of the geographical distribution of $^3He/^4He$ ratios, and revealed a relatively high contribution of mantle helium in tectonically active areas. It has also been recognized that helium isotope ratios in volcanic gases and in hot spring gases around active volcanoes are close to those found in MORBs (e.g., Sano and Wakita, 1985). In this contribution, we focus on the possible tectonic causes of upward release of mantle helium through the crust, with the exception of mantle helium carried by magmatic intrusion. In the following sections, we review two manifestations of mantle helium discovered in non-volcanic regions in Japan.

The first occurrence reviewed is in the Iide Mountains, composed of Mesozoic sedimentary and Late Cretaceous granitic rocks in Northeast Japan. This occurrence is attributed to newly generated magmas ascending in the present-day subduction system (Umeda et al., 2007). The other occurrence is in the source region of the 2000 western Tottori earthquake (M_j = 7.3), southwestern Japan. It is associated with a seismically active source fault acting as a preferential conduit for the transfer of mantle fluids through the crust (Umeda and Ninomiya, 2009). In the last section of this contribution, we present new helium isotope data on hot spring gas samples along the Itoigawa-Shizuoka Tectonic Line (ITSL), Central Japan, an onshore plate boundary, and also elucidate spatial variations in the $^3He/^4He$ ratios along the ISTL. Additionally, we propose an interpretation for the relationship between the helium isotope variations and geotectonic environments.

2. Release of Mantle Helium from the Mesozoic, Crystalline Iide Mountains

The Iide Mountains are located approximately 60 km west of the Quaternary volcanic front of Northeast Japan. The mountains began to uplift in the late Pliocene and the uplift rate increased in the Quaternary. The geology of the Iide Mountains is mainly composed of Mesozoic sedimentary rocks and Late Cretaceous to Paleogene granitic rocks (Figure 1). Miocene sediments, unconformably overlying the pre-Neogene basement, are distributed around the Mountains. They are comprised of terrestrial and marine sedimentary rocks, with frequent intercalations of volcaniclastic rocks and lava. Basaltic to andesitic intrusive rocks, K-Ar ages of which were reported to be 12.1 to 12.6 Ma (Umeda et al., 2007), intruded into the Late Cretaceous to Paleogene granitic rocks in the Iide Mountains. It follows that Miocene volcanism continued until around 12 Ma in this region. Although there is no indication of volcanism during the Pliocene and Quaternary, several high temperature hot springs, such as the Iide (55°C), Awanoyu (41°C) and Yunohira (56°C) hot springs have been recognized. On the basis of their chemical composition, they are either Na・Ca-Cl・SO_4・HCO_3 type or CO_2-Na-Cl・SO_4 type (Kimbara, 1992), thus similar to those hot springs in volcanic regions.

Seismic tomography and magnetotelluric soundings provide useful geophysical information on some sort of heat source causing hydrothermal activity in the crust. The updated seismic network in Japan has provided detailed, high resolution information on P and S wave velocity structures. P and S wave velocity structures in and around the Iide Mountains have been estimated at a higher spatial resolution. Figure 2 shows vertical cross-sections of V_p, V_s and V_p/V_s along the line running across the Iide Mountains in an east-west direction (A-A') (Umeda, 2009). The velocity patterns of P- and S-waves are generally similar except for the portion at depths of about 12 km. It appears that low velocity anomalies are located at depths greater than 15 km beneath the Iide Mountains for both the P- and S-waves. High V_p/V_s anomalies are also distributed at depths of more than 15 km. Seismic wave velocity varies depending on several physical conditions. Variations in fluid contents and temperature are likely to be the principal causes of the velocity anomalies observed in volcanic or geothermal regions (Nakajima and Hasegawa, 2003). On the grounds that hydrothermal activity occurs in the Iide Mountains, the seismic velocity anomaly detected is likely due to the higher temperature and the existence of magma- or a fluid-filled lower crust.

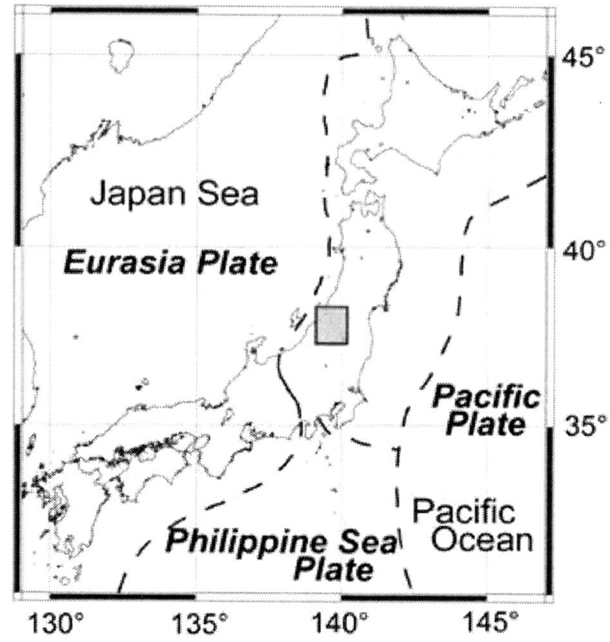

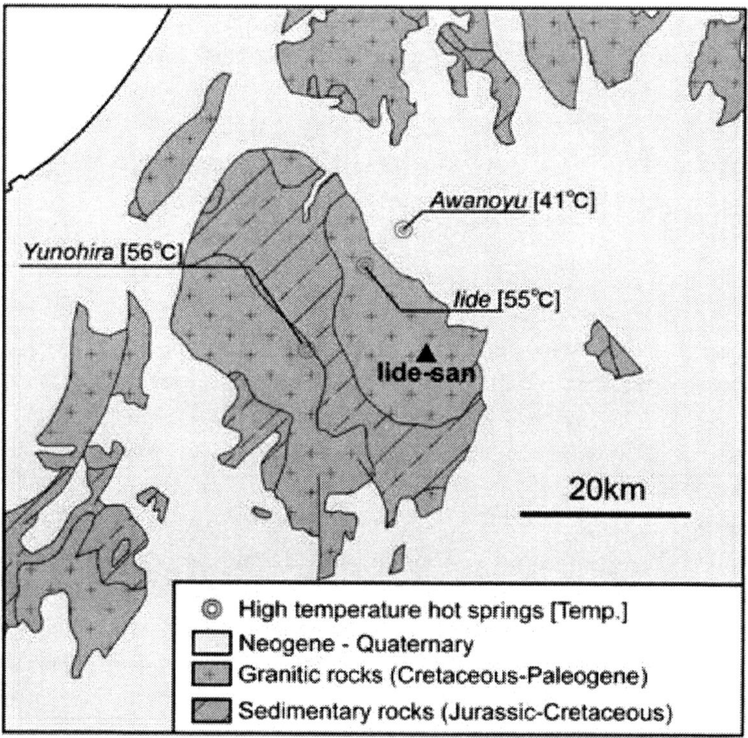

Figure 1. Simplified geological map around the Iide Mountains based on the map of Japan from the Geological Survey of Japan (1995)

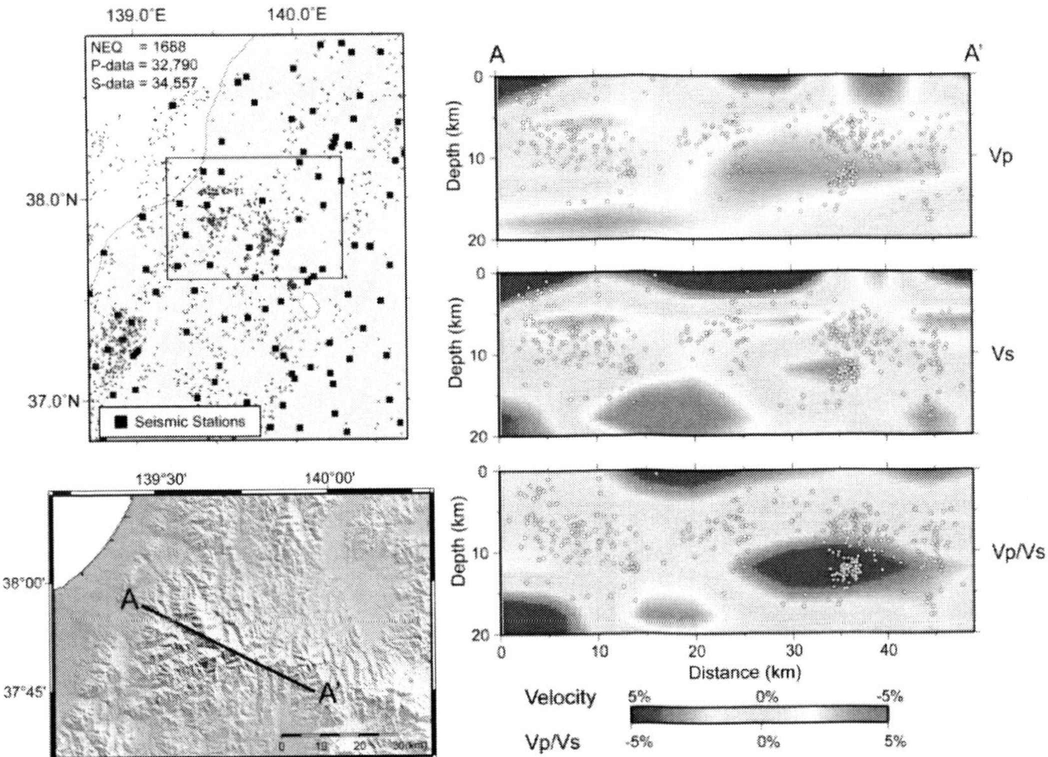

Figure 2. Vertical cross-sections of (a) P-wave, (b) S-wave and (c) V_p/V_s structures along the line A-A' across the Iide Mountains. Also shown are the locations of seismic stations and hypocenter distribution of earthquakes from the JMA catalog for this analytical area during the period from January 2002 to December 2008

Wide-band magnetotelluric soundings also detected an anomalously conductive body (< 10 Ωm) at a depth of 10-30 km beneath the Iide Mountains (Umeda et al., 2006). The conductor widens with increasing depth, and extends from the near-surface down to the base of the crust and perhaps into the upper mantle (Figure 3). There is a clear correlation between the low velocity anomaly and the conductor. The cut-off depth of crustal earthquakes is also in good agreement with the upper boundary of the conductor. It follows that the temperature of the conductor is higher than about 400 °C, because the cut-off depth is considered indicative of the boundary between the brittle and ductile zone (Hasegawa et al., 1991; Ito, 1993). The region of thinning of the brittle seismogenic layer is concordant with the distribution of the high-temperature hot springs. However, although it is clear that high-temperature materials lie in the middle-lower crust beneath the Iide Mountains, the origin of the heat source, such as hydrothermal fluids related to late remnant magmatism or a recently active shallow crustal magma system, cannot be inferred from the above geophysical data.

Table 1. Chemical and isotopic compositions of the hot spring gases and waters sampled along the ISTL. The analytical error for 4He/20Ne is ~15% of the values given. Abbreviation: ND = not determined, nd = not detected

No.	SiteName	Latitude	Longitude	Feature	$^3He/^4He(x10^{-6})$ [±1σ]			$^4He/^{20}Ne$	CO_2 (%)	N_2 (%)	CH_4 (%)	He (ppm)	Ar (%)	δD (‰)	$\delta^{18}O$ (‰)	Temp. (°C)	pH	EC (mS/m)
1	Kanakuma	36.489	137.912	Dissolved	4.18	±	0.07	1.06	ND	ND	ND	ND	ND	-79	-11.7	11.6	8.4	60
2	Sashikiri	36.448	137.981	Dissolved	1.26	±	0.02	0.67	ND	ND	ND	ND	ND	-78	-11.3	11.7	7.9	72
3	Kusayu	36.437	138.074	Dissolved	2.41	±	0.04	0.52	ND	ND	ND	ND	ND	-78	-11.8	17.5	9.1	18
4	Azuminochogatake	36.311	137.826	Dissolved	4.36	±	0.06	0.87	ND	ND	ND	ND	ND	-79	-11.4	13.7	6.5	10
5	Anazawa	36.333	138.002	Bubbles	0.97	±	0.02	11.4	6.2	1.6	92.2	24	0.03	-73	-10.1	29.7	7.6	494
6	Yamabe	36.248	138.001	Bubbles	3.44	±	0.06	2.04	nd	98.1	0.1	38	1.38	-88	-12.6	44.9	8.8	64
7	Tobira	36.185	138.085	Dissolved	1.16	±	0.02	0.75	ND	ND	ND	ND	ND	-81	-12.1	4.0	9.1	91
8	Tentoku	36.141	138.006	Dissolved	1.44	±	0.03	0.33	ND	ND	ND	ND	ND	-84	-12.1	5.8	2.7	72
9	Tyojya	36.165	137.948	Dissolved	2.27	±	0.06	4.14	ND	ND	ND	ND	ND	-80	-11.5	13.8	6.1	32
10	Ikenotaira	36.110	138.244	Dissolved	5.74	±	0.08	1.03	ND	ND	ND	ND	ND	-88	-12.8	25.0	9.2	13
11	Shimosuwa	36.077	138.091	Bubbles	6.62	±	0.10	16.3	nd	97.4	0.6	233	1.31	-83	-11.6	59.5	8.6	193
12	Kamisuwa	36.052	138.114	Bubbles	7.44	±	0.12	20.6	nd	60.1	38.5	171	0.74	-79	-11.4	60.1	8.1	83
13	Tatsunokojinyama	35.966	137.992	Bubbles	1.11	±	0.02	93.3	nd	68.3	30.7	797	0.80	-78	-11.1	35.2	7.7	156
14	Kaioizumi	35.887	138.409	Bubbles	7.68	±	0.10	58.7	35.5	63.3	0.3	286	0.57	-93	-13.3	53.1	6.8	268
15	Kobuchisawa	35.878	138.317	Bubbles	6.50	±	0.10	297	12.0	87.1	0.1	606	0.35	-86	-11.6	58.4	6.9	562
16	shiozawa	35.877	138.257	Bubbles	8.44	±	0.11	184	nd	85.3	14.1	873	0.36	-90	-12.7	33.8	9.8	274
17	Ojira	35.808	138.340	Bubbles	6.88	±	0.11	2150	nd	73.6	25.8	3110	0.18	-54	-3.9	33.6	7.7	47200
18	Mukawa	35.784	138.379	Bubbles	5.40	±	0.08	419	nd	56.5	43.1	354	0.14	-69	-8.6	39.3	7.4	1260
19	Kuromori	35.900	138.538	Bubbles	7.19	±	0.16	3.22	ND	ND	ND	ND	ND	-78	-10.6	5.2	6.8	902
20	Oyabu	35.762	138.320	Dissolved	7.39	±	0.13	11.8	ND	ND	ND	ND	ND	-91	-12.3	9.6	9.2	935
21	Kofu	35.662	138.570	Dissolved	5.64	±	0.08	33	ND	ND	ND	ND	ND	-82	-11.7	44.7	8.0	213
22	Togentenkeisen	35.651	138.410	Dissolved	0.96	±	0.03	0.52	ND	ND	ND	ND	ND	-77	-11.2	29.8	10.4	27.3
23	Nishiyama	35.554	138.307	Bubbles	4.05	±	0.05	17.6	nd	98.2	0.4	218	1.16	-83	-12.1	48.3	9.2	173
24	Momonoki	35.624	138.357	Bubbles	3.55	±	0.06	1.82	ND	ND	ND	ND	ND	-87	-12.6	41.4	9.2	106
25	Jikkoku	35.505	138.394	Dissolved	9.54	±	0.12	235	ND	ND	ND	ND	ND	-65	-9.1	28.2	9.4	790
26	Funayama	35.294	138.421	Dissolved	1.84	±	0.05	0.34	ND	ND	ND	ND	ND	-53	-8.3	9.1	7.9	38.4
27	Jyaoh	35.933	138.105	Dissolved	1.65	±	0.04	0.33	ND	ND	ND	ND	ND	-77	-11.5	9.7	7.6	11

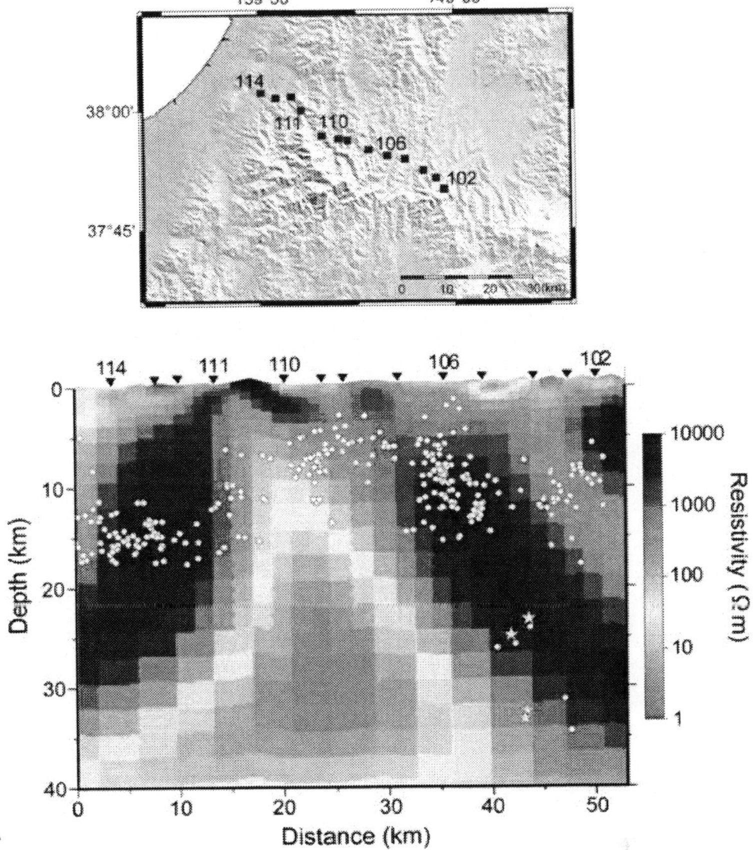

Figure 3. Two-dimensional resistivity model estimated by reanalyzing the magnetotelluric data acquired by Umeda et al. (2006). Also shown are hypocenters of earthquakes (white dots) and deep low-frequency earthquakes (stars). MT site locations are shown on the topographic map of the Iide Mountains and the surrounding regions

In order to examine the spatial relationship between the geophysical anomaly and helium isotope variations, we collected a total of 11 free gas and water samples from hot springs in and around the Iide Mountains. The $^3He/^4He$ ratios of the samples ranged from 0.22 to 7.9 RA. Most of the hot spring gases have higher $^3He/^4He$ ratios than the atmospheric ratio, that is, with remarkable differences in $^3He/^4He$ ratios compared with other regions consisting of Mesozoic sediments and Cretaceous granite away from active volcanoes. The $^3He/^4He$ ratio of arc-related volcanism in Northeast Japan rangs from 1.7 to 8.4 RA with a mean value of 5.1±1.7 RA (Hilton et al. 2002), consistent with the range of observed values in and around the Iide Mountains. Figure 4 shows the geographic distribution of the $^3He/^4He$ ratios. The hot springs having anomalously high $^3He/^4He$ ratios are located toward the central part of the Iide Mountains. In contrast, the hot springs and the gas wells having rather low $^3He/^4He$ ratios are distributed in the foothills and outside of the mountains. Consequently, the location of the geophysical anomaly correlates with the geographic distribution of hot springs with high $^3He/^4He$ ratios similar to those of mantle helium, indicating that the heat source is due to high-temperature fluid and/or melt derived from mantle materials.

The emanation of hot spring gases with high $^{3}He/^{4}He$ ratios requires the effective movement of mantle helium to the Earth's surface. Considering the apparent lack of volcanism since around 12 Ma, the potential ^{3}He sources are limited to the following: (1) fluid circulation through either the erupted Miocene basalts (~ 12 Ma) or an ancient and non-active magma chamber originally charged with mantle helium; or (2) direct addition of mantle helium from newly ascending magma in the present-day subduction system.

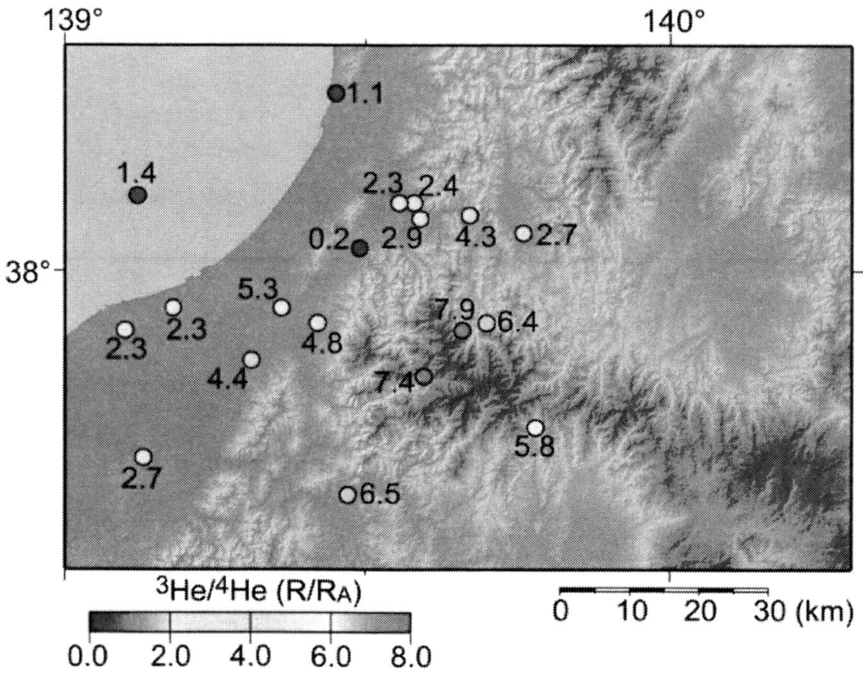

Figure 4. Geographic distribution of $^{3}He/^{4}He$ ratios of gases from hot springs and gas wells. Data reported in Umeda et al. (2007) and unpublished data are shown in this figure. Numbers indicate $^{3}He/^{4}He$ ratios as RA units

Once the basalt has erupted and/or the chamber is isolated and its volatile inventory is no longer being replenished by a mantle source, the helium isotope ratio would decrease through time due to the addition of radiogenic ^{4}He from U and Th decay, resulting in much lower ratios in the present-day (Craig and Lupton, 1976). For example, given U and Th abundances of 0.5 ppm and 1.0 ppm, respectively (Ebihara et al., 1984), and an initial He content of 1×10^{-6} cm^{3} STP/g (Mamyrin and Tolstikhin, 1984), we estimate that the present-day $^{3}He/^{4}He$ ratios in the basaltic magmas erupted and/or intruded around 12 Ma should be 5.3×10^{-6} (3.8 RA) at the present time. Therefore, $^{3}He/^{4}He$ ratios exceeding 3.8 RA observed around the Iide Mountains indicate a significant contribution of primordial ^{3}He carried by ascending magmas from the upper mantle in the present-day subduction system. In conclusion, helium isotope analyses can provide additional constraints on the origin of geophysical anomalies, such as active magma storage or a solidified high-temperature magma plug deep underground.

3. Release of Mantle Helium from the Source Region of the 2000 Western Tottori Earthquake

A magnitude (*Mj*) 7.3 crustal earthquake occurred in the western Tottori area, Southwest Japan, on 6 October 2000, where very few large earthquakes have occurred since the 1943 Tottori earthquake ($M_j = 7.2$) (Kanamori, 1972). Although this was a large, shallow, intraplate earthquake, there appear to be three characteristics that distinguish it from other common large earthquakes: (1) surface expression of a fault rupture is not clearly observed, (2) active faults are absent or unknown, and (3) the maximum shear strain rate, determined using the nationwide GPS network, is lower than in other areas. Thus, the earthquake fault is considered to be an immature fault; a fault at an early stage of evolution, deduced from the geomorphological, geological and seismological evidence (e.g., Inoue et al., 2002). From the viewpoint of seismic hazard and geological risk reduction, it is imperative to recognize such active faults lacking clear surface expression, especially in areas with a low density of active faults but where there is potential for lager magnitude earthquakes than would be expected because of the lack of any geomorphological signature.

The enrichment with a wide variety of terrestrial gases along active faults suggests that active faults may be major pathways in the crust for the gases, where materials are more porous and the permeability is relatively high (King, 1986). Walia et al. (2005) suggested that spatial variations of radon and helium concentrations in soil-gas can act as a powerful tool for the detection and mapping of active fault zones. Several researchers have made efforts to elucidate the geographic distribution of the $^3He/^4He$ ratios around active faults and their relationship to seismically active areas. Kennedy et al. (1997) attributed elevated $^3He/^4He$ ratios of up to 4 RA to transport of mantle helium through the fault structure along the San Andreas fault system. Kulongoski et al. (2005) insisted that mantle helium found in groundwaters of the Morongo Basin moved via deeply penetrating active faults, and that episodic seismicity and associated hydrofracturing drive transfer of volatiles from the mantle to the crust. In this section, the characteristics of helium isotope ratios around the source region of the 2000 western Tottori earthquake using both the data reported by Umeda and Ninomiya (2009) and unpublished data are discussed.

Figure 5 shows the spatial distribution of the air-corrected $^3He/^4He$ ratios of gases from water samples around the source region of the 2000 Western Tottori Earthquake. As mentioned in the previous section, generally, the crustal helium component is expected to be dominant in water samples from wells in regions away from active volcanoes (e.g., Sano and Wakita, 1985; Umeda et al., 2009). Figure 5 reveals that water samples collected in the aftershock zone following the 2000 earthquake are characterized by $^3He/^4He$ ratios higher than the atmospheric value. It should be noted that the sample with $^3He/^4He$ ratios about four times higher than the atmospheric value came from the well, the nearest well to the epicenter of the mainshock. Although the Yokota volcanic rocks of early Pleistocene age are found near wells far from the aftershock distribution, the $^3He/^4He$ ratios are markedly lower than the atmospheric value, indicating an insignificant contribution of mantle helium released from basaltic volcanic rocks. The $^3He/^4He$ ratios plotted as a function of distance from the main trace of the source fault has a clear trend of decreasing $^3He/^4He$ ratios with distance from the source fault of the 2000 earthquake.

Geochemical findings of elevated $^3He/^4He$ ratios observed along active faults indicate that faults may play an important role not only as major pathways but also may play a role in the triggering of earthquakes by lessening the shear strength of the source fault. However, the transfer of mantle fluids through the lower crust to the Earth's surface remains obscure because the ductile lower crust is usually considered a barrier to mantle volatiles (e.g., Hilton, 2007). Geological studies of natural ductile shear zones formed beneath brittle shear zones offer the key to an understanding of the transfer of mantle fluids in the lower crust. There appears to be a clear tendency for fluid to flow into and along shear zones (e.g., Kerrich et al., 1984). For example, hydrous minerals are localized in the vicinity of ductile shear zones developed in the host anhydrous lithology (Yoshino, 2002). Numerical experiments studying pressure distributions in plastic and viscous media also demonstrated that brittle shear zones are at lower pressures compared to the surrounding rocks, whereas ductile zones are at higher pressures (Mancktelow, 2006). These results lead to the conclusion that ductile shear zones, connected to deep zones with near-lithostatic, pressurized fluids facilitate migration of fluids upward to the base of the seismogenic regime. Accordingly, the transfer of mantle helium to the shallow aquifer is considered to be efficient along highly permeable rupture zones. Near the Earth's surface, these fluids then mixed with groundwater derived from meteoric water, but without significant dilution by crustal radiogenic helium, resulting in the emanation of groundwaters with high $^3He/^4He$ ratios along the trace of the source fault. In addition, we suggest that helium isotopes can be regarded as a tool for investigating and/or mapping concealed active faults with no surface expression.

4. Variations in ^{3}He/^{4}He Ratios Along the ISTL and its Tectonic Implications

4.1. Geological Background

The ISTL, a fault zone about 250 km long (Figure 6), is the most active among these faults in terms of long-term slip rates (Ikeda and Yonekura, 1986) and cyclicity of faulting (Okumura, 2001). Observational data on the surface deformation field obtained from the nationwide GPS earth observation network revealed ongoing crustal activity around the ISTL, with horizontal shortening at rates of $\sim 0.3\times10^{-6}$ strain/year (Sagiya et al., 2002). The ISTL is considered to be an onshore plate boundary between the North American (or Okhotsuk) and the Eurasian (Amurian) plates (Seno et al., 1996). The ISTL fault system can be divided into three distinct segments; northern, middle, and southern. The northern segment of the ISTL north of Suwa lake is recognized to be composed of east-dipping reverse faults extending about 60 km in length. The middle ISTL around Suwa lake consists of left-lateral strike-slip faults about 50 km long, trending NW-SE, and the southern segment of the ISTL is characterized by west-dipping reverse faults extending over a length of 50 km.

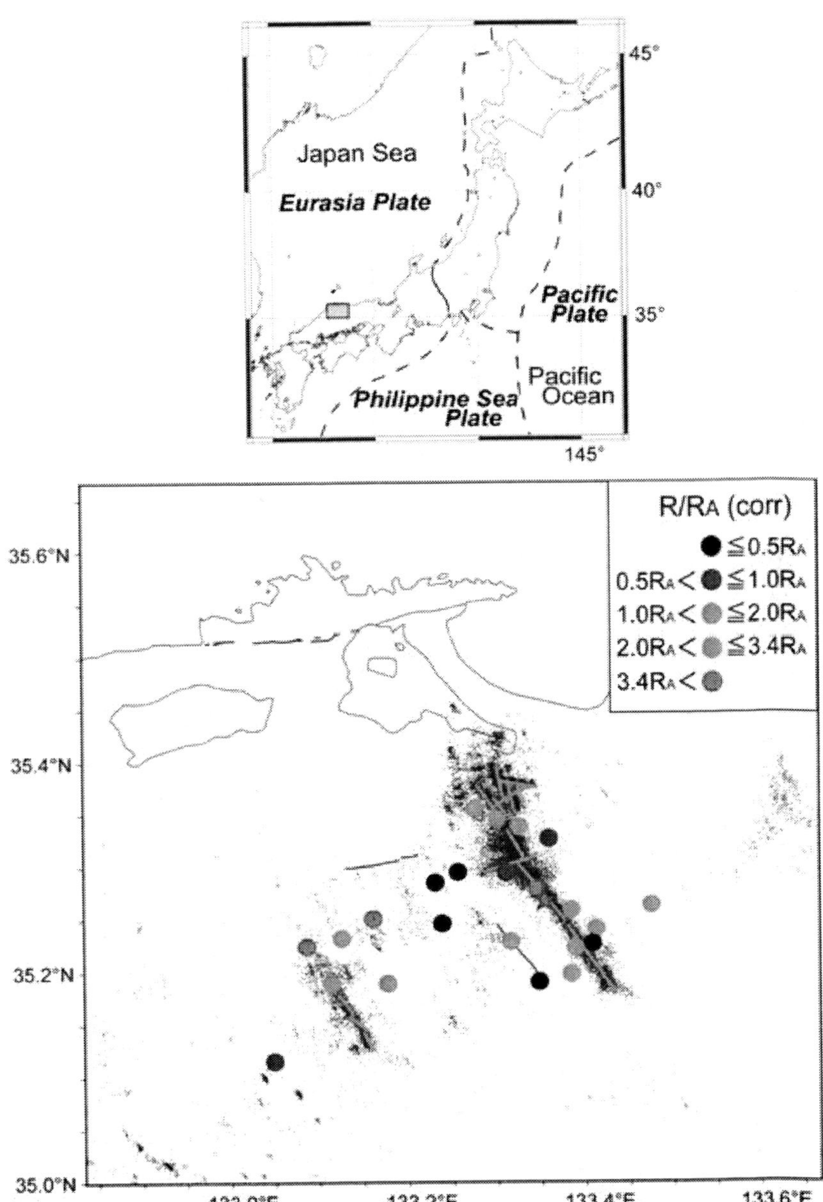

Figure 5. Geographic distribution of the air-corrected $^3He/^4He$ ratios in groundwaters around the seismic source region. Data reported in Umeda and Ninomiya (2009) and unpublished data are shown in this figure. Red stars and small black dots denote the epicenters of the mainshock and aftershock events (January 2000 to December 2003), respectively. Also shown in this figure are source fault segments (orange lines) activated by the mainshock, which was resolved by Fukuyama et al. (2003). Dark blue lines denote active faults identified by Nakata and Imaizumi (2002)

The ISTL represents the southwestern boundary the northern Fossa Magna rift basin to the north, and is the boundary between the accretionary prism of southwest Japan and the Izu-Bonin volcanic ridge to the south. The northern Fossa Magna is a rifted basin (half graben) formed as a back-arc spreading of the Japan Sea in Early to Middle Miocene. The master fault that formed the northern Fossa Magna was estimated to be an east-dipping, low-angle normal fault system developed in an E-W extensional stress field (e.g., Sato et al., 2004). This rifted basin is filled with a thick pile of sediments and volcanic rocks with a thickness of more than 6 km (Kato, 1992). Since the early Pleistocene, this low-angle fault has reactivated as an reverse fault under an E-W compressional stress regime, so that the basin sediments are on the hanging wall side (Sato et al., 2004). In the southern ISTL, the fore-arc sediments of Honshu arc and the sediments deposited on the Izu-Bonin volcanic ridge have been strongly deformed by the collision of the two arcs. Thus, the southern ISTL was produced by collisional processes (Kano et al., 1990) and its nature and geometry are very different from the northern one. The hanging wall of the main active strand of the southern ISTL lies horizontally over Pliocene to Quaternary basin fill, forming a west-dipping, low-angle active nappe (Ikeda et al., 2009).

4.2. Analytical Procedures

A total of 27 gas and water samples were collected from hot springs wells along the ISTL (Figure 6). Gas samples were collected in glass sample containers with vacuum cocks at both ends. The gas was introduced into the evacuated container by water displacement using an inverted funnel and an injection syringe. Details of the sample collection methods can be found in Nagao et al. (1981). When there were no visible bubbles, water samples were also collected to measure the isotopic ratios of dissolved gases in the water. The dissolved gases were expelled from the solution by ultrasonic agitation and collected in a glass bottle.

Major components of the gas samples were determined using gas chromatography with Ar and O_2 as carrier gases. In order to evaluate the concentration of He, Ar, O_2 and N_2, a TCD detector was used, while CH_4 and CO_2 concentrations were determined using a FID detector coupled with a methanizer. Estimated measurement uncertainty is $\pm$ 0.01 % for major components, and $\pm$ 0.001 % for minor components. Stable isotope ratios (D/H and $^{18}O/^{16}O$) of sampled hot spring waters and gases were determined with a Micromass Optima isotope ratio mass spectrometer (IRMS), using the zinc metal reduction method of Coleman et al. (1982) and the CO_2-H_2O equilibrium method of Yoshida and Mizutani (1986). Values of hydrogen and oxygen isotopes in samples are expressed in δ ‰ vs. SMOW (standard mean ocean water). Estimated total uncertainty of measurement is $\pm$ 1.0 ‰ for δD, and $\pm$ 0.1 ‰ for $\delta^{18}O$.

The isotopic ratios of He and Ne were measured using a modified VG5400 mass spectrometer at the Laboratory for Earthquake Chemistry, University of Tokyo. The ^3He and ^4He ion beams were detected on a double collector system: ^3He by axial counting and ^4He by the high Faraday collector, (feedback resistor = 10 GΩ). A resolving power of 600 allowed the complete separation of the $^3He^+$ beam from the H^{3+} and HD^+ beams. Measured $^3He/^4He$ ratios are normalized to a standard $^3He/^4He$ gas ($^3He/^4He=1.71\times10^{-4}$) prepared and stored in a stainless steel container on the inlet line. Neon was released from a cryogenic trap at 45 °K, and blank levels of ^4He and ^{20}Ne determined were about 2×10^{-10} cm^3 STP and 3×10^{-10} cm^3 STP, respectively. These blank levels represent less than 0.1 % of the amount of sampled gases, so blank correction was not required (e.g., Aka et al., 2001).

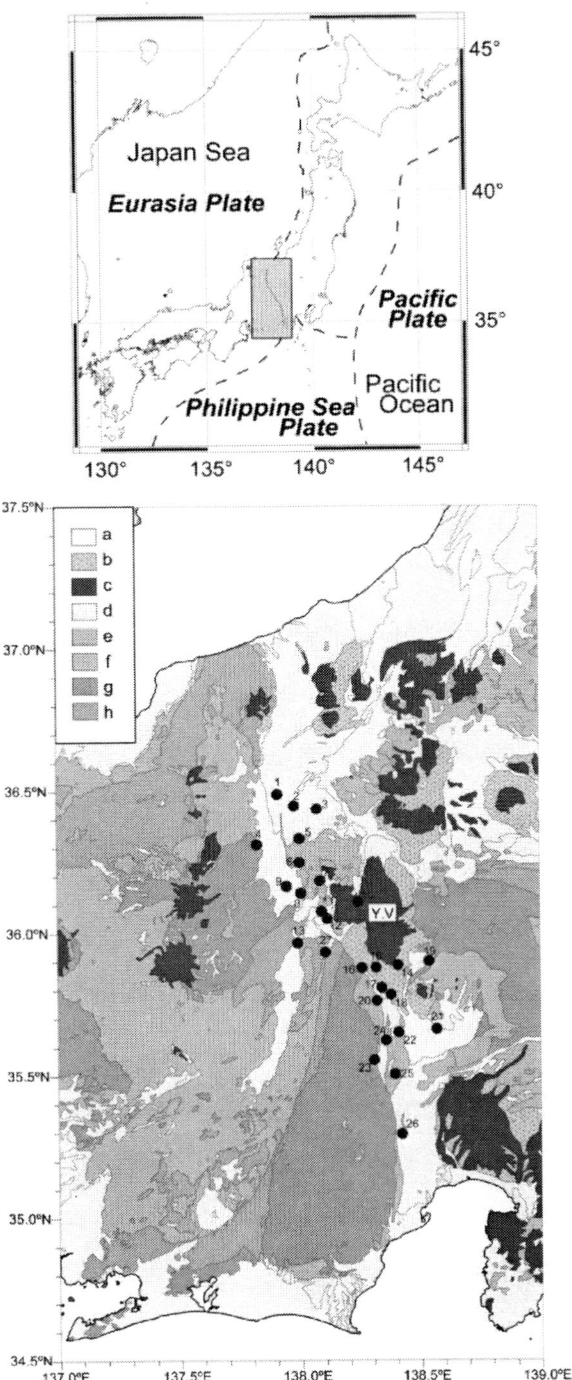

Figure 6. Simplified geological map along the ISTL based on Geological Survey of Japan (1995) and localities of hot springs sampled in this study indicated with sample numbers that correlate with Table 1. a, Quaternary sediments; b, debris; c, Quaternary volcanic rocks; d, Middle Miocene to Pliocene volcanic and sedimentary rocks; e, Middle to Late Miocene volcanic and sedimentary rocks; f, Miocene to Late Cretaceous igneous rocks; g, Mesozoic sedimentary rocks; h, Mesozoic metamorphic rocks; Y.V., Yatsugatake volcano group.

4.3. Chemical and Isotopic Composition

The analytical results for the hot spring samples are presented in Table 1. Oxygen and hydrogen isotopes can be used to estimate the origin of the hot spring waters. The δD - $\delta^{18}O$ relationship in the hot spring waters is shown in Figure 7. The isotopic trend of hot springs is similar to the meteoric water line ($\delta D = 8 \times \delta^{18}O + 10$) (Craig, 1961). This suggests that either the water had a short residence time and little chance to interact with the local geology, or that the waters have not reached high enough temperatures to initiate significant water/rock interaction. The water sample from the Ojira well (Table 1, Site 17) is the farthest off the meteoric water line, with δD = -54 ‰ and $\delta^{18}O$ = -3.9 ‰. The plot of δD and $\delta^{18}O$ from the Ojira Site connect to the meteoric water line on a trending through the Mukawa (Site 18) and the Kuromori (Site 19) wells, which are located near the Quaternary Yatsugatake volcano group (Kaneoka et al., 1980). This trend could reveal mixing between meteoric water and a magmatic vapor component (Giggenbach, 1992).

Along the ISTL, N_2 is the predominant gas constituent in the springs, followed by CH_4. Note also, that the He concentrations in hot spring gases from the Ojira well are extremely high, exceeding 3000 ppm. In Figure 8, a more meaningful Ar-He(x10)-N_2(/100) triangular diagram (Giggenbach et al., 1983) is shown. This figure shows the relative Ar, He and N_2 content of end-member compositions of gases derived from: (i) pure air (N_2/Ar = 83), (ii) air-saturated waters (ASW) at 20 °C (N_2/Ar = 38), (iii) gases directly deriving from arc-type magma chambers (N_2/Ar = 800-10,000) (Giggenbach et al., 1996) and (iv) gases typically

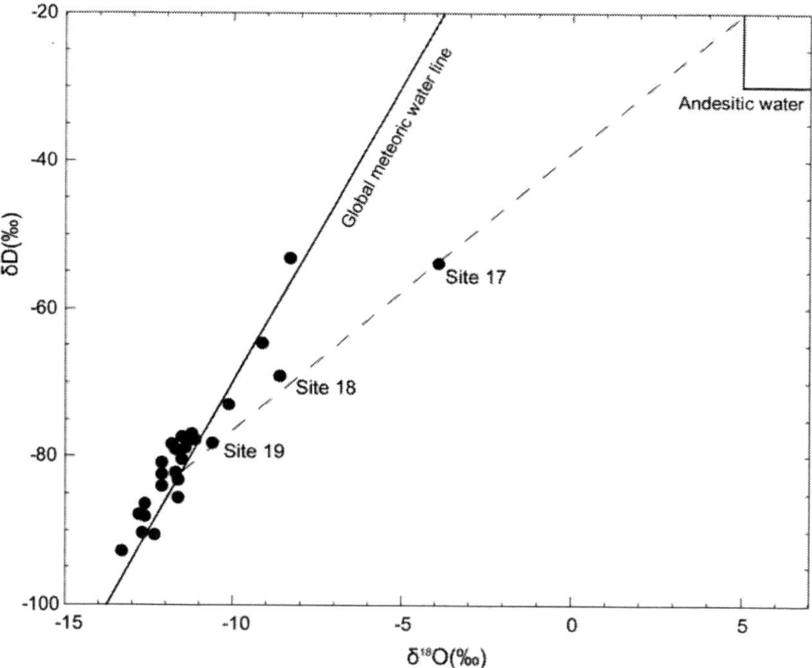

Figure 7. δD - $\delta^{18}O$ relationship of groundwaters around the seismic source region. The global meteoric water line (Craig, 1961) is shown in this figure

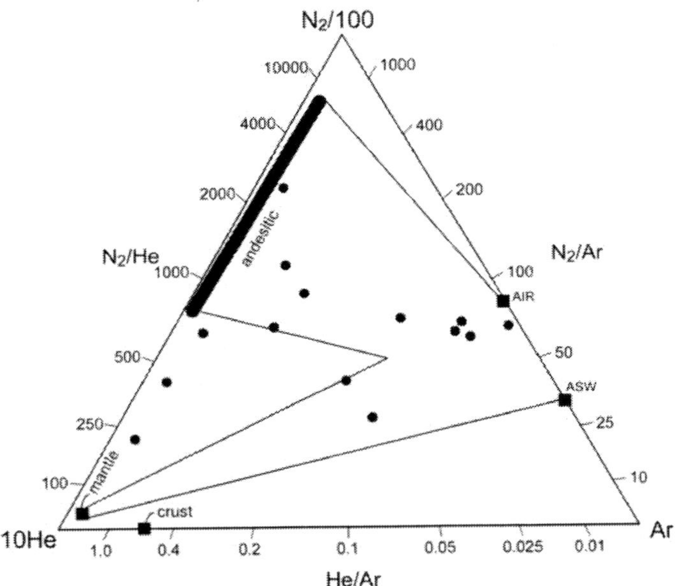

Figure 8. Relative N_2-He-Ar composition of gases from hot springs along the ISTL. Solid squares represent air, air-saturated water (ASW), crust and mantle composition. Data on mantle- and crust-derived gases and typical arc-related gases are from Giggenbach (1996).

generated in the crust and/or the mantle. The N_2/He and He/Ar values of these samples indicate either arc- or mantle and/or crust derived gases that have shifted toward ASW or air. Gas samples collected from the Kobuchisawa (Site 15), Shiozawa (Site 16) and Mukawa (Site 18) wells are relatively enriched in arc-type gas. This indicates that the N_2/Ar ratios are consistent with the δD - $\delta^{18}O$ relationship.

The $^3He/^4He$ ratios of the samples range from 0.96×10^{-6} to 9.54×10^{-6}, independent of the helium concentration in each gas sample. Figure 9 shows the relationship between the measured $^3He/^4He$ and $^4He/^{20}Ne$ ratios. The contribution of relative mantle helium can be estimated by a simple three-component mixing model (Sano and Wakita, 1985). Assuming that helium in the gas samples is composed of mantle, crustal and atmospheric helium component, we calculated the relative contribution of these three types of helium using the following equations:

$$(^3He/^4He) = (^3He/^4He)_m \times M + (^3He/^4He)_c \times C + (^3He/^4He)_a \times A \qquad (1)$$

$$1/(^4He/^{20}Ne) = M/(^4He/^{20}Ne)_m + C/(^4He/^{20}Ne)_c + A/(^4He/^{20}Ne)_a \qquad (2)$$

$$\text{where; } M + C + A = 1 \qquad (3)$$

and with M, C and A representing the contributions of mantle, crust and atmospheric helium components, respectively, and subscripts m, c and a denote mantle, crust and atmospheric ratios, respectively. In this calculation, we used $(^3He/^4He)_m = 11\times10^{-6}$, $(^3He/^4He)_c = 1.5\times10^{-8}$, $(^3He/^4He)_a = 1.4\times10^{-6}$, $(^4He/^{20}Ne)_m = 1000$, $(^4He/^{20}Ne)_c = 1000$, and $(^4He/^{20}Ne)_a = 0.32$, respectively. The mantle helium is estimated to be 2.4% to 77% of the total helium content of the gas samples.

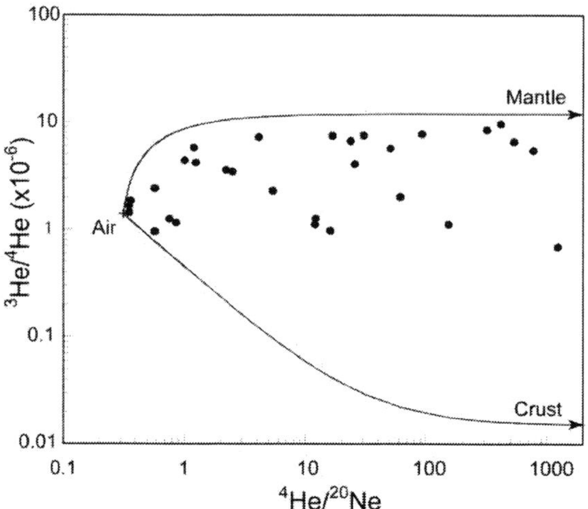

Figure 9. $^3He/^4He$ versus $^4He/^{20}Ne$ diagram for the hot spring gases along the ISTL

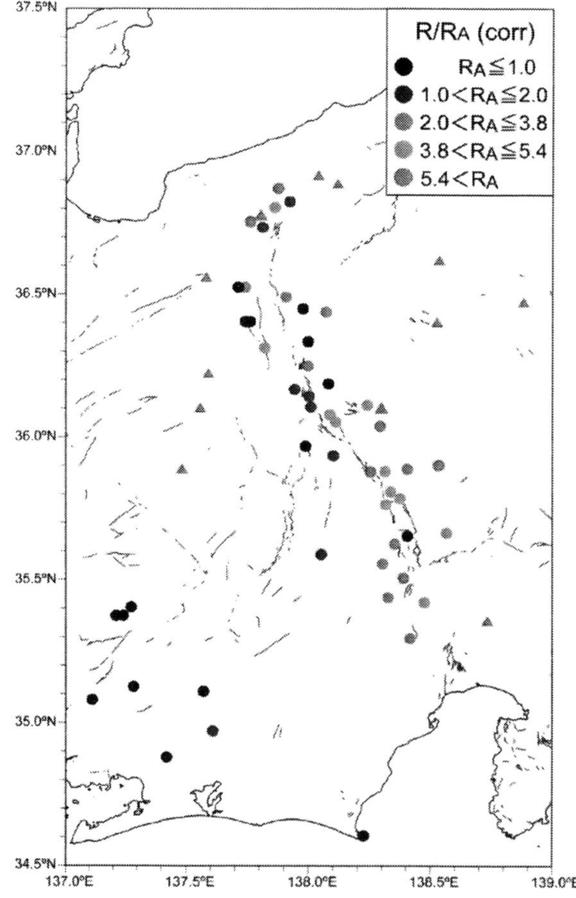

Figure 10. Geographic distribution of the air-corrected $^3He/^4He$ ratios in hot spring gases around the ISTL. Also shown are active volcanoes (red triangles) and active faults (dark blue lines) (Nakata and Imaizumi, 2002)

4.4. Geographic Distribution of $^{3}He/^{4}He$ Ratios and Their Relationship to Tectonics

The geographic distribution of $^{3}He/^{4}He$ ratios in terrestrial gases or waters may reveal mantle-crust interactions in different tectonic provinces, and the occurrence and distribution of conduits for the mantle volatiles within the crust (e.g., Hilton, 2007). Figure 10 shows the spatial distribution of the air-corrected $^{3}He/^{4}He$ ratios along the ISTL. Data from hot springs around the ISTL reported in Sano et al. (1982), Sano and Wakita (1985), Kusakabe et al. (2003), and Umeda et al. (unpublished data) are also shown in this figure. In the case of dissolved water samples, air-corrected values are obtained by assuming that all the neon in the samples is of atmospheric origin and dissolved in air-saturated water ($^{4}He/^{20}Ne = 0.254$ at 15 °C).

It should be noted that hot spring gases collected around the middle ISTL are characterized by $^{3}He/^{4}He$ ratios higher than the mean $^{3}He/^{4}He$ ratio of 5.4±1.9 RA of arc-related volcanism around the world (Hilton et al., 2002). As shown in Figure 6, the Quaternary age Yatsugatake volcano group are bordered to the west by the middle ISTL. Consequently, anomalously high $^{3}He/^{4}He$ ratios near Quaternary volcanoes indicate a significant contribution of primordial ^{3}He carried by ascending magmas from the upper mantle in the present-day subduction system. Recent seismic tomography studies revealed the existence of a low velocity zone beneath the Yatsugatake volcano group interpreted as a magmatic conduit by Panayotopoulos et al. (2008). Similarly, high $^{3}He/^{4}He$ ratios in hot spring samples collected in the northernmost ISTL, where the Quaternary age Shirouma-Oike volcano is located (Sakuyama, 1979), also indicate a relatively high contribution of magmatic helium. These results are in concordance with insight from analysis of stable isotope ratios in water samples and major components of gas samples. The magmatic influence on the $^{3}He/^{4}He$ ratio in groundwater could be limited in the range of several kilometers to 10 or so kilometers radius from the centers of Quaternary volcanoes (e.g., Sano et al., 1990; Sakamoto et al., 1992).

As discussed above, the crustal helium component is expected to be dominant in samples from wells in regions away from active volcanoes (e.g., Sano and Wakita, 1985). That is, the $^{3}He/^{4}He$ ratios would be lower than the atmospheric value owing to the addition to groundwater with radiogenic helium from the decay of uranium and thorium in crustal rocks over geologic time. Along the northern ISTL, most of samples have $^{3}He/^{4}He$ ratios similar or even slightly lower than the atmospheric value. Rather high $^{3}He/^{4}He$ ratios from hot spring wells, such as the Kanakuma (Site 1), Kusayu (Site 3), and Yamabe (Site 6) wells, dug into the Middle Miocene to Pliocene volcanic and sedimentary formations of the northern Fossa Magna might be explained by the admixture of primordial helium in mafic igneous rocks. However, this model cannot explain the elevated $^{3}He/^{4}He$ ratio of up to 4 RA collected from the Azuminochogatake well (Site 4) dug into Mesozoic sedimentary rocks. Along the southern ISTL, the observed $^{3}He/^{4}He$ ratios of samples are significantly higher than those from the northern ISTL. Although the Jikkoku well (Site 25) is located away from Quaternary volcanoes, the $^{3}He/^{4}He$ ratio is much higher than 3.8 RA. This value is the expected $^{3}He/^{4}He$ ratio at the present time for basaltic intrusions of pre-Quaternary age in northeast Japan, allowing for post-intrusive radiogenic ingrowth of ^{4}He by decay of uranium and thorium in the ancient magma (see Section 2).

The potential interpretation of significant ^{3}He enrichment is thought to be due to either nucleogenic ^{3}He produced by the decay of higher than average concentrations of ^{6}Li, or contamination of groundwater by the decay of tritium from nuclear bomb tests. The ^{3}He/^{4}He production rate in the crust is expected to vary according to the abundance of the α-producing isotopes (^{232}Th, ^{235}U, ^{238}U) and those nuclides involved in ^{6}Li (n, α), ^{3}H (β^-) and ^{3}He reactions. The production rate was estimated to be 10^{-7} (~0.1 RA) for average granitic crust (Morrison and Pine, 1955). Consequently, nucleogenic ^{3}He from the decay of ^{6}Li cannot account for the high ^{3}He/^{4}He ratios in the groundwaters. With respect to tritiogenic helium, assuming an average tritium concentration of 5.7 T.U. from the precipitation in Japan between 1970's and the 2000's (Yabusaki et al., 2003), the accumulated amount of tritiogenic helium by complete decay of 5.7 T.U. would produce 1.4×10^{-14} cm^3 STP /g H_2O (e.g., Schlosser et al., 1989). Thus, tritiogenic helium could not be a significant contributor to the total ^{3}He of hot spring gases, considering that ^{4}He concentration is 4.9×10^{-8} cm^3 STP /g H_2O in air-saturated water (Ozima and Podosek, 2002). As mentioned previously, it has been recognized that elevated ^{3}He/^{4}He ratios observed in seismically active areas are associated with faulting that might trigger upward transport of mantle helium from the subcrustal lithosphere. Therefore, the observations described above are indications that both the northern and the southern segments of the ISTL may be associated with leakage of mantle helium within the crust to the Earth's surface.

The slip-rates of the ISTL have been estimated based on seismic reflection profiles, drill core data and analysis of tectonic geomorphology. The northern segments of ISTL, with dip angles of ~30°, exhibits constant slip-rates in the range of 2.0 to 5.8 m/kyr since the middle Pleistocene (Matsuta et al., 2004). In contrast, the southern ISTL has been estimated to have slip rates as high as 7.5 to 11 m/kyr during the late Quaternary time, based on the deformation of fluvial terraces (Ikeda et al., 2009). Earthquakes in the crust are generally thought to arise from the frictional instability of existing faults induced by coupling of increased hydraulic pressure and tectonic shear stress (Sibson et al., 1992). In particular, low-angle thrusting requiring extremely low shear resistance is thought to be associated with the occurrence of highly overpressured fluids (Hubbert and Rubey, 1959). "Fault-valve" behavior proposed by Sibson et al. (1992) depends on the likelihood that fault permeability is highest immediately after faulting, so that an earthquake rupture transecting overpressurized fluids in the lower crust induces postseismic discharge along the rupture zone. If the interval of fault recurrence in the southern ISTL are shorter than those in the northern ISTL, the reason for rather high ^{3}He/^{4}He ratios observed in the southern ISTL compared to the northern part may be ascribed to the relatively more effective transfer of mantle fluid to the surface due to more frequent development of higher permeability pathways.

In addition, the southern ISTL is a collision zone where the Izu volcanic ridge is impinging into Central Japan. It has been recognized that there is a large seismic gap associated with the Philippine sea slab in the region north to northeast of the Izu Peninsula. Ishida (1992) proposed that the gap reflects the splitting of the Philippine Sea slab into two. However, although recent seismic tomography by Nakajima and Hasegawa (2007) confirmed the existence of a continuous, high-velocity seismic zone corresponding to the Philippine Sea plate from west to east across the north of the Izu Peninsula at depths of 70 to 80 km, the amplitude of the high-velocity zone appears to be weak below the Izu volcanic ridge. Nakajima and Hasegawa (2007) argued that fluids derived from the Pacific Sea slab could migrate upward through fracture zone and/or conduits produced within the fossil volcanic ridge, and thus contributed to the lower the seismic velocity. If so, an extremely high flux of aqueous fluids through the Izu-Bonin volcanic ridge may result in a large contribution of mantle helium in the southern ISTL, a possible reason for the rather high $^3He/^4He$ ratios observed there.

CONCLUDING REMARKS

A great deal of effort has been made on obtaining information about $^3He/^4He$ ratios in gas samples from fumaroles, natural gas wells, hot spring and drinking wells in Japan. Generally, fumarolic and hot spring gases associated with active volcanoes have $^3He/^4He$ ratios higher than the atmospheric value of 1.4×10^{-6}, indicating a significant contribution of primordial ^3He carried by ascending magmas from the upper mantle. However, anomalously elevated $^3He/^4He$ ratios are observed in the Iide Mountains and the source region of 2000 western Tottori earthquake away from active volcanoes. The former plausible ^3He source is attributed to mantle helium derived from the latent crustal magma storage associated with newly ascending magmas in the present-day subduction system. The latter source can be interpreted as aqueous fluids with mantle helium migrating upward along active faults. Newly developed helium isotope data along the ISTL also reveals probable mantle degassing through the onshore plate boundary between the North American and the Eurasian plates. Accordingly, the spatial distribution of $^3He/^4He$ ratios in gas samples from crustal fluids are considered to provide potentially useful information for determining not only latent magmatic activity but also potential pathways for mantle volatiles, such as in tectonically active zones.

ACKNOWLEDGMENTS

We would like to thank K. Nagao, University of Tokyo, for helping with the helium isotope analyses. We also appreciate the constructive comments given by G. F. McCrank, an ex-JAEA international Fellow.

REFERENCES

Aka, FT; Kusakabe, M; Nagao, K; Tanyileke, G. *Appl. Geochem.,* 2001, 16, 323-338.

Clarke, WB; Beg, MA; Craig, H. *Earth Planetary Sci. Lett.*, 1969, 6, 213-220.

Coleman, ML; Shepherd, TJ; Durham, JJ; Rouse, JE; Moore, GR. *Anal. Chem.,* 1982, 54, 993-995.

Craig, H. *Science,* 1961, 133, 1833-1834.

Craig, H; Lupton, JE. *Earth Planet. Sci. Lett.,* 1976, 31, 369-385.

Craig, H; Clarke, WB; Beg, MA. *Earth Planet. Sci. Lett.,* 1975, 26, 125-132.

Ebihara, M; Nakamura, Y; Wakita, H; Konda, T. *Geochem. J.,* 1984, 18, 287-295.

Fukuyama, E; Ellsworth, WL; Waldhauser, F; Kubo, A. *Bull. Seismol. Soc. Am.,* 2003, 93, 1468-1478.

Geological Survey of Japan, *Geological Map of Japan, 1, 1,000,000 3rd Edition CD-ROM Version, Digital Geoscience Map G-1*; Geol. Surv. Jpn.: Tsukuba, 1995.

Giggenbach, WF. *Earth Planet. Sci. Lett.,* 1992, 113, 495-510.

Giggenbach, WF. Monitoring and mitigation of Volcano Hazards; Scarpa, R; Tilling, R. I; Ed; *Springer Verlag: Berlin,* 1996, 222-256.

Giggenbach, WF; Gonfiantini, R; Jangi, BL; Truesdell, AH. *Geothermics,* 1983, 12, 199-222.

Güleç, N; Hilton, DR; Mutlu, H. *Chem. Geol.,* 2002, 187, 129-142.

Hasegawa, A; Zhao, D; Hori, S; Yamamoto, A; Horiuchi. S. *Nature*, 1991, 352, 683-689.

Hilton, DR. *Science*, 2007, 318, 1389-1390.

Hilton, DR; Fischer, TP; Marty, B. *Noble Gases in Cosmochemistry and Geochemistry*; In: D; Porcelli, CJ; Ballentine, R; Wieler, Ed; Reviews in Mineralogy and Geochemistry 47; Mineral. Soc. Am: Washington DC., 2002, 319-370.

Hubbert, MK; Rubey, WW. *Bull. Geol. Soc. Am.,* 1959, 70, 115-205.

Ikeda, Y; Yonekura, N. *Bull. Dep. Geogr., Univ. Tokyo,* 1986, 18, 49-63.

Ikeda, Y; Iwasaki, T; Kano, K; Ito, T; Sato, H; Tajikara, M; Kikuchi, S; Higashinaka, M; Kozawa, T; Kawanaka, T. *Tectonophys.,* 2009, 472, 72-85.

Inoue, D; Miyakoshi, K; Ueta, K; Miyawaki, A; Matsuura, K. *J. Seismol. Soc. Jpn.*, 2002, 54, 557-573.

Ishida, M. *J. Geophys. Res.,* 1992, 97(B1), 489-513.

Ito, K. *Tectonophys*., 1993, 217, 11-21.

Kanamori, H. *Phys. Earth Planet. Interiors,* 1972, 5, 426-434.

Kaneoka, I; Mehnert, H; Zashu, S; Kawachi, S. *Geochem. J.,* 1980, 14, 249-257.

Kano, K; Kosaka, K; Murata, M; Yanai, S. Tectonophys., *1990*, 176, 333-354.

Kato, H. *Bull. Geol. Surv. Jpn.,* 1992, 43, 1-30.

Kennedy, BM; van Soest, MC. *Science*, 2007, 318, 1433-1436.

Kennedy, BM; Kharaka, YK; Evans, WC; Ellwood, A; DePaolo, DJ; Thordsen, J; Ambats, G; Mariner, RH. *Science,* 1997, 278, 1278-1281.

Kerrich, R; Latour, TE; Willmore L. *J. Geophys. Res.*, 1984, 89(B6), 4331-4343.

King, CY. *J. Geophys. Res.,* 1986, 91(B12), 12269-12281.

Kimbara, K. *Distribution map and catalogue of hot and mineral springs in Japan,* ; Geol. Surv. Jpn.: Tsukuba, 1992, 394.

Kulongoski, JT; Hilton, DR; Izbicki, JA. *Geochimica et Cosmochimica Acta,* 2005, 69, 3857-3872.

Kusakabe, M; Ohwada, M; Satake, H; Nagao,K; Kawasaki, I. In *Volcanic, Geothermal, and Ore-Forming Fluids: Rulers and Witnesses of Processes within the Earth*; SF; Simmons, I; Graham, Ed; Society of Economic Geologists Special Publication 10; Economic Geol. Pub. Co.: Littleton, (CO), 2003, 75-89.

Kurz, MD; Jenkins, WJ; Hart, SR. *Nature,* 1982, 297, 43-47.

Kurz, MD; Jenkins, WJ; Hart SR; Clague, D. *Earth Planet. Sci. Lett.,* 1983, 66, 388-406.

Lupton, JE. *Annu. Rev. Earth Planet. Sci.,* 1983, 11, 371-414.

Mamyrin, BA;Tolstikhin, IN. *Helium isotopes in nature*, Elsevier: Amsterdam, 1984, 273.

Mancktelow, NS. *Geology,* 2006, 34, 345-348.

Matsuta, N; Ikeda, Y; Sato, H. *Earth Planet. Space,* 2004, 56, 1323-1330.

Marty, B; O'Nions, RK; Oxburgh, ER; Martel, D; Lombardi, S. *Tectonophys.,* 1992, 206, 71-78.

Morrison, P; Pine J. *Ann. N. Y. Acad. Sci.,* 1955, 62, 69-92.

Nagao, K; Takaoka, N; Matsubayashi, O. *Earth Planet. Sci. Lett.,* 1981, 53, 175-188.

Nakajima, J; Hasegawa, A. *J. Volcanol. Geotherm. Res.*, 2003, 127, 1-18.

Nakajima, J; Hasegawa, A. *J. Geophys. Res.,* 2007, 112, B08306, doi:10.1029/2006JB004770.

Nakata, T; Imaizumi, T. *Digital active fault map of Japan (DVD-ROM)*; University of Tokyo Press: Tokyo, 2002, 60.

Okumura, K. *J. Seismol.,* 2001, 5, 411-431.

O'Nions, RK; Oxburgh, ER. *Earth Planet. Sci. Lett.,* 1988, 90, 331-347.

Ozima M; Podosek FA. *Noble Gas Geochemistry (Second Edition)*; Cambridge University Press: Cambridge, 2002, 286.

Panayotopoulos, Y; Hirata, N; Sato, H; Iwasaki, T; Kato, A; Imanishi, K; Kuwahara, Y; Cho, I. *Eos Trans. AGU,* 2008, 89(53), Fall Meet. Suppl., T13B-1937.

Porcelli, D; Ballentine, CJ. *Rev. Mineral. Geochem.,* 2002, 47, 411-480.

Sagiya, T; Nishimura, T; Iio, Y; Tada, T. *Earth Planet. Space*, 2002, 54, 1059-1063.

Sakamoto, M; Sano, Y; Wakita, H. *Geochem. J.,* 1992, 26, 189-195.

Sakuyama, M. *J.Volcanol.Geotherm. Res.,* 1979, 5, 179-208.

Sano, Y; Wakita, H. *J. Geophys. Res.*, 1985, 90, 8729-8741.

Sano, Y; Wakita, H; Williams, S. N. *J. Volcanol. Geotherm. Res.,* 1990, 42, 41-52.

Sano, Y; Takahata, N; Igarashi, G; Koizumi, N; Sturchio, N. *Chem. Geol.,* 1998, 150, 171-179.

Sato, H; Iwasaki, T; Kawasaki, S; Ikeda, Y; Matsuta, N; Takeda, T; Hirata, N; Kawanaka, T. *Tectonophys.,* 2004, 388, 47-58.

Schlosser, P; Stute, M; Sonntag, C; Münnich, KO. *Earth Planet. Sci. Lett.,* 1989, 94, 245-256.

Seno, T; Sakurai, T; Stein S. *J. Geophys. Res.,* 1996, 101(B5), 11305-11315.

Shimoike, Y; Notsu, K. *J. Volcanol. Geotherm. Res.,* 2000, 101, 211-221.

Sibson, RH. *Tectonophys.,* 1992, 211, 283-293.

Torgersen, T. *J. Geophys. Res.,* 1993, 98, 16257-16269.

Torgersen T; Clarke W. *Geochim. Cosmochim. Acta,* 1985, 49, 1211-1218.

Umeda, K. *Stability and Buffering Capacity of the Geosphere for Long-term Isolation of Radioactive Waste: Application to Crystalline Rock*; A; Hooper, J; Andersson, B; Forinash, R; Munier, K; Umeda, Ed; Nuclear Energy Agency, Organization for Economic Co-operation and Development: Paris, 2009, 289-301.

Umeda, K; Ninomiya, A. *Geochem. Geophys. Geosyst.,* 2009 10, Q08010, doi:10.1029/2009GC002501.

Umeda, K; Ninomiya, A; Negi, T. *J. Geophys. Res.,* 2009, 114, B01202, doi:10.1029/2008JB005812.

Umeda, K; Asamori, K; Negi, T; Ogawa, Y. *Geochem. Geophys. Geosys.,* 2006, 7, Q08005, doi:10.1042/2006GC001247.

Umeda, K; Asamori, K; Ninomiya, A; Kanazawa, S; Oikawa, T. *J. Geophys. Res.,* 2007, 112, B05207, doi:10.1029/2006JB004590.

Walia, V; Su, TC; Fu, CC; Yang, TF. *Radiat. Meas.,* 2005, 40, 513-516.

Yabusaki, S; Tsujimura, M; Tase, N. *Bull. Terrestri. Environ. Res. Center. Univ. Tsukuba,* 2003, 4, 119-124.

Yoshida, N; Mizutani, Y. *Anal. Chem.,* 1986, 58, 1273-1275.

Yoshino, T. *Bull. Earthq. Res. Inst. Univ. Tokyo,* 2002, 76, 479-500.

In: Helium: Characteristics, Compounds…
Editors: Lucas A. Becker

ISBN: 978-1-61761-213-8

Chapter 4

ELASTIC AND INELASTIC PROCESSES WITH SPIN-POLARIZED METASTABLE HELIUM ATOMS IN GAS DISCHARGE

V. A. Kartoshkin [*] ***and G. V. Klementiev***

Ioffe Physico-Technical Institute of the Russian Academy of Sciences, St. Petersburg, Russian Federation

ABSTRACT

At the interaction between the spin-polarized excited atom and paramagnetic ground state atom or molecule in gas discharge, elastic and inelastic processes can take place simultaneously. It means that besides the chemo-ionization of the atom or molecule at the expense of atom's excitation energy (inelastic process), an exchange of electrons is possible without a great depolarization (elastic process, or spin exchange). In such a case these two processes give rise to a remarkable spin polarization transfer between colliding particles. Influencing each other, these two processes result in a change in the spin exchange and frequency shift cross section values.

The helium metastable atoms (He^*), having a large store of internal energy (19.8 eV), are capable to ionize the molecule or atom (A) even at thermal energies of relative motion

$$He^* + A \to (HeA)^* \to \text{ionization products} \tag{1}$$

Usually investigations are performed with unpolarized particles, and the cross sections for decay of metastable's states due to collisions with molecules and atoms are determined. Experiments with polarized particles give one a possibility to obtain an information about elastic processes. Due to the paramagnetism of the collision partners (the ground state alkali atoms, paramagnetic molecules, and hydrogen atoms etc.) the spin-exchange process

[*] Corresponding author: e-mail: victor.kart@mail.ioffe.ru

$$He^*(m_1) + A(m_2) \rightarrow (HeA)^* \rightarrow He^*(m_1^{'}) + A(m_2^{'}), \tag{2}$$

$$m_1 + m_2 = m_1^{'} + m_2^{'} ,$$

is possible simultaneously with the chemi-ionization (1).

The kinetic equations, describing such a situation in the optical pumping experiment, were obtained. The expression for the spin-exchange and frequency resonance shifts cross sections of processes, taking into account the mutual influence, were derived. It was shown that the complex elastic cross sections do significantly change because of the chemi-ionization process. It gave us a possibility to get much information about spin-exchange and chemi-ionization cross sections from the optical orientation's and magnetic resonance's experiments. In the frames of the theory of complex interaction potentials energy dependences of the chemi-ionization, spin-exchange and frequency shift cross section were obtained for the systems "polarized helium metastable atom - hydrogen atom, ground state alkali atom, or paramagnetic molecule".

INTRODUCTION

In helium gas discharge there are a great number of neutral and charged particles of which an investigation is of interest for many branches of physics and applications in technique and biology. After introducing the optical pumping (optical orientation, or polarization, of paramagnetic atoms in their ground or excited state), this technique was used to study many kinds of interatomic processes, among which for precision experiments are of great importance the spin exchange collisions, being not only the strong relaxation mechanism but also shifting the transition frequency. Moreover, they allow one to produce the" indirect" polarization of those atoms (such as the H and N atoms), of which the direct optical orientation is impossible up to date due to the fact that the first atomic transition from the ground state lies in the far ultraviolet [1,2]. On the subject of spin exchange there are many theoretical and experimental papers (see [3,4,] and therein).

Since the direct optical orientation (optical pumping) of helium atoms in the metastable triplet state [5], many processes with spin-polarized atomic particles were investigated in a helium gas discharge in a wide temperature range [4, 6-12]. Due to high excitation energy of the 2^3S_1 helium state, for almost all atoms there is a great probability to be ionized even at thermal energies and often these atoms may "change" their electron coordinates with those of the triplet helium atoms. Such spin-exchange processes, accompanied by the chemi-ionization at the expense of the internal energy of the other collision partner, are the subject of this paper of ours. We shall describe below an expanding of the theory of spin exchange on such "double" processes [13- 17]. In explaining the spin exchange process by interference of the scattering amplitudes for different electron terms of the quasi-molecule formed during the collision of two atomic particles (atoms or molecules) we took into account an influence of inelastic chemi-ionization process on spin exchange by introducing complex scattering phase angles instead of real ones having been used previously in the case of pure spin exchange [17-19]. Also will be quoted results of experimental efforts in investigating the pair systems in which one partner is the 2^3S_1 helium atom and the other is the hydrogen atom, or alkali atom, or oxygen molecule, or the 2^3S_1 helium atom [20-25]. The indirect spin polarization of hydrogen atoms was carried out in the He-H_2 gas mixture due to simultaneously proceeding chemi-ionization and spin exchange, the contribution of each process being determined [15].

Such a division was possible in optical pumping experiments by analyzing as amplitudes of magnetic resonance of two collision partners as line widths of orientation and alignment signals [26,27]. Also was considered the question about the fulfillment of the Wigner rule, according to which the total electron spin does not change as a result of collision. This question is of considerable interest for physics of atomic collisions and due to its correctness was possible indirect spin-polarization of atomic particles in binary collisions, such as chemi-ionization and spin exchange, when one of the collision partners was preliminary optically oriented and both collision partners were in the S - state [1,2, 9,10,15] .

The total spin conservation rule was shown to be correct in several systems considered below and even in a not obvious situation as in the system He(2^3 S_1)-O_2($^3\Sigma_g^-$) [27]. In the case of chemi-ionization and spin exchange processes taking place simultaneously one can get spin polarized electrons, ions, atoms, molecules, or molecular ions. Of interest is a possible redistribution of the spin-polarization between electrons, nucleus and rotational motion.

THE THEORY OF SPIN EXCHANGE

Since the direct optical orientation (optical pumping) of helium triplet metastable atoms a great many processes were investigated in gas discharge in a wide temperature range. Here we shall be interested in the spin exchange, taking place during the collision of two atomic particles, which is accompanied by the ionization of one of collision partners at the expense of the internal energy of other particle. The systems under study have atoms or molecules with the zero orbital electron moments and the electron spins S_1, S_2 = 1/2, 1. The examples are the collision pairs He(2^3S_1)-H(1 $^2S_{1/2}$) [15] or He(2^3 S_1)-O_2($^3\Sigma_g^-$) [27,30] where one particle (the helium atom), due to high excitation energy of the metastable state, can ionize hydrogen atom or oxygen molecule even at low energy of relative motion. We could derive the expressions for the cross sections of such processes in supposing the Wigner total spin conservation rule to be valid. The fulfillment of the Wigner rule is not obvious, but we notice that there are some examples of systems, not in all cases but often with zero orbital moment, in which the Wigner rule was confirmed in experiments on optical orientation of atoms [9].

The process of pure spin exchange, i.e. without chemi-ionization due to high internal energy of one of collision partners, was studied theoretically in many papers (see for example [3,4]). It explains by interference of the elastic scattering amplitudes for different electron terms of the quasi-molecule formed during the collision of two atomic particles (atoms or molecules). The influence of inelastic chemi-ionization process on spin exchange could be taken into account by making use of complex scattering phase angles instead of real ones having been used previously in the case of pure spin exchange [31].

Consider the collision system with electron spins S_1=1/2 and S_2 = 1, for instance the system He(2 $^3S_1$)-A(n $^2S_{1/2}$), in which the chemi-ionization

$$\mathrm{He}(2^3S_1) - \mathrm{A}(^2S_{1/2}) \rightarrow \begin{cases} \mathrm{HeA}^+ + e^- \\ \mathrm{He}(1^1S_0) + \mathrm{A}^+ + e^- \end{cases} \quad (3)$$

takes place simultaneously with the spin exchange

$$He(2^3S_1, m_1) + A(^2S_{½}, m_2) \rightarrow He(2^{-3}S_1, m_2') + A(^2S_{½}, m_2')\ , \tag{4}$$

where $m_1 + m_2 = m_1' + m_2'$, m_i (i =1,2) being the projection of electron spins on the quantization axe.

The quasi-molecule $(HeA)^*$, formed during the collision, time of which is of the order of 10^{-12}s, might evolve on two terms - the doublet term V_d with the total spin S = 1/2 and the quartet term V_q with the total spin S = 3/2.

The corresponding scattering amplitudes are

$$f_{d,q} = \frac{1}{2ik}\sum_{l=0}^{\infty}(2l+1)\left[\exp\left(2i\eta^{d,q}\right)-1\right]P_l(\cos\theta), \tag{5}$$

where the Legendre polynomial ($P_l(\cos\theta)$) expansion is used and scattering phase (η) angles may be calculated from the partial wave equation according to [31].

For the term with the greater multiplicity V_q, the ionization process (3) is forbidden in accordance with the Wigner spin conservation rule since the total spin of initial species is 3/2, whereas that of products of reaction (3) would be 1/2 Also, due too low collision time, are insufficient transitions between doublet and quartet terms of the quasi-molecule $(HeA)^*$, that allows one to ignore an influence of spin exchange process on chemi-ionization. Hence we might put the phase shifts to be real for the quartet term (η^q) and complex for the doublet term ($\eta^d = \chi^d + i\lambda^d$).

Let Φ_{HeA} be the initial (just before collision) wave function of the (He^*A) system. The asymptotic wave function (after scattering) will be as

$$\Psi_{HeA}(S_1m_1;S_2m_2) \sim \Phi_{HeA}(S_1m_1;S_2m_2)\cdot\exp(i\,\mathbf{kr}) + \frac{\exp(ikr)}{r}\sum_{m_1m_2}\Phi_{HeA}(S_1m_1;S_2m_2)\cdot f(m_1m_2;m_1'm_2';\theta), \tag{6}$$

where **k** being the initial wave vector, the vector **r** describing the relative position of nuclei, and $f(m_1m_2;\, m_1'm_2';\, \theta')$ being the amplitude of scattering through the angle θ in the system of center of mass.

According to the another angular moments' coupling scheme, the spins S_1 and S_2 should give the total spin S with the values 1/2 or 3/2 and the magnetic quantum number M_S, to which corresponds the molecular wave function $\Xi(S,M_S)$, that yields

$$\Psi_{HeA}(S_1m_1;S_2m_2) \sim \sum_S C^{SM_S}_{S_1m_1S_2m_2}\,\Xi(S,M_S)\cdot\left[\exp(i\,\mathrm{kr}) + f_S(\theta)\frac{\exp(ikr)}{r}\right]. \tag{7}$$

where $M_S = m_1 + m_2$, $C^{SM_S}_{S_1 m_1 S_2 m_2}$ being the Clebsch - Gordan coefficients, and $f_S(\theta)$ being the amplitude of elastic scattering for the doublet or quartet term.

From (6) and (7) one can conclude that the amplitude of changing the atoms' spin coordinates should be connected with the scattering amplitudes on molecular terms through the vector coupling coefficients as

$$f\left(m_1 m_2; m_1' m_2'; \theta\right) = \sum_S C^{SM_S}_{S_1 m_1 S_2 m_2} C^{S'M_S'}_{Sm_1' S_2 m_2'} f_S(\theta) \tag{8}$$

The cross section of the spin exchange process $m_1 m_2 \rightarrow m_1' m_2'$ then is given by the expression

$$\sigma\left(m_1 m_2; m_1' m_2'; \theta\right) = \left|f\left(m_1 m_2; m_1' m_2'; \theta\right)\right|^2 \tag{9}$$

The use of the the Clebsch - Gordan coefficients' values for the angular moments S_1=1/2 and S_2=1 leads to the conclusion that in such a case all cross sections for spin change of collision partners should be identical:

$$\sigma(1/2,0;-1/2,1) = \sigma(1/2,-1;-1/2,0) = \sigma(-1/2,0;1/2,-1) = \sigma(-1/2,1;1/2,0) = \frac{2}{9}\left|f_q - f_d\right|^2 \tag{10}$$

The integration of (10) with respect to θ yields the desired cross section for the process (4) in the form

$$\sigma^{tr} = \frac{2\pi}{9k^2}\sum_{l=0}^{\infty}(2l+1)\left[1 - 2\exp(-2\lambda_l^d)\cos 2\left(\chi_l^d - \eta_l^q\right) + \exp(-4\lambda_l^d\right]. \tag{11}$$

The cross section for the ionization process (1) can be calculated in a manner given in [31] as

$$\sigma^{abs} = \frac{\pi}{k^2}\sum_{l=0}^{\infty}(2l+1)\left[1 - \exp(-4\lambda_l^d)\right] \tag{12}$$

The similar ideas may be used when considered the case of colliding particles with the equal spins S_1, S_2 =1 , as for the pair He(2^3S_1)-O_2($^3\Sigma_g^-$) or He(2^3S_1)- He(2^3S_1). Here at the collision of two particles is formed the quasi-molecule with the total spin S = 0, 1, or 2. The corresponding molecular terms are singlet (V_s), triplet (V_t), and quintet (V_q), scattering amplitudes being f_s, f_t, and f_q. For the singlet and triplet terms are possible both chemi-ionization and spin exchange, whereas for the term with the greatest multiplicity (V_q) chemi-ionization process is forbidden in accordance with the Wigner rule. Therefore we must put the scattering phase angle to be real (η_l^q) for the quintet term and complex ($\eta_l^{s,t} = \chi_l^{s,t} + i\,\lambda_l^{s,t}$) for two other terms.

The cross section of the spin change process $m_1m_2 \rightarrow m_1'm_2'$ being equal at the scattering on the angle θ to the square of (9), the cross sections $\sigma_i(\theta)$, i=1, 2, 3 become

$$\begin{aligned}
&\sigma_1(\theta) = \sigma(1,0;0,1) = \sigma(0,1;1,0) = \sigma(-1,0;0,-1) = \sigma(0,-1;-1,0) = \frac{1}{4}\left|f_t - f_s\right|^2, \\
&\sigma_2(\theta) = \sigma(1,-1;0,0) = \sigma(0,0;1,-1) = \sigma(-1,1;0,0) = \sigma(0,0;-1,1) = \frac{1}{9}\left|f_s - f_q\right|^2, \\
&\sigma_3(\theta) = \sigma(1,-1;-1,1) = \sigma(-1,1;1,-1) = \left|\frac{1}{3}f_s - \frac{1}{2}f_t + \frac{1}{6}f_q\right|^2,
\end{aligned} \tag{13}$$

from which it follows that $\sigma_1(\theta)$ is due to the interference of scattering amplitudes on the triplet and quintet terms, $\sigma_2(\theta)$ is due to that on the singlet and triplet terms, and $\sigma_3(\theta)$ is due to the scattering amplitudes' interference on all the three terms of the quasi-molecule.

The integration of (13) with respect to θ yields

$$\begin{aligned}
&\sigma_1^{tr} = \frac{\pi}{4k^2}\sum_{l=0}^{\infty}(2l+1)\left[1 - 2\exp(-2\lambda_l^t)\cos 2(\chi_l^t - \eta_l^q) + \exp(-4\lambda_l^t)\right] \\
&\sigma_2^{tr} = \frac{\pi}{9k^2}\sum_{l=0}^{\infty}(2l+1)\left[1 - 2\exp(-2\lambda_l^s)\cos 2(\chi_l^s - \eta_l^q) + \exp(-4\lambda_l^s)\right] \\
&\sigma_3^{tr} = \frac{\pi}{36k^2}\sum_{l=0}^{\infty}(2l+1)\left[\,1 - 12\exp(-2\lambda_l^s - 2\lambda_l^t)\cos 2(\chi_l^s - \chi_l^t) + 4\exp(-2\lambda_l^s)\cos 2(\eta_l^q - \chi_l^s) - \right. \\
&\left. - 6\exp(-2\lambda_l^t)\cos 2(\eta_l^q - \chi_l^t) + 4\exp(-4\lambda_l^s) + 9\exp(-4\lambda_l^t)\,\right]
\end{aligned} \tag{14}$$

Frequency Shifts due to Spin Exchange with Chemi-Ionization

It is well known [4] that at the collision of two atomic particles A and B with the spins S_1 and S_2 and the magnetic quantum numbers m_1 and m_2, the spin exchange processes of the type

$$A\,(m_1) + B(m_2) \rightarrow A\,(m_1') + B(m_2'), \tag{15}$$

where $m_1 + m_2 = m_1' + m_2'$, often result in a significant and well measured frequency shift of the Larmor precession of atomic spins in a magnetic field that affects the work of important quantum electronic devices such as gyroscopes, magnetometers, frequency standards, etc.[4]. Theoretical aspects of this problem were considered in [4,32-36]. For two-level atomic system, the magnetic resonance frequency shift ($\Delta\omega$) due to spin exchange (1) is given by the expression

$$\Delta\omega = \alpha\, N\, v\, \sigma_{sh}\, P \quad , \tag{16}$$

where

$$\sigma_{sh} = \frac{\pi}{k^2}\sum_{l=0}^{\infty}(2l+1)\sin 2(\eta_l^2 - \eta_l^1) \tag{17}$$

is the frequency shift cross section , N and P are the density and the orientation of the other particles to which the spin orientation can be transferred, v is the relative velocity of colliding particles, η_l^1 and η_l^2 are the scattering phases angles for terms with different multiplicity, and the factor α depends on the spins of two atoms.

The formulae (16) and (17) are valid for the spin exchange when inelastic processes are unimportant, i.e. for the case of pure, or simple, spin exchange (15). The transfer of orientation in spin exchange from the particle A to the particle B in [13, 17] was accompanied also by the effective inelastic process (chemi-ionization) , due to high internal energy of the particle A

$$A^* + B \rightarrow A + B^+ + e^- . \tag{18}$$

As in the preceding paragraph, consider the case of atoms with the spins $S_A = 1$, $S_B = 1/2$ and zero orbital moments. At the collision of such particles the quasi-molecule $(AB)^*$ is produced with the total electron spins S = 1/2 or 3/2, to which correspond the doublet or quartet molecular terms. The spin exchange process (15) is possible for both molecular terms, but the chemi-ionization (18) is prohibited for the quartet term in accordance with the Wigner spin conservation rule. The scattering phase angles is therefore real (η_l^q) for the quartet term, but complex ($\eta_l^d = \chi_l^d + i\lambda_l^d$) for the doublet term.

The calculations carried out in [19] give us the frequency shift cross section as

$$\sigma_{sh} = \frac{\pi}{k^2}\sum_{l=0}^{\infty}(2l+1)\exp\left(-2\lambda_l^d\right)\sin 2(\eta_l^q - \eta_l^d), \tag{19}$$

and the spin factor (see (16)) in the expression for the frequency shift will be as

$$\alpha = \frac{2}{3} . \tag{20}$$

Comparing the formulae for the frequency shift cross sections (17) and (19), one can reveal , unlike simple spin exchange, the appearance of the exponential factor in the case of two competition processes, thus in general, excluding possible oscillations at relatively low ionization probabilities, reduces noticeable the spin exchange cross section. If the ionization probability for the doublet term approaches unity for the partial waves which contribute to σ_{sh}, i.e. $\lambda_l^d \rightarrow \infty$, the frequency shift tends to zero.

Kinetics of Optical Orientation of Metastable (2^3S_1) Helium Atoms

At the interaction between spin-polarized metastable helium atom and paramagnetic atom or molecule in gas discharge elastic and inelastic processes take place simultaneously. In such a case these two processes influence each other giving rise to a change of the cross section's value for the elastic process. It means, that besides the chemi-ionization of the particle at the expense of the atom's excitation energy (inelastic process), an exchange of electrons is possible without a great depolarization (s.c. spin exchange, or elastic process).

The kinetics of the optical orientation in the "excited atom - paramagnetic particle" mixture was studied taking into account the simultaneously occurring processes of spin-exchange and chemi-ionization. The equations, describing such a situation, are obtained. It was shown how the widths of magnetic resonance lines depend on the chemi-ionization and spin-exchange cross sections. It gave us a possibility to get information about spin-exchange and chemi-ionization cross sections from the optical orientation and magnetic resonance experiment for the following systems: He(2^3S_1) - Li($2^2S_{1/2}$)[37], Na($3^2S_{1/2}$)[38], K($4^2S_{1/2}$)[39], Rb($5^2S_{1/2}$)[40], Cs($6^2S_{1/2}$)[41], H($1^2S_{1/2}$)[15], D($1^2S_{1/2}$)[42] and He(2^3S_1) - O_2($^3\Sigma^-_g$)[27], He(2^3S_1)[43].

a. Interaction between Particles with Electron Spins S=1/2 and S=1

Consider the interaction of two atoms, one of which denoted below as A has the electron spin S=1/2 and the other is the helium isotope ^{4}He, which is diamagnetic in the ground state (1^1S_0) and paramagnetic in the metastable state (2^3S_1) where its electron spin is S=1. The metastable helium atom in the 2^3S_1 state we will denote below as B. As noted above, at the interaction between metastable helium atoms and paramagnetic particles elastic (spin exchange) and inelastic (chemi-ionization) processes take place simultaneously. Each of these processes influence observables in the experiment quantities - orientations and alignments. The kinetics equations describing the evolutions of the observable will be derived.

In view of the fact that the particle collision time (t_{coll}~10^{-12} s) in much smaller than the hyperfine interaction time in the alkali atom (for example, hyperfine splitting is $\Delta\nu=1771\cdot10^6$ Hz for ^{23}Na, $\Delta\nu=1420\cdot10^6$ Hz for ^{1}H, $\Delta\nu=3035\cdot10^6$ Hz for ^{85}Rb, etc. [44]), we shall assume that the electron orientation evolves during the collision, whereas the nuclear orientation evolves in the time between collisions.

Begin our consideration with the spin exchange process. Since the electronic spin of the helium atom is S_1=1 and the electronic spin of the alkali atom is S_2=1/2, the particles under study can be described by the corresponding density matrices P and ρ of 3×3 and 2×2 dimensionality. In this case the P and ρ matrices have the form

$$P = \begin{pmatrix} P_{11} & 0 & 0 \\ 0 & P_{22} & 0 \\ 0 & 0 & P_{33} \end{pmatrix}, \quad \rho = \begin{pmatrix} \rho_{11} & \rho_{12} \\ \rho_{21} & \rho_{22} \end{pmatrix} \tag{21}$$

In neglecting the correlation between the ρ and P matrices, the interacting particles comprise a system that can be described by the density matrix Φ as

$$\Phi = \rho \otimes \mathrm{P} \,, \tag{22}$$

where the sign $\otimes$ denotes the direct matrix product.

The transition to the electron-spin-coupled representation is given by the transformation $\Phi^c = \Gamma\Phi\Gamma'$, explicit form of the matrix Γ being in the [19].

The evolution of the elements of the density matrix Φ due to spin exchange in the coupled moment representation Φ^c [32,36] is given by the equation

$$\left(\frac{d\Phi^c_{pq}}{dt}\right)_{coll} = -(w - id)\Phi^c_{pq}(1-\delta_{pq}) \,, \tag{23}$$

where w = - Im H_{pq} , d = Re H_{pq}, δ_{pq} is the Kronecker symbol..

In order to get kinetics equations describing the evolution of the density matrices P and ρ we need to compute the trace of the eq.(23), as for example,

$$\frac{d\rho}{dt} = Tr_{B_e}\left[\left(\frac{d\Phi}{dt}\right)_{coll}\right]. \tag{24}$$

Using the relations (21)-(24), we can write out the expression for the longitudinal orientation of particle A as

$$\frac{d\langle S_A\rangle^z}{dt} = -\frac{4}{9}w[\rho_{22}(P_{11}+P_{22}) - \rho_{11}(P_{22}+P_{33})] = -\frac{4}{9}w\left(\frac{4}{3}\langle S_A\rangle^z - \frac{1}{2}\langle S_B\rangle^z - \frac{1}{3}\langle Q_B\rangle^{zz}\langle S_A\rangle^z\right), \tag{25}$$

where $\langle S_A\rangle_z$ is the longitudinal orientation of the particle A $[\langle S_A\rangle^z = (\rho_{11}-\rho_{22})/2)]$, $\langle S_B\rangle^z$ is the longitudinal orientation of the particle B $[(\langle S_B\rangle^z = (P_{11}-P_{33})]$, $\langle Q_B\rangle^{zz}$ is the longitudinal components of the electronic alignment of the particle B $[\langle Q_B\rangle^{zz} = \frac{1}{2}(P_{11}+P_{33}) - P_{22})]$.

The above-mentioned considerations can be used to obtain the equations for the evolution of the observable quantities in the spin exchange process as

$$\frac{d\langle S_A\rangle^z}{dt} = -\frac{1}{\tau_1}\left(\frac{4}{3}\langle S_A\rangle^z - \frac{1}{2}\langle S_B\rangle^z - \frac{1}{3}\langle Q_B\rangle^{zz}\langle S_A\rangle^z\right), \frac{1}{\tau_1} = N_B C_{se} \tag{26}$$

$$\frac{d\langle S_B\rangle^z}{dt} = -\frac{1}{\tau_1^{'}}\left(\frac{1}{2}\langle S_B\rangle^z - \frac{4}{3}\langle S_A\rangle^z + \frac{1}{3}\langle Q_B\rangle^{zz}\langle S_A\rangle^z\right), \frac{1}{\tau_1^{'}} = N_A C_{se} \tag{27}$$

$$\frac{d\langle Q_B\rangle^{zz}}{dt} = -\frac{1}{\tau_1^{'}}\left(\frac{3}{2}\langle Q_B\rangle^{zz} - \langle S_B\rangle^z\langle S_A\rangle^z\right), \frac{1}{\tau_1^{'}} = N_A C_{se} \tag{28}$$

$N_{A,B}$ being the concentrations of the A and B particles and C_{se} being the spin exchange rate constant.

Taking into account that spin exchange takes place simultaneously with chemi-ionization, we can write the kinetics equations describing the chemi-ionization process in the form

$$\frac{d\langle S_A\rangle^z}{dt} = -\frac{1}{3\tau_2}\left(\langle S_A\rangle^z - \frac{1}{2}\langle S_B\rangle^z\right), \frac{1}{\tau_2} = N_B C_{ci}, \tag{29}$$

$$\frac{d\langle S_B\rangle^z}{dt} = -\frac{1}{3\tau_2^{'}}\left(\langle S_B\rangle^z - \frac{4}{3}\langle S_A\rangle^z - \frac{2}{3}\langle Q_B\rangle^{zz}\langle S_A\rangle^z\right), \frac{1}{\tau_2^{'}} = N_A C_{ci} \tag{30}$$

$$\frac{d\langle Q_B\rangle^{zz}}{dt} = -\frac{1}{\tau_2^{'}}\left(\frac{1}{3}\langle Q_B\rangle^{zz} - \frac{2}{3}\langle S_B\rangle^z\langle S_A\rangle^z\right), \frac{1}{\tau_2^{'}} = N_A C_{ci}. \tag{31}$$

Chemi-ionization collisions lead not only to evolution of orientation and alignment, but also to a change of the metastable helium atoms concentration according to the equation

$$\frac{dN_B}{N_B dt} = -\frac{1}{3\tau_2}\left(1 - 2\langle S_A\rangle^z\langle S_B\rangle^z\right), \frac{1}{\tau_2^{'}} = N_A C_{ci}, \tag{32}$$

$N_{A,B}$ being the concentrations of the A and B particles and C_{ci} being the chemi-ionization rate constant.

The effect of optical pumping can be taken into account by introducing the corresponding terms in equations (26)-(32), namely φ_p, $\varphi_p^{'}$, and $\varphi_p^{''}$ ($\varphi_p = L_p/\tau_p$, where τ_p is the pumping time for orientation or alignment and L_p characterizes the light used for pumping. Further, one must take into account in the equations the contributions due to diffusion (τ_p) and other relaxation processes (τ_r), which influence the evolution of the corresponding spin polarization moments.

Thus, the system of equations for the orientation and the alignment in the matrix form is of the form

$$\frac{d}{dt}\begin{bmatrix}\langle S_A\rangle^z \\ \langle S_B\rangle^z \\ \langle Q_B\rangle^{zz}\end{bmatrix}=[A]\cdot\begin{bmatrix}\langle S_A\rangle^z \\ \langle S_B\rangle^z \\ \langle Q_B\rangle^{zz}\end{bmatrix}+[B]\cdot\begin{bmatrix}\langle S_A\rangle^z\langle Q_B\rangle^{zz} \\ \langle S_A\rangle^z\langle Q_B\rangle^{zz} \\ \langle S_A\rangle^z\langle S_B\rangle^z\end{bmatrix}+[\varphi_p] \tag{33}$$

the matrices $[A]$, $[B]$, and $\lfloor\varphi_p\rfloor$ given by the expressions

$$[A]=\begin{bmatrix}-\left(\frac{4}{3\tau_1}+\frac{1}{3\tau_2}\right)-\frac{1}{\tau_d}-\frac{1}{\tau_r} & \left(\frac{1}{6\tau_2}+\frac{1}{2\tau_1}\right) & 0 \\ \left(\frac{4}{9\tau_2^{'}}+\frac{4}{3\tau_1^{'}}\right) & -\left(\frac{1}{3\tau_2^{'}}+\frac{1}{2\tau_1^{'}}\right)-\frac{1}{\tau_d^{'}}-\frac{1}{\tau_r^{'}} & 0 \\ 0 & 0 & -\left(\frac{1}{3\tau_2^{'}}+\frac{3}{2\tau_1^{'}}\right)-\frac{1}{\tau_d^{''}}-\frac{1}{\tau_r^{//}}\end{bmatrix}= \tag{34}$$

$$=\begin{bmatrix}-p_1 & r_1 & 0 \\ r_2 & -p_2 & 0 \\ 0 & 0 & -p_3\end{bmatrix},$$

$$[B]=\begin{bmatrix}\frac{1}{3\tau_1} \\ \frac{2}{9\tau_2^{'}}-\frac{1}{3\tau_1^{'}} \\ \frac{2}{3\tau_2^{'}}+\frac{1}{\tau_1}\end{bmatrix}, \tag{35}$$

$$\varphi_p=\begin{bmatrix}\varphi_p \\ \varphi_p^{'} \\ \varphi_p^{''}\end{bmatrix}. \tag{36}$$

The matrix elements entering in (36) show themselves under certain pumping conditions. In the case of optical pumping of the metastable helium atoms by the linearly polarized light, there is the alignment alone. In the case of pumping by circularly polarized light, there are both orientation and alignment of the helium atoms. In the case of optical pumping of alkali atoms, the electronic orientation in the system of Zeeman sublevels of alkali atoms is produced. Further, due to the hyperfine interaction arises both orientation and alignment in the system of hyperfine sublevels of the alkali atoms. Unfortunately, the direct optical pumping of hydrogen and deuterium atoms is impossible up to date, because resonance lines of these atoms are in the far ultraviolet range.

We are interested here in the optical orientation of the metastable He atoms. As a result of collisions with partners in the ground state, the polarization will be redistributed between the collision partners. From (33)-(36), it follows that the magnetic resonance line widths of metastable helium atoms, oriented (Δf_{or}) and aligned (Δf_{al}), caused by collisions with alkali, hydrogen and deuterium atoms in the ground state are given by the expressions:

$$\pi\Delta f_{or} = \frac{1}{2\tau_1'} + \frac{1}{3\tau_2'}, \tag{37}$$

$$\pi\Delta f_{al} = \frac{3}{2\tau_1'} + \frac{1}{3\tau_2'}$$

From which it fellows that $\pi(\Delta f_{al} - \Delta f_{or})$, i.e. the difference of magnetic-resonance line widths within the multiplicative constant, is equal to the rate constants for spin exchange between helium and alkali, hydrogen, or deuterium atoms

$$1/\tau_1' = N_A C_{se} \tag{38}$$

B. Interactios between Particles with Electron Spins S=1 and S=1

Considering the evolution of the density matrices

$$P = \begin{pmatrix} P_{11} & 0 & 0 \\ 0 & P_{22} & 0 \\ 0 & 0 & P_{33} \end{pmatrix} \text{ and } \rho = \begin{pmatrix} \rho_{11} & \rho_{12} & \rho_{13} \\ \rho_{21} & \rho_{22} & \rho_{23} \\ \rho_{31} & \rho_{32} & \rho_{33} \end{pmatrix} \tag{39}$$

like to that carried out above, we could get the kinetics equations which describe the influence of spin exchange and chemi-ionization processes on the observable quantities. The system of equations for longitudinal component, the particle A being the oxygen molecule or helium atom in the triplet metastable state and the particle B being the metastable helium atom, is of the form

$$\begin{aligned} \frac{d\langle S_A\rangle^z}{dt} &= -N_B v\left[\left(\frac{1}{3}\sigma_1^{tr} + \frac{1}{3}\sigma_2^{tr} + \frac{2}{3}\sigma_3^{tr}\right) + \left(\frac{1}{9}\sigma_s^{abs} + \frac{1}{3}\sigma_t^{abs}\right)\right]\langle S_A\rangle^z + N_B v\left[\left(\frac{1}{3}\sigma_1^{tr} + \frac{1}{3}\sigma_2^{tr} + \frac{2}{3}\sigma_3^{tr}\right) + \right. \\ &\left. + \left(\frac{1}{9}\sigma_s^{abs} + \frac{1}{6}\sigma_t^{abs}\right)\right]\langle S_B\rangle^z + N_B v\left[\left(\frac{1}{3}\sigma_1^{tr} - \frac{1}{6}\sigma_2^{tr} - \frac{1}{3}\sigma_3^{tr}\right) + \left(-\frac{1}{18}\sigma_s^{abs} + \frac{1}{12}\sigma_t^{abs}\right)\right]\langle S_A\rangle^z \langle Q_B\rangle^{zz} + \\ &+ N_B v\left[\left(-\frac{1}{3}\sigma_1^{tr} + \frac{1}{6}\sigma_2^{tr} + \frac{1}{3}\sigma_3^{tr}\right) + \left(\frac{1}{18}\sigma_s^{abs} + \frac{1}{12}\sigma_t^{abs}\right)\right]\langle S_B\rangle^z \langle Q_A\rangle^{zz}, \end{aligned} \tag{40}$$

$$\begin{aligned} \frac{d\langle Q_A\rangle^{zz}}{dt} &= -N_B v\left[\left(\sigma_1^{tr} + \sigma_2^{tr}\right) + \left(\frac{1}{9}\sigma_s^{abs} + \frac{1}{3}\sigma_t^{abs}\right)\right]\langle Q_A\rangle^{zz} + N_B v\left[\left(\sigma_1^{tr} + \sigma_2^{tr}\right) + \left(-\frac{1}{9}\sigma_s^{abs} + \frac{1}{6}\sigma_t^{abs}\right)\right]\langle Q_B\rangle^{zz} + \\ &+ N_B v\left[\frac{3}{2}\sigma_2^{tr} + \left(\frac{1}{6}\sigma_s^{abs} + \frac{1}{4}\sigma_t^{abs}\right)\right]\langle S_A\rangle^z \langle S_B\rangle^z + N_B v\left[\frac{1}{2}\sigma_2^{tr} + \left(\frac{1}{18}\sigma_s^{abs} - \frac{1}{12}\sigma_t^{abs}\right)\right]\langle Q_A\rangle^{zz} \langle Q_B\rangle^{zz}, \end{aligned} \tag{41}$$

The equations describing the evolution of orientation and alignment of the particle B can be obtained from Eqs. (40) and (41) by interchanging indices A and B of the quantities N, $\langle S \rangle$, and $\langle Q \rangle$.

Thus we can introduce the cross sections σ^{abs} and σ^{tr}, which characterize the decay of orientation and alignment of colliding particles due to chemi-ionization and spin exchange, as

$$\sigma_{or}^{tr} = \frac{1}{3}\sigma_1^{tr} + \frac{1}{3}\sigma_2^{tr} + \frac{2}{3}\sigma_3^{tr}, \tag{42}$$

$$\sigma_{or}^{abs} = \frac{1}{9}\sigma_s^{abs} + \frac{1}{3}\sigma_t^{abs}, \tag{43}$$

$$\sigma_{al}^{tr} = \sigma_1^{tr} + \sigma_2^{tr}, \tag{44}$$

$$\sigma_{al}^{abs} = \frac{1}{9}\sigma_s^{abs} + \frac{1}{3}\sigma_t^{abs} \tag{45}$$

The system of equations for the orientation and the alignment has the form

$$\frac{d}{dt}\begin{bmatrix} \langle S_A \rangle^z \\ \langle S_B \rangle^z \\ \langle Q_A \rangle^{zz} \\ \langle Q_B \rangle^{zz} \end{bmatrix} = [A]\cdot\begin{bmatrix} \langle S_A \rangle^z \\ \langle S_B \rangle^z \\ \langle Q_A \rangle^{zz} \\ \langle Q_B \rangle^{zz} \end{bmatrix} + [B]\cdot\begin{bmatrix} \langle S_A \rangle^z \langle Q_B \rangle^{zz} \\ \langle S_B \rangle^z \langle Q_A \rangle^{zz} \\ \langle S_A \rangle^z \langle S_B \rangle^z \\ \langle Q_A \rangle^{zz} \langle Q_B \rangle^{zz} \end{bmatrix} + \begin{bmatrix} L_A^p / \tau_A^p \\ L_B^p / \tau_B^p \\ L_A^{p'} / \tau_A^{p'} \\ L_B^{p'} / \tau_B^{p'} \end{bmatrix} \tag{46}$$

From (40) - (45) it follows that the magnetic resonance line widths of helium metastable atoms, oriented (Δf_{or}) or aligned (Δf_{al}) , caused by collisions with oxygen molecules in the ground state, are given by the expressions:

$$\pi\Delta f_{or} = N_A v(\sigma_{or}^{abs} + \sigma_{or}^{tr}), \;\; \pi\Delta f_{or} = N_A v(\sigma_{al}^{abs} + \sigma_{al}^{tr}) \tag{47}$$

taking into account equations (42)-(45). Further, from (47), (42)-(45) it follows that the difference between the line widths for the orientation and alignment is according to the relation

$$\pi(\Delta f_{al} - \Delta f_{or}) = \frac{2}{3} N_A v(\sigma_1^{tr} + \sigma_2^{tr} - \sigma_3^{tr}) \tag{48}$$

Thus, it becomes really possible to determine the cross sections for spin exchange and the chemi-ionization from the experiments on optical orientation of atoms, taking into account the fact that these cross sections are given by the relation

$$\langle\sigma(T)\rangle = \frac{C(T)}{\langle v\rangle} = \frac{\langle\sigma(E)v\rangle}{\langle v\rangle} \tag{49}$$

where <v> is the average relative velocity of colliding particles.

EXPERIMENT

A. Magnetic Resonance in a System of Optically Polarized Metastable Helium Atoms and Atoms with Electron Spin S=1/2 in Ground State

Below we have studied the interaction of optically polarized metastable helim atoms with alkali atoms in the ground state and experimentally determined the rate constant of spin exchange during the collisions of the spin-polarized He ($2\,^3S_1$) atoms with alkali atoms. As it was shown above, the decay rates of the orientation $\langle S_B\rangle^z$ and the alignment $\langle Q_B\rangle^{zz}$ of the metastable helium atoms depend on collision processes (1) and (2) in accordance with the equations (33). As it can be seen from (37), the contributions to magnetic resonance (MR) line width for alignment and oriented atoms should differ from one another. Just this difference permits one to determine the rate constant of the two spin- dependent processes taking place simultaneously.

Figure 1 shows the standard experimental setup for investigating the optical pumping of atoms. The 2^3S_1 metastable helium atoms were optically, whether oriented with circularly polarized light or aligned with linearly polarized light, by the radiation (λ=1.08 μm) propagating along a static magnetic field $H_0 \approx 35$ mOe induced by Helmholtz solenoid (2) within the magnetic shield (1). A helium capillary lamp (3), in which a high frequency (hf) discharge was excited, was used as the pumping source. The metastable state of helium atoms was populated by exciting the high frequency discharge within absorption chamber (6) containing gaseous isotope of ^{4}He (p=1 Torr at T=300 K) and metallic alkali. The gas discharge chamber was placed in a thermostat. The working temperature was determined by using a thermistor glued onto the coolest part of the absorption chamber. Variation of the temperature in the working range for each alkali metal permitted one to regulate the concentration of alkali atom from 10^{10} cm^{-3} to 10^{12} cm^{-3}.

Magnetic resonance was excited in the system of Zeeman sublevels of the 2^3S_1 helium atoms by applying an amplitude modulated radio-frequency magnetic field H_1 = h sinΩt sinωt, where $\Omega/2\pi$=250 Hz, and $\omega/2\pi$ is the frequency of the radio-frequency field. The H_1 was perpendicular to the static magnetic field H_0. In the experiments, variations (at the modulation frequency Ω) of the intensity of the pumping radiation passing through the absorption chamber were recorded with scanning the frequency ω in the region of the magnetic-resonance frequency $\omega_0 = \gamma H_0$, where $\gamma/2\pi = 2.8$ MHz/Oe and γ is the gyromagnetic ratio of the 2^3S_1 helium atoms.

The pumping by circularly polarized light can produce both orientation and alignment in the system of Zeeman sublevels of the 2^3S_1 state of the helium metastable atoms. The relative part of alignment in the signal depends on the emission spectrum of the pumping lamp and on the length of the optical path in the absorption camber. The estimates due to this contribution can be done in accordance with [45]. In order to avoid an admixture of the alignment signal to the orientation signal, the shape of the magnetic resonance line of the metastable helium atoms was analyzed for the case of a large resonance radio-frequency magnetic field H_1 following the method suggested in [45]. In our experiment this part did not exceed 10%. At small values of the amplitude of the radio-frequency field $h \leq 0.1$ mOe, the magnetic resonance line shape was nearly Lorentzian, and the magnetic resonance line width depended mainly on the spin exchange, chemi-ionization, and other collisional relaxation processes in the absorption chamber in compliance with the equations (33)-(36).

The line widths were determined by measuring widths of the magnetic-resonance line at the half of the amplitude of the resonance signal.

To find the spin exchange and chemi-ionization rate constants sought, the increment $\delta f_{or,al}$ of magnetic resonance line widths (both for oriented and aligned metastable helium atoms) was determined appearing in response to heating the absorption chambers

$$\delta f_{or,al} = \Delta f_{or,al}(N) - \Delta f_{or,al}(N_0) \,, \tag{50}$$

where $\Delta f_{or,al}(N_0)$ is the line width at the temperature T when the concentration of alkali atoms is negligible and therefore it does not influence the line width, and $\Delta f_{or,al}(N)$ is the line width of the magnetic resonance line at the concentration of alkali atoms equal to N. To get the plots of $\Delta f_{or,al}$ versus the concentration of alkali atoms, the temperature was "converted" into the value N in accordance with the data in [46].

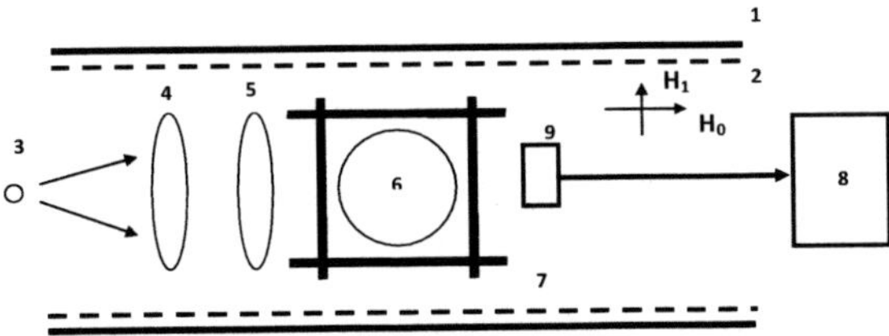

1 - Five layers magnetic shield, 2 - Solenoid,
3 - Resonance lamp,
4 - Polaroid,
5 - λ/4 - mice plate,
6 - Absorption chamber
7 - Helmholtz coils
8 - System of registration
9 -. Photo-detector
H_1 is the resonance radio-frequency magnetic field,
H_0 is the static magnetic field

Figure 1. Experimental setup

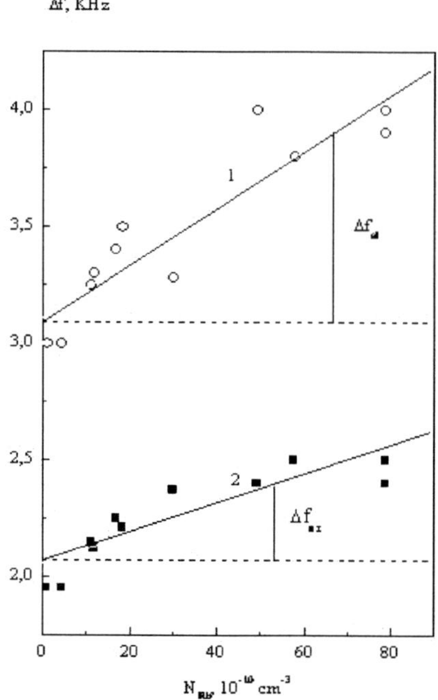

Figure 2. Dependence of the magnetic resonance line width of aligned (1) and oriented (2) metastable 2^3S_1 helium atoms on the concentration of ground ($5^2S_{1/2}$) state rubidium atoms (N_{Rb}) [40]

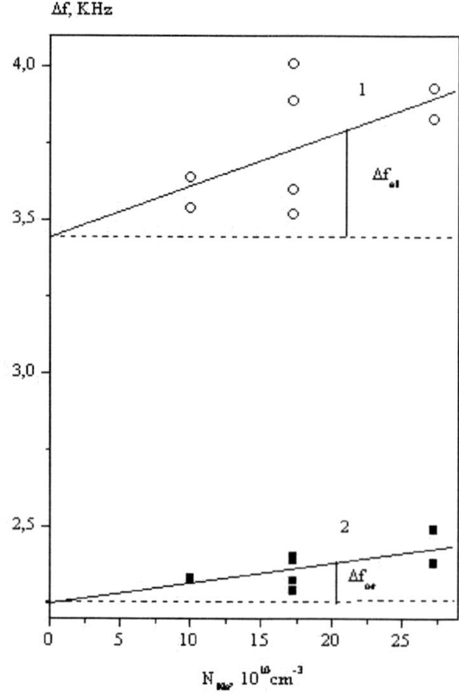

Figure 3. Dependence of the magnetic resonance line width of aligned (1) and oriented (2) metastable 2^3S_1 helium atoms on the concentration of ground ($3^2S_{1/2}$) state sodium atoms (N_{Na}) [38]

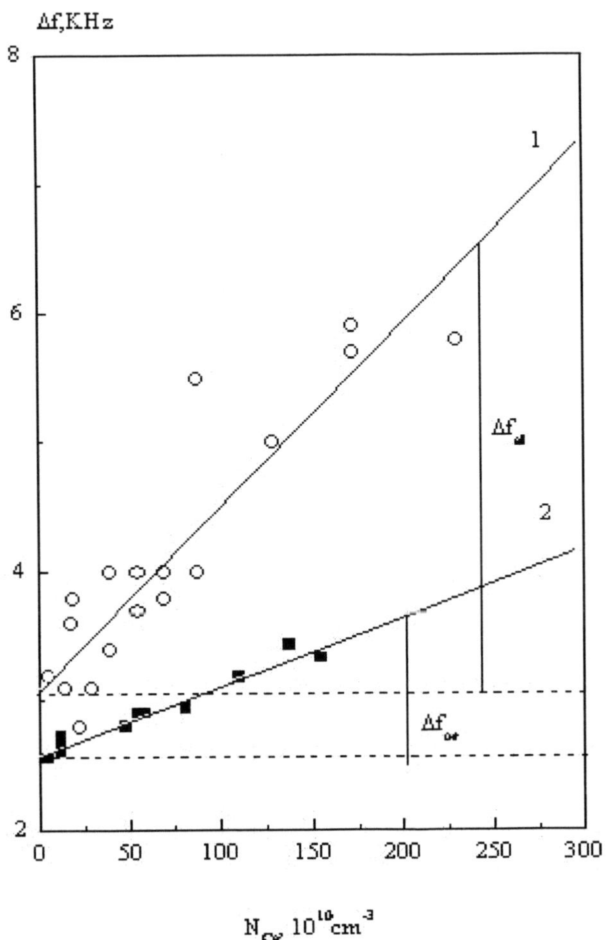

Figure 4. Dependence of the magnetic resonance line width of aligned (1) and oriented (2) metastable 2^3S_1 helium atoms on the concentration of ground ($6^2S_{1/2}$) state cesium atoms (N_{Cs}) [41]

Figure 2, Figure 3, and Figure 4 show some examples of the plots $\Delta f_{or,al}(N)$ for the Rb, Cs, and Na atoms. As it can be seen from these figures, the contributions due to the oriented (2) and aligned (1) atoms to the magnetic-resonance line width differ significantly. It should be noted that the width of the magnetic resonance line of polarized helium atoms also depends on the rate of diffusion, spin exchange with electrons, relaxation in the volume of the chamber, etc. (see equations (33)-(36)). In the experiments described variations (with increasing temperature) of the magnetic resonance line width due to these processes do not exceed the measurement error. Contribution due to these processes to the magnetic-resonance line was taken into account via extrapolation of the magnetic resonance lines to the zero concentration of the alkali atoms. The results of such extrapolation are presented in the Figure 2-Figure 4. Using Eqs. (37), (50) and Figs. 2-4, we can get the desirable rate constants of spin exchange and chemi-ionization processes. Below are the experimental results on chemi-ionization and spin exchange rate constants which were received using method of optical orientation as described above. The rate constants are found to be as follows:

for the $He(2^3S_1)$-$Cs(6^2S_{1/2})$ system - $C_{se}= (2.8\pm0.8)\times10^{-9}$ cm^3s^{-1} and $C_{ci}= (1.0\pm0.3)\times10^{-9}$ cm^3s^{-1} at T=320 K [41], for the $He(2^3S_1)$ - $Na(3^2S_{1/2})$ system - $C_{se}= (2.3\pm1.1)\times10^{-9}$ cm^3s^{-1} and $C_{ci}= (2.9\pm1.4)\times 10^{-9}$ cm^3s^{-1} at T=420 K [38], and for the $He(2^3S_1)$ - $Rb(5^2S_{1/2})$ system-C_{se} = $(3.1\pm0.6)\times 10^{-9}$ cm^3s^{-1} and $C_{ci}= (1.8\pm0.4)\times10^{-9}$ cm^3s^{-1} at T=320 K [40].

In conclusion, it should be noted that the spin exchange rate constants have been determined for the first time in our papers, whereas the chemi-ionization rate constants have been determined earlier by different authors. The experimental data presented above have been compared with experimental and theoretical data of other authors, the comparison being done in [38, 40, 41].

B. Magnetic Resonance in a System of Optically Polarized Particles with Electron Spins S=1 [($He(2^3s_1)$ - $O_2(^3\Sigma^-_G)$]

The experiments on optical orientation of helium atoms in the 2^3S_1 metastable state were carried out in the helium-oxygen gas mixture placed in the absorption chamber (6) (see Figure 1).The experimental procedure was as described above and analogous to that with the helium-alkali mixture. The experiments were run at temperature T=300 K. In order to maintain a constant concentration of molecular oxygen in the absorption chamber during the experiment, the experimental setup was equipped with ballast volume. This was necessary because of the relatively low concentration of the O_2 molecules (of the order of 10^{13} - 10^{14} cm^{-3}) and because of the possibility of molecules diffusing towards the wall of the absorption chamber. Experimentally, the magnetic resonance line widths were determined for the oriented and aligned He^* atoms at different partial concentrations of the molecular oxygen in the absorption chamber. This made it possible to separate the part of magnetic resonance (MR) line width caused by the addition of the molecular oxygen to helium. Figure 5 shows the MR line width of the metastable helium atoms as a function of the concentration of molecular oxygen. One can see from the figure that the MR line widths of the oriented (2) and alignment (1) atoms are markedly different. According to (48), using the difference between the line widths for the orientation and alignmet signals we have obtained as a result the following relations between three cross sections:

$$\sigma_1^{tr} + \sigma_2^{tr} - \sigma_3^{tr} = (13 \pm 3)\cdot 10^{-16}\, cm^2 \qquad (51)$$

Although this quantity is rather difficult to analyze, because it includes three cross sections which are dependent on interference between the scattering amplitudes for all three terms of the ($He(2^3S_1)$ - $O_2(^3\Sigma^-_g)$) quasi-molecule, one can obtaine an important information concerning the interaction by using some model potentials.

INTERACTION POTENTIALS

During the collision of metastable helium atoms, possessing the electron spin $S_1=1$, with alkali atoms in the ground state (with the electron spin $S_2=1/2$), a quasi-molecule is formed that can be described by two terms with the total spins $S_s=1/2$ and $S_q=3/2$ with the corresponding doublet $V_d(R)$ and quartet $V_q(R)$ potentials. The spin exchange process is due to both terms, whereas the chemi-ionization is only due to the term of the lower multiplicity, since for the term of higher multiplicity the total spin of the system would not be conserved in the chemi-ionization process and hence the Wigner spin conservation rule would not be fulfilled [47]. The presence of the inelastic process for the doublet term can be taken into account by introducing a complex interaction potential

$$V_d(R) = U_d(R) + \frac{i}{2}\Gamma_d(R) \quad , \qquad (52)$$

whose imaginary part (the auto-ionization width $\Gamma(R)$) is responsible for a decrease in the concentration of particles due to the process of the ionization [31].

Below we are discussing the $He^*(2^3S_1)$ - $Na(3^2S_{1/2})$, $K(4^2S_{1/2})$, or $Li(2^2S_{1/2})$ systems. The parameters of the doublet term and the imaginary part of the interaction potential were calculated in [48-50]. It is known that the difference between the doublet and the quartet terms is determined by the exchange interaction in the $He^*(2^3S_1)$ -B $(n^2S_{1/2})$ quasi-molecule, where B is the alkali atom, i.e.

$$V_q = V_d(R) + V_{ex}(R) \quad . \qquad (53)$$

Therefore, in order to construct the quartet interaction potentials one has to determine the exchange interaction $V_{ex}(R)$ for all three systems under study. To do this, we have used the results from [51]. In accordance with [51], the exchange interaction is of the form

$$\begin{gathered} V_{ex}(R) = R^{\frac{2}{\alpha}+\frac{2}{\beta}-\frac{1}{\alpha+\beta}-1} \cdot \exp[-(\alpha+\beta)\cdot R]\cdot[J_{\alpha\beta}(R) + J_{\beta\alpha}(R)], \\ J_{\alpha\beta}(R) = A^2 \cdot B^2 \cdot 2^{-2-\frac{2}{\alpha+\beta}} \cdot \left(\frac{2}{\alpha+\beta}\right)^{2+\frac{1}{\alpha+\beta}} \cdot \Gamma\left(\frac{1}{\alpha+\beta}\right)\cdot\left(\frac{\alpha+\beta}{2\beta}\right)^{\frac{2}{\alpha}-\frac{2}{\alpha+\beta}} \cdot \\ \int_0^1 \exp\left[\frac{y-1}{\beta} + (\beta-\alpha)Ry\right](1-y)^{2/\beta-1/(\alpha+\beta)}(1+y)^{2/\alpha-2/(\alpha+\beta)}\left[1+\frac{\beta-\alpha}{\beta+\alpha}y\right]^{-2-1/(\alpha+\beta)} dy \end{gathered} \qquad (54)$$

where $\alpha^2/2$ and $\beta^2/2$ are the binding energy of electrons for the metastable helium atom and for the alkali atom in the ground state, R is the internuclear distance, and Γ is the Euler gamma- function.

As it was noted above, the parameters of the doublet term and the imaginary part of the interaction potential for $He(2^3S_1)$ - $Na(3^2S_{1/2})$, $K(4^2S_{1/2})$, $Li(2^2S_{1/2})$ were calculated in [48-50]. The exchange interaction for these systems and quartet terms were determined in [38,39,37].

At the interaction between particles with electron spins $S_1 = S_2 = 1$ (in our case such are the $He(2^3S_1)$ - $O_2(^3\Sigma^-_g)$, $He(2^3S_1)$ systems), the situation is similar to that discussed above. The main difference is now that the quasi-molecule should be described by three potentials : V_q (for the total spin $S_q = 2$), V_t (for the total spin $S_t = 1$), and V_s(for the total spin $S_s = 0$) and by two auto-ionization widths (for the total spins $S_s = 0$, $S_t = 1$). In order to calculate the spin-exchange and chemi-ionization cross sections, we have to know the interaction potentials of the systems under investigation:

$$V_{s,t}(R) = U_{s,t}(R) + \frac{i}{2}\Gamma_{s,t}(R) \quad , \tag{55}$$

$$V_q(R) = U_q(R) \quad . \tag{56}$$

In [52] the interaction potentials were calculated for the $He(2^3S_1)$-$O_2(^3\Sigma_g)$ system. However, the main attention was devoted to calculating the real part of the potentials ($U_{s,t,q}(R)$). In order to get imaginary part of the potentials, the experimental data on the chemi-ionization cross sections were used. Complex potentials describing the quasi-molecule are given in the [30]. In the region of interest the singlet and triplet potentials of the He^*-O_2 quasi-molecule differ insignificantly (as for the He^*-He^* system [53]), so we can suppose that $V_s(R)=V_t(R)$ and $\Gamma_s(R)= \Gamma_t(R)$. The cross sections for these systems are determined by Eqs. (28) and (42)-(45) and taking into account the small difference between singlet and triplet terms we became

$$\sigma_s^{abs} = \sigma_t^{abs} = \sigma^{abs} \quad , \tag{57}$$

$$4\sigma_1^{tr} = 9\sigma_2^{tr} = 36\sigma_3^{tr} \quad . \tag{58}$$

In such case the Eq (47) has the form

$$\pi\Delta f_{or} = N_A v\left(\frac{4}{9}\sigma^{abs} + \frac{5}{36}\sigma^{tr}\right) \tag{59}$$

$$\pi\Delta f_{al} = N_A v\left(\frac{4}{9}\sigma^{abs} + \frac{13}{36}\sigma^{tr}\right)$$

Taking into account Eq. (59) and experimental data (Figure 5) one can get experimental values of spin exchange and chemi-ionization cross sections for the $He(2^3S_1)$-$O_2(^3\Sigma_g)$ system. At the temperature of T=300 K these cross sections are [30] as

$$\sigma^{abs} = 27 \cdot 10^{16}\ cm^2 \pm 20\% \tag{60}$$

$$\sigma^{tr} = 39 \cdot 10^{16}\ cm^2 \pm 20\%$$

CALCULATIONS OF THE SPIN EXCHANGE AND CHEMIIONIZATION CROSS SECTIONS

The complex potentials obtained make it possible to describe completely the spin exchange and chemi-ionization processes occurring in the interaction of a metastable helium atom with atoms o molecules with a non zero electron spin. As already noted above, a system consisting of a helium atom in the metastable state and atom or molecule with unpaired electron spin evolves in several terms. Some of these terms are complex and one is real. Therefore, it should be taken into account that the phase shifts of scattering from the complex potentials are complex ($\eta = \chi + i\lambda$) and that for the real term is real (η).

The scattering phases were determined in quasi-classical approximation using the Jeffries approximation [31]

$$\eta_l = \int_{R_0}^{\infty} F_1(R)dR - \int_{R_0'}^{\infty} F_0(R)dR \; , \tag{61}$$

where

$$\begin{aligned} &F_1^S(R) = \left[2\mu(E - V_S(R) - \frac{(l+1/2)^2}{2\mu R^2}) \right], \\ &S = s,t,q, \\ &F_0(R) = \left[2\mu E - \frac{(l+1/2)^2}{R^2} \right]. \end{aligned} \tag{62}$$

Here, E is the kinetic energy, R_0 and R_0' are the roots of equations $F_1^S(R) = 0$ and $F_0(R) = 0$ (for F_1^S, the largest root should be taken), and $V_S(R)$ is the atomic interaction potential corresponding to the total spin S of the system.

To analyze the chemi-ionization and spin exchange cross sections and compare them with the experimental values, one has to turn from energy dependence of cross sections to their dependence on temperature. To this aim, the calculated cross sections were averaged over the Maxwellian distribution of velocities. Using (11), (12), (14), (61), and (62) it is possible to calculate the interesting us cross sections. The results of such calculations are given in the following papers: $He(2^3S_1)$ - $Li(2^2S_{1/2})$ [37], $Na(3^2S_{1/2})$[38], $K(4^2S_{1/2})$[39], $H(1^2S_{1/2})$[17] and $He(2^3S_1)$ - $O_2(^3\Sigma^-_g)$[30], $He(2^3S_1)$[43].

In the case of optical orientation of atoms in an helium-alkali atom (or molecular) plasma, the orientation transfer between atomic particles is caused by concurrent elastic and inelastic processes - spin exchange and chemi-ionization. It is known that collisions with spin exchange result in a frequency shift of the magnetic resonance.

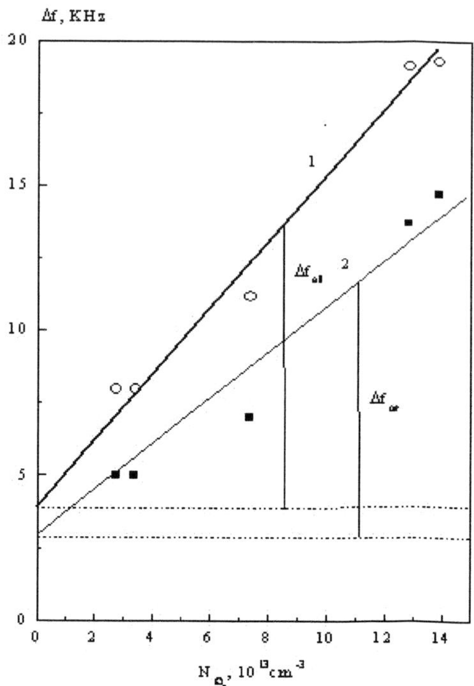

Figure 5. Dependence of the magnetic resonance line width of aligned (1) and oriented (2) metastable 2^3S_1 helium atoms on the concentration of the O_2 ($^3\Sigma^-_g$) molecules [20]

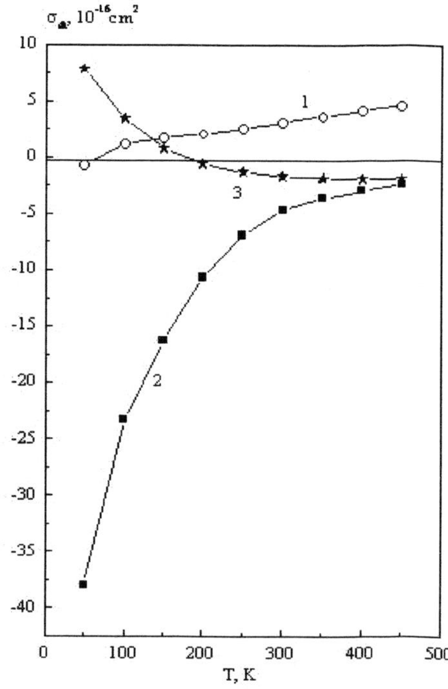

Figure 6. Temperature dependence of the cross section of magnetic resonance frequency shift. for the $He(2^3S_1)$-$K(4^2S_{1/2})$ (1), $He(2^3S_1)$-$Li(2^2S_{1/2})$ (2) and $He(2^3S_1)$-$Na(3^2S_{1/2})$ (3) systems

The shifts due to spin exchange have a substantially effect, in particular, on the performance of alkaline-helium magnetometers [22]. It was found possible to eliminate frequency shifts of the magnetic resonance in these instruments by choosing the optimal temperature of an absorption chamber [57]. This situation points to the presence of a rather sharp temperature dependence of the frequency shifts and to the fact that different collisional systems may differ in the shift sign.

Figure 6 shows the temperature dependences of the cross section of magnetic resonance frequency shifts for the $He(2^3S_1)$-$K(4^2S_{1/2})$ (1), $He(2^3S_1)$-$Li(2^2S_{1/2})$ (2), and $He(2^3S_1)$-$Na(3^2S_{1/2})$ (3) systems. These shift cross sections and shifts of magnetic resonance lines were calculated for the systems: $He(2^3S_1)$ - $Li(2^2S_{1/2})$ [54], $Na(3^2S_{1/2}$)[55], $K(4^2S_{1/2})$[56], $H(1^2S_{1/2})$[19], and $He(2^3S_1)$ - $O_2(^3\ \Sigma^-_g)$[30]. As we can see from Figure 6, the temperature dependences for $He(2^3S_1)$-$Li(2^2S_{1/2})$ (2) and $He(2^3S_1)$-$Na(3^2S_{1/2})$ (3) systems reverse signs.

Conclusion

At the interaction between the spin-polarized excited atom and paramagnetic ground state atom or molecule in gas discharge, elastic and inelastic processes can take place simultaneously. It means that besides the chemo-ionization of the atom or molecule at the expense of atom's excitation energy (inelastic process), an exchange of electrons is possible without a great depolarization (elastic process, or spin exchange). In such a case these two processes give rise to a remarkable spin polarization transfer between colliding particles. Influencing each other, these two processes result in a change in the spin exchange and frequency shift cross section values.

Spin dependent processes, proceeding simultaneously, result in a redistribution of electron spin polarization between the collision partners and also in transfer of the polarization to reaction products such as electrons, atomic and molecular particles, including ions. Further, the polarization attained by molecular particles can be redistributed between angular moments of molecular particles (electron and nuclear spins and rotation degrees of freedom) [58,59].

Acknowledgments

The authors thank Dr. S.P.Dmitriev, Dr. N.A.Dovator, Dr. V.D.Melnikov, and Dr A.I.Okunevich of the Ioffe Physico -Technical Institute of the Russian Academy of Sciences for useful collaboration.

References

[1] Anderson, LW; Pipkin, FM; Baird, JC. *Phys. Rev. Lett.*, 1958, *1*, 229-230.

[2] Anderson, LW; Pipkin, FM; Baird, JC. *Phys. Rev.*, 1959, *116*, 87-98.

[3] Walker, TG; Happer, W. *Rev. Mod. Phys.*, 1997, *69*, 629-642.

[4] Happer, W. *Rev. Mod. Phys.*, 1972, *44*, 169-249.
[5] Walters, GK; Colegrove, FD; Schearer, LD. *Phys. Rev. Lett.*, 1962, *8*, 439-442.
[6] Dupont-Roc, J; Leduc, M; Laloe, F. *J. de Phys.*, 1973, *34*, 961-976.
[7] Dupont-Roc, J; Leduc, M; Laloe, F. *J. de Phys.*, 1973, *34*, 977-987.
[8] Schearer, LD. *Phys.Rev.Lett.*, 1969, *22*, 629-631.
[9] Sevast'yanov, BN; Zhitnikov, RA. *Zh. Eksp. Teor. Fiz. (in Russian)* 1969, *56*, 1508-1518.
[10] Dmitriev, SP; Zhitnikov, RA; Okunevich, AI. *Zh. Eksp. Teor. Fiz. (in Russian),* 1976, *70*, 69-75.
[11] Blinov, EV; Zhitnikov, RA; Kuleshov, PP. *Tech. Phys. Lett.*, 1976, *2*, 117-118.
[12] McCusker, MV; Hatfield, LL; Walters, GK. *Phys. Rev. Lett.*, 1969, *22*, 817- 820.
[13] Okunevich, AI. *Zh. Eksp. Teor. Fiz. (in Russian)* 1976, *70*, 899-907.
[14] Okunevich, AI. *Opt. Spektrosk.(in Russian),* 1982, ***52*, 7-10.**
[15] Dmitriev, SP; Zhitnikov, RA; Kartoshkin, VA; Klement'ev, GV; Melnikov, VD. *JETP,* 1983, *58*, 485-491.
[16] Kartoshkin, VA; Klementiev, GV; Melnikov, VD. *JETP Lett.*, 1984, *39*, 156-159.
[17] Klementiev, GV; Melnikov, VD; Kartoshkin, VA. *KHIMICHESKAYA FIZIKA (in Russian),* 1985, ***4*,** 37-41
[18] Hickman, AP; Morgner, H. *J.Phys.B- Atomic and Molecular Physics* 1976, *9*, 1765-1787.
[19] Kartoshkin, VA; Klementiev, GV; Melnikov, VD. *Zh. Tekh. Fiz.(in Russian),* 1985, 55, 131-136.
[20] Kartoshkin, VA; Klementiev, GV**.** *Optik. Spektrosk. (in Russian),* 1987, *63*, 465-467.
[21] Klementiev, GV; Kartoshkin, VA; Dmitriev, SP; Melnikov, VD. *Physica Scripta,* 1996, *54*, 174-178.
[22] Blinov, EV; Zhitnikov, RA; Kuleshov, PP. *Zh. Tekh. Fiz. (in Russian)* 1979, 49, 588-596.
[23] Dmitriev, SP; Zhitnikov, RA; Okunevich, AI. *Zh. Tekh. Fiz. (in Russian)* 1982, *52***,** 1235-1236**.**
[24] Dmitriev, SP; Okunevich, AI**.** *Optik. Spektrosk.(in Russian),* 1982, *52,* 5-7.
[25] Okunevich, AI**.** *Optik. Spektrosk. (in Russian)* 1983, *54,* 787-794.
[26] Dmitriev, SP; Dovator, NA; Zhitnikov, RA; Kartoshkin, VA; Melnikov, VD**.** *JETP Lett.,* 1997, *65*, 399-401.
[27] Kartoshkin, VA; Klement'ev, GV. *Optik. Spektrosk. (in Russian),* 1996, *80,* 372-377.
[28] Blinov, EV; Ginzburg, BI; Zhitnikov, RA; Kuleshov, PP. *Zh. Tekh. Fiz.(in Russian)***,** 1984, *54***,** 287-292**.**
[29] Blinov, EV; Zhitnikov, RA; Kuleshov, PP. *Zh. Tekhnich. Fiz. (in Russian),* 1982, *52,* 904-908**.**
[30] Kartoshkin, VA; Klementiev, GV. *Optik.Spektrosk.(in Russian),* 1996, *80*, 608-613.
[31] Mott, NF; Massey, HSW. *The Theory of Atomic Collisions;* 3 ed;Clarendon Press; Oxford, UK, 1965, 858.
[32] Bender, PL. *Phys.Rev.A,* 1964, *134*, A1174-A1180.
[33] Dalgarno, A; Rudge, MJ. *Proc. Roy. Soc., (London)* 1965, *A286*, 519-524
[34] Balling, LC; Hanson, RJ; Pipkin, FM. *Phys.Rev.A,* 1964, *133*, 607-626.
[35] Grossetete, F. *J. de Phys.,* 1964, *25*, 383-396.
[36] Baranger, M. In Atomic and Molecular Processes; Bates, DR; Academic Press; Spectral

line broadening in plasma; New York; NY,1962; pp.493-549
[37] Kartoshkin, VA. *Tech. Phys. Lett.,* 2007, *33*, 1047-1049.
[38] Dmitriev, SP; Dovator, NA; Kartoshkin, VA. *Tech. Phys*., 2009, *54*, 1557-1559.
[39] Kartoshkin, VA; Klementiev, GV. *Opt. Spectrosc*., 2007, *102*, 510-513.
[40] Dmitriev, SP; Dovator, NA; Kartoshkin, VA. *Tech. Phys. Lett*., 2008, *34*, 693-695.
[41] Dmitriev, SP; Dovator, NA; Kartoshkin, VA. *JETP Lett., 1997, 66, 151-154.*
[42] Dmitriev, SP; Kartoshkin, VA; Klement'ev, GV. Tech. Phys,.1997, 42, 18-21.
[43] Kartoshkin, VA. *Opt. Spectrosc.,* 2008, *105*, 657-662.
[44] Pomerantsev, NM; Ryzhkov, VM; Skrotskii, GV. Fizicheskie osnovy kvantovoi magnitometrii (in Russian) (Physical Foundation of Quantum Magnetometry) ; Nauka: Moskow, SU,1972, 448.
[45] Klementiev, GV; Melnikov, VD; Kartoshkin, VA. *Opt. Spectrosc.,* 1988, 65, 468-470.
[46] Nesmeyanov, AN. Vapor Pressure of the chemical elements; Gary R; Elsevier Pub. Co., Amstermdam- New York, 1963, 462.
[47] Wigner, E. *Gotting.Nachr*., 1927, *4*, 375.
[48] Cohen, JS; Martin, RL; Lane, NF. *Phys.Rev.A,* 1985, *31*, 152-160.
[49] Scheibner, KF; Cohen, JS; Martin, RL; Lane, NF. *Phys.Rev.A,* 1897, *36*, 2633-2643.
[50] Kimura, M; Lanc, NF. *Phys.Rcv.A,* 1990, *41*, 5938-5942.
[51] Smirnov, BM. *Asimptoticheskie metody v teorii atomnih stolknovenii* (in Russian) (Assimpthotical methods in the theory of atomic collisions) Atomizdat: Moscow, SU, 1973, 296.
[52] Leisin, O; Morgner, H; Müller, WZ. *Physik A.-Atoms and Nuclei,* 1982, 304, 23-30.
[53] Muller, MW; Merz, A; Ruf, MW; Hotop, H; Meyer, W; Movre, M. *Z.Phys.D- Atoms, Molecules and Clusters,* 1991, 21, 89-112.
[54] Kartoshkin, VA. *Tech. Phys*., 2009, *54*, 737-739.
[55] Kartoshkin, VA. *Opt. Spectrosc.,* 1999, 87, 178-180.
[56] Kartoshkin, VA. *Tech. Phys*., 2007, *52*, 123-125.
[57] Blinov, EV; Ginzburg, BI; Zhitnikov, RA; Kuleshov, PP. *Zh. Tekh.Fiz. (in Russian),* 1984, *54,* 2315-2322.
[58] Kartoshkin, VA; Klementiev, GV; Melnikov, VD. *Opt. Spectrosc*., 1997, *82*, 365-367.

In: Helium: Characteristics, Compounds…
Editors: Lucas A. Becker
ISBN: 978-1-61761-213-8

Chapter 5

EFFECT OF HE PLASMA TREATMENT ON THE REACTIVITY OF POROUS METHYL-DOPED SILICON DIOXIDE LAYERS

F. N. Dultsev
Institute of Semiconductor Physics SB RAS, Novosibirsk, Russia

1. ABSTRACT

Helium in the state of plasma has the properties strikingly different from those of atomic helium. Using helium plasma, one may obtain VUV (vacuum ultraviolet) radiation. At present, He VUV radiation is widely used to modify porous layers based on silicon dioxide. In the present work, we demonstrate the experimental data on the action of He plasma treatment on methyl-doped silicon dioxide layers. It is shown that the He VUV radiation affects the materials with different pore sizes in principally different ways. We propose a model based on the experimental data. The reasonable character of the model was confirmed by quantum chemical calculations. According to our model, the action of He plasma is considered mainly as the VUV radiation (21 eV) causing rupture of chemical bonds. Detachment of hydrogen atom from the methyl groups causes the appearance of the positive charge on silicon atoms, which leads to an increase in dπ-pπ overlapping, which, in turn, decreases the reactivity of the layers. This causes passivation of the layers.

2. INTRODUCTION

Helium is widely used at present in plasmachemical processes in semiconductor industry. Many unique properties of helium contribute into its diverse application areas. Due to the high thermal conductivity and fluidity of helium, it is used in electronic industry to provide thermal contact between substrates and holders in reactors. Another important feature of helium is the possibility to generate the vacuum ultraviolet (VUV) radiation in helium plasma; this property is widely used in the synthesis of dielectric layers based on silicon dioxide. The addition of helium into the reactor during plasma-enhanced chemical vapour deposition (PECVD) processes causes an increase in the contribution from the photochemical reaction, which allows one to decrease the RF power necessary for deposition, and to obtain more uniform films at lower temperatures [1-5]. Recently, PECVD processes have won application for obtaining carbon nanotubes (CNTs) and nanofibers (CNFs), and the use of He allowed decreasing the synthesis temperature [6].

Recently developed new class of dielectrics with low permittivity, so-called low-K dielectrics, are widely used in modern microelectronics to decrease the capacity of interlayer insulators, which is necessary to increase the switching rate in integrated circuits. Low-k materials should possess sufficient modulus of elasticity, hardness, rupture strength, thermal conductivity. In this situation, the main method of achieving low permittivity is to obtain a porous material. Silicon dioxide layers are suitable for this purpose. So, not only the search for new methods of obtaining low-k materials with sufficient mechanical strength but also better understanding of the chemical nature of processes is important.

However, the use of highly porous materials brings about the necessity to take into account the higher chemical activity of their surface, which causes complications of technological application. To decrease the reactivity of porous materials, methods of their stabilization and passivation are being developed. Treatment in helium plasma is used most frequently [7-15]. Some examples show that plasma treatment [14, 15] causes pore closure, which results in a decrease in the chemical activity of these layers. Quite contrary, the authors of [16, 17] showed that the treatment of SiO_2 films in the plasma of inert gases caused an increase in the rate of etching these layers in 1% HF solution. Choosing a proper inert gas and excluding the ion component in surface treatment, the authors demonstrated that the film reactivity changes depending on the energy of UV radiation of the plasma.

Understanding of the dependence of chemical properties on film structure will allow one to modify the surface in the desired direction. The key to such understanding is in investigation of the electron structure of surface atoms.

The effect of porosity on chemical activity may be considered in two aspects: first, due to the developed surface, with the reaction proceeding not only at the interface but in volume; this effect may be called macrostructural. Second, changes of the electron structure of surface atoms, which leads to changes in their chemical reactivity; this effect may be called microstructural. A decrease in pore radius (to 1 nm and less) causes changes in the electron structure of surface atoms. This is accompanied by the changes in chemical properties. So, the electron structure of surface atoms should be necessarily taken into account in the case of the films with small pores (1 nm in radius and less).

In the present work, on the basis of experimental data, we propose a mechanism describing passivation of the surface of the films based on silicon dioxide under the action of UV radiation that was generated in helium plasma. The effect of the treatment of porous films in plasma on the reactivity of the films was simulated. Theoretical consideration is useful also due to the fact that the instrumental analysis of these systems is hindered because the concentration of surface groups may be small, often below the detection limit of direct spectroscopic methods. In this case, modeling may be the basic method to predict the properties of thus treated surfaces.

3. EXPERIMENTAL

The reactivity of layers was evaluated on the basis of etching rate, so experiments included the following sequence of operations: at first, the preliminary UV-treatment of the samples was carried out, then etching. Plasmachemical pretreatment and etching were performed with the Oxford PlasmaLab-80 RIE chamber, modified in order to exclude the direct effect of the plasma ions on the sample.

3.1. Preliminary treatment

The Oxford PlasmaLab-80 set-up has two sources: RIE and ICP. Preliminary treatment was performed using ICP (remote plasma) alone. Pressure in the chamber was 3 mTorr, the ICP power was varied from 400 to 600 W. The sample was placed so that the effect of the ions and the action of the UV radiation from the ICP source on the sample could be excluded. For this purpose, a net and a mirror were built into the chamber (the sample was placed in the shade from the ICP source); similar design was described in [17,18]. Helium and argon were used as plasma-forming gases. Three types of UV radiation sources were used in the work: helium plasma (21 eV), argon plasma (10 eV), medium-pressure mercury lamp (4 eV).

3.2. Etching

After irradiation, the samples were transferred into a similar chamber for etching; this chamber was not equipped with any additional nets or screens. Etching was carried out in the Standard Plasma Processing mode. Etching parameters were: pressure in the chamber, 60 mtorr, power of RIE and ICP generators was 100 and 60 W, respectively. Films were etched using the standard gas mixtures NF_3:Ar=1:2, CF_2Cl_2:Ar = 1:2. Etching conditions were chosen so that the formation of a polymer film was excluded.

3.3. Methods

Changes in film thickness were monitored with a LEF-3M ellipsometer (λ = 632.8 nm, light incidence angle 55-75 deg, measurements were carried out at 3-5 angles). Porosity was measured using ellipsometric adsorption porosimetry (EAP) with the help of an automatic ellipsometer (λ=632.8 nm, light incidence angle 70 deg). The procedure was described in more detail in [19]. Measurement and calculation of micropores were carried out similarly to [20].

3.4. Materials

Porous layers used in the present investigation were obtained with the help of sol-gel procedure or chemical vapour deposition (CVD). The samples with the pore radius 5 nm and larger were obtained using sol-gel technology. The initial porous methyl-doped oxide (MDO) films were characterized by about 25% porosity and the dielectric constant of 2.5, film density of 1.25 g/cm^3 and refractive index of 1.33 at the wavelength λ = 632.8 nm, film thickness 600-640 nm. Pore size distribution for these layers is shown in Figure 1, curve 1.

The layers with pore radius smaller than 5 nm were obtained by CVD. Layer porosity was about 25%, too; thickness was 600-650 nm. The fraction of methyl groups in the films of both types was 25% with respect to silicon. Pore size distribution for the samples with the average pore radius of 1.1 nm is shown in Figure 1, curve 2.

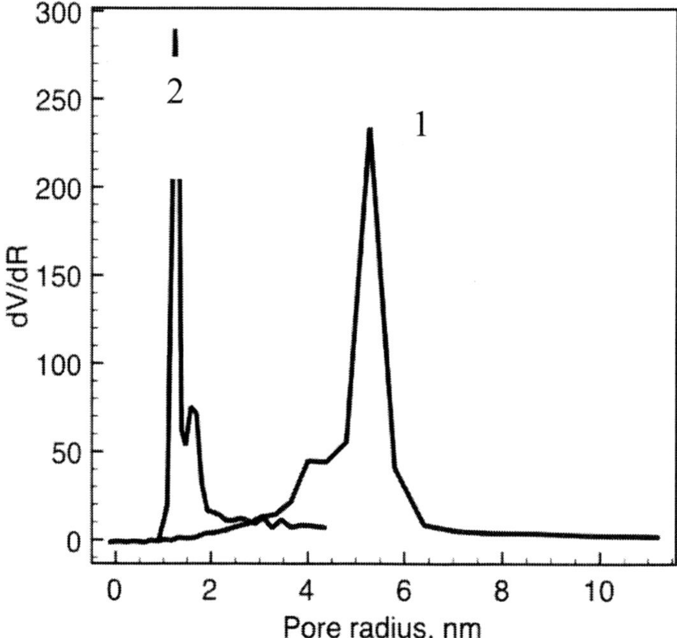

Figure 1. Pore size distribution for the initial layers.

In order to check the plasma power and thus to decrease scattering of measured etching rates, we monitored the etching rate of thermal silicon dioxide layers. The films of thermal silicon dioxide were obtained by the oxidation of the silicon substrate at a temperature of 1000 °C; their thickness was 100 nm, refractive index 1.46 at λ=632.8 nm.

4. Results and Discussion

The effect of the UV radiation on the chemical properties of the layers is illustrated in Figure 2 showing the dependence of the initial etching rate on pore size after pretreatment in He plasma. Etching was carried out in the gas mixture NF_3 + Ar. The effect of the UV pretreatment on the chemical properties of layers depends on the energy of radiation; with an increase in energy, the effect of preliminary treatment increases. The largest effect is observed in the case of He plasma (21 eV), while there is almost zero effect for the radiation of the mercury lamp (4.3 eV), see Figure 3 showing changes in etching rate depending on the energy of UV-pretreatment for the layers with pore radius 1 nm. So, it may be stated that changes of the reactivity of layers are connected exactly with the porous structure of layers. The chemical composition of the layers obtained using different procedures (sol-gel or CVD) was approximately the same with respect to methyl group content (the fraction of methyl groups was about 25%), the films had the same refractive index (n = 1.23 - 1.30) and total porosity (about 25-30%). The ellipsometric investigation of the porous structure showed that the main difference is in pore size. The results obtained using EAP are shown in Figure 1. Pore size distribution is shown for pore radius 5.5 nm (this is the maximal radius for the layers studied in the present work) and 1.1 nm (minimal). Calculation of pore size distribution was carried out using the desorption branch of the adsorption-desorption isotherm. Toluene was used as an adsorbate.

The mechanism of the influence of UV radiation on porous layers may be considered in the following manner. On the basis of the results shown in Figure 2, we chose two samples with pore radii 1 nm and 5.5 nm that exhibited the most clearly pronounced effects of the action of UV radiation but with opposite signs: in the first case, film passivation occurs, while in the second case etching rate increases, that is, surface activation occurs. After the UV treatment, the samples were transferred into the etching chamber. Etching was carried out in various gas mixtures but the observed regularities were the same for different gas compositions. In our experiments we carried out plasmachemical etching in CF_2Cl_2:Ar mixture. One can see from the experimental dependencies of the rate of etching in CF_2Cl_2 on the time of preliminary irradiation in helium plasma that the etching rate is essentially dependent on pore size.

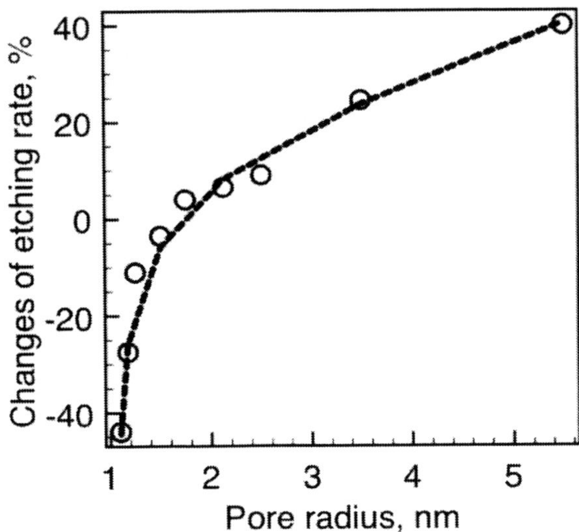

Figure 2. Dependence of the relative etching rate on pore radius after irradiation (21 eV). Etching was carried out in the plasma NF_3 : Ar = 1:2, total pressure in the chamber was 60 mtorr. Etch time: 40 s, irradiation time: 1 min.

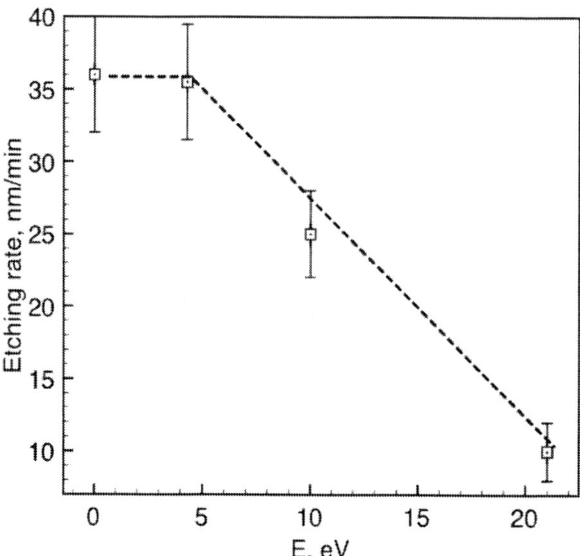

Figure 3. Dependence of etching rate on the energy of preliminary irradiation (pretreatment) for microporous layers.

For instance, for the samples with pore radius 1.1 nm, and for thermal silicon dioxide, etching rate decreases with irradiation time, while for the samples with pore radius 4-6 nm etching rate increases. Changes in the etching rate for the film are shown in Figure 4. One can see that the etching rate changes with an increase in irradiation time. Etching time was chosen constant. This was done in order to exclude the uncertainty connected with non-uniformity over the film thickness during irradiation. After irradiation, the refractive index did not change for all the film types. In parallel, during one experiment, etching of thermal silicon dioxide was carried out either after treatment in helium plasma or without such a treatment. After treatment in helium plasma, the average rate of etching of thermal silicon dioxide in CF_2Cl_2 plasma decreased in all the cases. The dependence of etching rate on irradiation time is shown in Figure 4 (curve 1). So, one may state that the treatment in helium plasma leads to a decrease in the reactivity of layer surface both for layers with small pores (1.1 nm) and for thermal silicon dioxide, while for the samples with larger pores (5.5 nm) we observe an increase in etching rate (curve 3 in Figure 4). It should also be mentioned that the etching rates for non-irradiated samples with small pores (1.1 nm in radius) and larger ones (4-6 nm) did not differ much; for instance, etching rate for the films with pore radius 5.5 nm was 35-36 nm/min, while for the films with pore radius 1.1 nm etching rate was 29-30 nm/min; however, after irradiation etching rates for these two types of films changed in quite different manners.

Changes of etching rate with etching time (depth) for the samples with smaller and larger pores are shown in Figure 5. This dependence points to the non-uniformity of plasma action over the film thickness: one can see what is the depth affected by the plasma during irradiation. We see that the irradiation affects only the surface layer; its thickness is only weakly dependent on irradiation time.

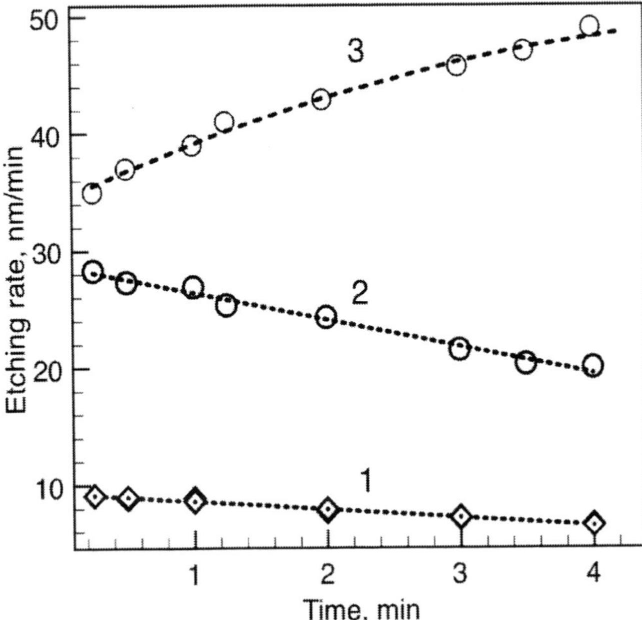

Figure 4. Dependence of etching rate on the time of irradiation in plasma, for the constant etching time of 6 min: 1 – microporous layers, 2 – mesoporous layers. Etching was performed in the plasma CF_2Cl_2:Ar=1:2, total pressure in the chamber was 60 mtorr.

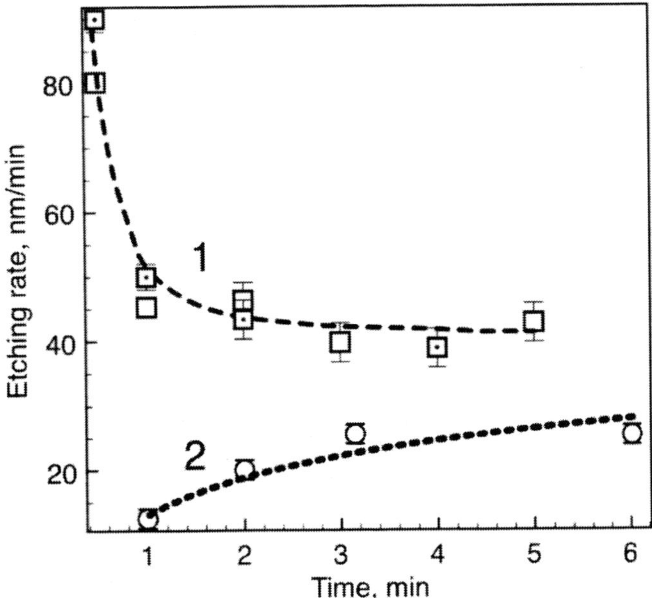

Figure 5. Dependence of etching rate on etch time for the constant time of UV irradiation: 1 – for the films with pore radius 5.5 nm, irradiation time 0.5 min (empty squares), 1 min (squares with dots inside), etching rate without irradiation: 35-36 nm/min; 2 – for films with pore radius 1.1 nm, irradiation time 1 min, etching rate without irradiation: 29-30 nm/min.

Let us consider possible reasons of changes in the reactivity of thin SiO_2-based films. First of all, in order to explain these reasons, one should consider the electronic structure of silicon dioxide surface. Due to the large charge of the nucleus of Si atom, the energy of the unoccupied 3d orbitals decreases, and they turn out to be situated closely to the 2p orbials of O. So, the possibility of additional overlapping of 3d orbitals of Si and 2p orbitals of O atoms arises, which strengthens the bond between silicon and oxygen. So, the diversity of chemical properties of silicon dioxide is connected with the lability of Si-O-Si bond. The introduction of additional groups (methyl groups in the case under consideration) can lead to rather large changes in the reactivity of this bond. To explain the regularities observed in our experiments, we performed modeling with the consideration of the electronic structure of Si-O-Si lattice fragment that forms a pore. The structure of such a pore and the method of its calculation are considered below.

4.1. Molecular structure

It is known that the basic topological units determining the middle-range order in vitreous SiO_2 and in amorphous silicon dioxide layers are polyhedrons (tridimite groups) composed mainly of four 6-membered rings (4-, 5- and 7-membered rings are also present). These rings are composed of $SiO_{4/2}$ tetrahedrons [21]. Because of this, for modeling, we used the structure formed by the rings composed of –Si–O–Si–O atoms. Each ring contains 6 silicon atoms. These small rings are arranged in the form of a larger ring with the inner radius equal to the pore radius (in the case under consideration, the number of small rings was chosen so that a pore 1 nm in radius was formed). The cylindrical pore is built of these large rings; such a ring is shown in Figure 6. The pore composed of these rings is shaped like a cylinder. The fraction of methyl groups is 20 – 25% of the number of silicon atoms. This range corresponds to the real films with the porosity about 25% and pore radius 1 nm, for example, as reported in [22]. This approach was described in more detail in [23]. Calculation of the adsorption complex (AC) was carried out by means of molecular mechanics (MM2), while the structure of the fragment shown in Figure 6 was optimized using MNDO/PM3. A detailed description of the approach and its applications was presented in [24, 25].

The modeling procedure starts with optimization of the structure shown in Figure 6 for the minimum of total energy; the accompanying changes in Si–O–Si angles were assessed; the distribution over the angle values was plotted. For this purpose, the range 135-153 ^{o} was divided into 10 intervals. Total number of Si-O-Si angle values in the structure under simulation was 58. Then we plotted the dependencies of the number of angles on angle value in the form of the diagram, see Figures 7-8. We made this to simplify the comparison of our calculation results with the experimental FTIR data obtained by other researchers. Calculation results suggest that the dependence of Si-O-Si angles on the presence of substituent starts to appear for a ring with the radius not larger than 1-2 nm. The calculated angle distribution shown in Figures 7-8 is directly connected with the frequency of Si–O–Si bond vibrations in the IR spectra within the range 1000 – 1200 cm^{-1}. Table 1 shows the experimentally determined vibrational frequencies of Si-O bond depending on the value of Si–O–Si angle, these data were determined from the experimental FTIR spectra reported in literature, in particular in [15].

Table 1. Wavenumbers of Si-O-Si bond vibrations determined from experimental FTIR spectra [15], depending on Si-O-Si bond angle.

Si–O–Si angle, deg	<140	144	155
Wavenumber, cm^{-1}	1000	1070	1150

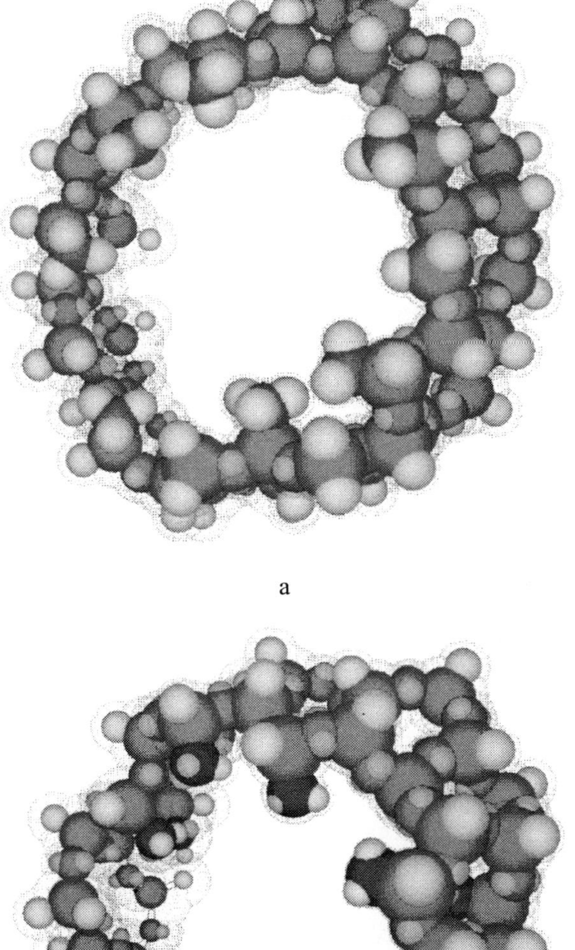

a

b

Figure 6. Modelled structure imitating a pore. a – initial structure, b – after substitution of methyl groups for NH_2 groups.

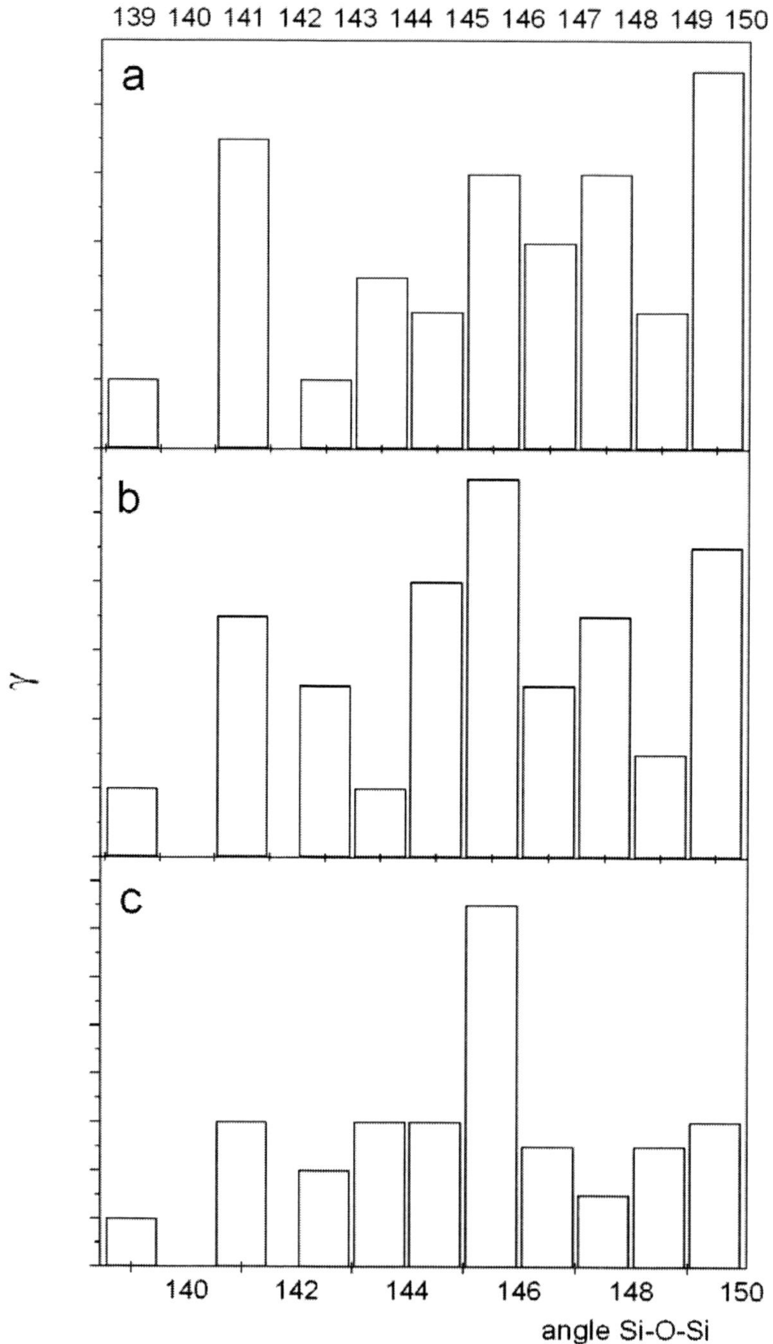

Figure 7. Calculated distribution of Si–O–Si angles: (a) initial structure, (b) during the formation of CH*-groups, which corresponds to the film after UV irradiation. The formation of bonds with the Si-O-Si angle within 144-146 deg. corresponds to the line at 1075 cm^{-1} in the experimental FTIR spectrum [15], (c) during the interaction with ammonia after He plasma treatment.

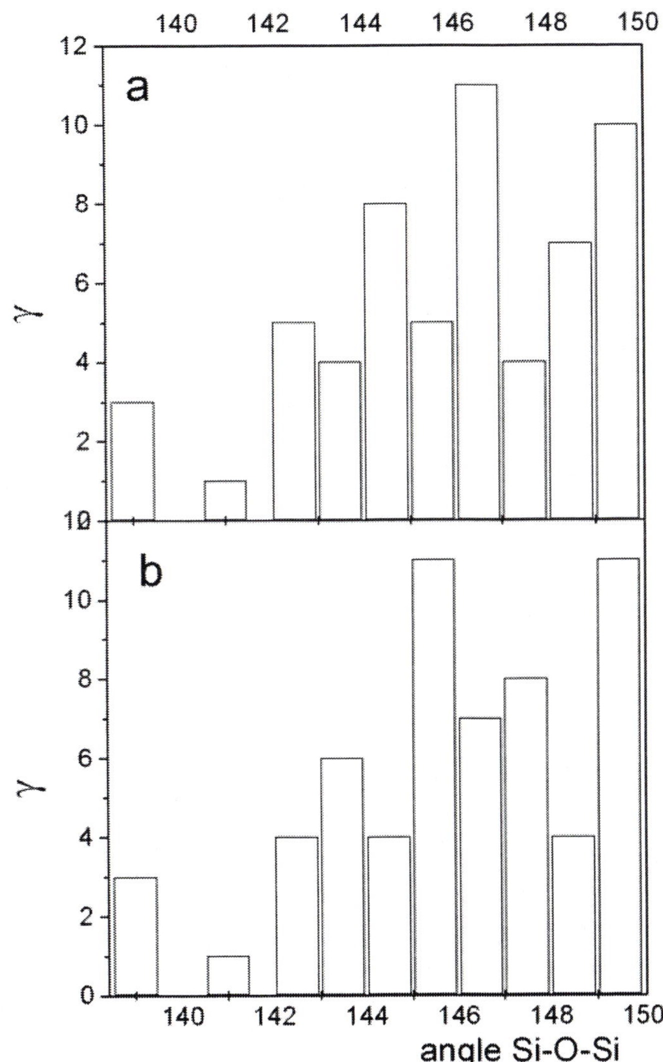

Figure 8. Distribution of Si–O–Si angles: (a) after substitution of methyl groups for NH2 groups; (b) with the formation of CH2–NH2 bonds.

Before we pass to discussing the results of experiments and modeling, the initial conditions and assumptions are to be formulated. Firstly, the treatment in He plasma is considered mainly as the UV irradiation. To remind, the energy of the UV radiation for helium is about 21 eV (480 kcal/mol), while for argon it is 10 eV (210 kcal/mol) [14]. So, the energy of UV radiation in plasma is sufficient to break not only C-H and Si-C bonds but also Si-O bonds. The UV radiation of He plasma is able to break Si-O bonds in the material, but in view of the impossibility to remove the formed fragments, the most probable process is recombination with the distribution of the excess energy over the lattice. It may be assumed that this process has no direct influence on the change of porosity. This assumption is not in conflict with the results reported in [15, 17, 26]. So, following the approach described by us, we do not consider the effect of the UV radiation of plasma causing rupture of the strong Si-O-Si bonds, though we do not exclude this possibility as well. It should also be noted that the appearing additional dπ-pπ overlapping causes strengthening of Si-O bonds.

Considering the energies of Si-C and C-H bonds we should stress that the energy of a C-H bond is strongly dependent on its surroundings. For instance, the energy of C-H bond in pentane is about 395 kJ/mol (94.5 kcal/mol), while in toluene it is 359 kJ/mol (85.7kcal/mol) [27]. In other words, the shift of the electron density causes changes in the acidity of H atom (the acidity is higher for smaller bond energy). Since the charge of the nucleus is larger for silicon than for carbon, the acidity of the H atom of a methyl group bound to silicon is higher than in hydrocarbons; the energy of the bond decreases (by about 5 kcal/mol, according to our calculations). However, it should be kept in mind that the energy of Si-C bond can increase due to π-conjugation. So, the energies of Si-C and C-H bonds can be considered to be comparable.

Below the effect of various kinds of treatment on the reactivity of methyl-doped layers will be considered; with the help of modeling, the mechanism explaining these effects will be proposed.

4.2. Modeling

The action of He plasma on silicon dioxide layers with pore size less than 2 nm was considered mainly as the interaction with CH_3 group because this group is the most accessible one from the viewpoints of energy and stoichiometry. It was assumed in calculations that the detachment of hydrogen atom or entire CH_3 group occurs. Calculation was carried out through the adsorption complex. This approach allowed us to determine the most probable reaction route. The calculated distribution of Si–O–Si angle values depending on the amount of the introduced CH_2* (CH_x) groups is shown in Figure 7a-b. One can see from this distribution that the removal of hydrogen atoms is accompanied by an increase in the number of bonds having the angle 144-146 degrees.

Considering such a surface treated in ammonia plasma, we may assume that NH_2* interacts with the CH_2* group giving the chain Si – CH_2 – NH_2. After the plasma is switched off, the unshared electron pair on nitrogen atom forms a bond (due to Van der Waals forces) with the silicon atom inside the pore. The possibility of the formation of such a bond in the pore is higher than on an open surface. It is most probable that the following reaction occurs: NH2* binds CHx groups with Si giving a chain like –CH– Si–O–Si–O–Si–O–Si–, and the end of this chain gets bound to the beginning through NH_2. Changes of the distribution of Si–O–Si angles in this situation are shown in Figure 7-c. To explain the IR spectrum observed after treatment in ammonia plasma without preliminary treatment in He, we calculated several reasonable reaction models. The model describing the experimental observations in the best manner is the following: the rupture of CH_3 groups, the formation of both Si–NH_2 and Si–CH_2–NH_2. Changes of the distribution of Si–O–Si angles as the layer interacts with ammonia are shown in Figure 8. It is clearly seen from this Figure that the most probable angle is about 144-146°. It should be noted that the number of angles 142 ^{o} and less has decreased.

4.3. Comparison of modeling results with experiment and model verification.

For treatment in He: as we do not observe an increase in the intensity of the line at 1000 cm^{-1}, it is rather improbable that the treatment in He plasma causes detachment of oxygen atom and its removal from silicon atom with the formation of dangling bond. Quite contrary, the observed decrease of the frequency of CH_3 group vibrations is connected either with the detachment of hydrogen or with the change of $d\pi$-$p\pi$ overlapping during adsorption at the silicon atom, which increases the rigidity of the Si–O–Si bond. In turn, this causes weakening of the bond holding CH_3 group, which correspondingly leads to a decrease in the frequency of CH_3 vibrations. An evidence in favour of increased overlapping is the fact that the intensity of the line at the frequency of 1070 cm^{-1} and above increases, which suggests the increase of the Si-O-Si angle. So, FTIR spectra show that the treatment in He causes a decrease of the intensity of the line at 1000 cm^{-1} and an increase in the intensities of lines near 1070-1150 cm^{-1}. These changes occur due to CH_3 groups. It is these groups that are able to render the electron density: with the detachment of a proton or a hydrogen atom, the order of the Si-O bond increases, that is, the active centre is formed at the CH_3 group: for instance, CH_2* or CH_2^-. In both cases, the electron density shifts from the CH_x group giving the effective negative charge δ on Si atom. This causes changes of the geometry, namely, an increase in the value of Si-O-Si angle. This model is in good agreement with our calculations. One can see in Figure 7a, b that indeed, if we assume that the detachment of proton from the CH_3 group occurs in He plasma, the shift of the distribution toward larger angles occurs. Subsequent treatment in NH_3 plasma (after treatment in He plasma) promotes grafting at these CHx* groups, as well as at Si atoms; NHx* may serve as a bridge. It is clear that the pore size decreases or the pores become closed because the reaction occurs at the very entrance of the pore. So, if porosity is determined on the basis of refractive index (using ellipsometer), we would not see any essential changes. It is also necessary to stress that the number of built-in NH_2 groups may be not large because building-in will occur on the surface; this may be the reason why they are not seen in the IR spectra. Modeling shows that the distribution should not much differ from the initial state. It may be noted that the angle distribution has become more uniform, with the prevalence of one angle value (Figure 7c). This is in agreement with the experimental FTIR spectra in which we actually observe some narrowing of the lines.

In the case of treatment with NH3 (without preliminary treatment in He) there are no active CH_x centers, and the interaction proceeds mainly due to the formation of the bond between NHx* and Si atom. In this situation, Si-O-Si angle increases, which causes an increase of the frequencies to 1070 and to 1150 cm^{-1}. On the basis of the intensity of these bands, one may make conclusions concerning the amount of adsorbed molecules. The band of the CH_3 group gets shifted to smaller frequencies, which is one more evidence of a strong decrease of π-bonding in Si-O-Si system. As an increase in π-bonding causes an increase in Si-O-Si angle, a pore 1 nm in radius may increase by 0.4-0.5 Å (about 2 %), which is observed experimentally. The experimental data are well confirmed by modeling results. One can clearly see in the distributions shown in Figure 8 a,b that the number of bonds with the angle within the range 146-148 $^{\circ}$ increases, while the angles less than 142 degrees almost disappear. Such a behaviour is characteristic both of the case of the substitution of CHx groups for NH2 (8a) and for the case of attachment of NH2 groups at the silicon atom (Figure 8b).

So, the correctness of the chosen model may be stated.

Below we will discuss passivation, changes in porosity during treatment and etching.

The experimentally observed dependence of etching rate on the time of preliminary irradiation in plasma suggests that the thickness over which film activation (or passivation) occurs is weakly dependent on irradiation time. After irradiation for a long time (4-6 minutes), irreversible changes also occur in the film. For example, etching rate gets completely recovered for the layers with smaller pores (1.1 nm), while for the layers with larger pores (4-6 nm) we observe a small decrease in radius after long-term treatment in helium plasma. In our opinion, an increase in etching rate after VUV irradiation for the films with the pores larger than 2-3 nm in radius (see Figure 2) is connected with the fact that the surface atoms have somewhat higher degree of freedom in comparison with the surface atoms of smaller pores (where the effect of the surface potential is great). This freedom in larger pores results in the noticeable contribution from the processes involving rupture of Si-O and other groups. So, a more friable layer is formed; it is etched with a higher rate.

Figure 9 shows the pore size distribution for the layers with pore radius about 5 nm for different times of UV radiation. The pore size distribution after long-time plasma treatment is shown in Figure 9 (curve 2); for convenience, the data are normalized for layer thickness. One can see that the radius of pores in the film volume decreases; so does total porosity (by about 6-8%). Such behaviour is similar to thermal annealing; a decrease in pore radius can be explained by the action of Laplacian forces. The mechanism of this process was described in [28]. In turn, this may serve as a confirmation of the correctness of our assumption that the excess energy gets distributed over the lattice. Comparing the results obtained in the present work with the data reported in [28], we may say that at the depth of 200 nm the temperature of annealing, achieved due to UV irradiation and energy redistribution, is equivalent to 200 – 250 C.

The behaviour of the layers with pores about 1 nm in radius is quite different As it has already been mentioned above, refractive index is unchanged after UV irradiation, nor does the pore size distribution change. At the same time, we observe a decrease in etching rate. We also see that the pore size distribution changes after etching. The pore size distribution after plasma treatment and etching of the layers with small pores (1.1nm) is shown in Figure 10. It is noteworthy that after etching the pores of a larger radius (~ R=2 nm) appear, though the pores of this size were not detected in the initial film. Since plasma treatment and etching are accompanied by local heating, the appearance of the pores of a larger radius may be connected with the fact that additional strain caused by dπ-pπ overlapping arises in the film under irradiation; this strain may promote the rupture of Si-O bond during etching; it may also cause longer relaxation time. Brought together, these factors cause an increase in pore radius due to the rupture of chemical bonds and junction of pores. Bond rupture occurs along the common wall between two pores into the depth of the layer. Heating during etching stimulates such a rearrangement. Under etching without preliminary treatment in plasma, similarly to the case of treatment only in He plasma, pores 2 nm in radius do not appear. Internal strain that simplifies bond breakage serves as a stimulating factor for the junction of two pores; thus the strain is eliminated.

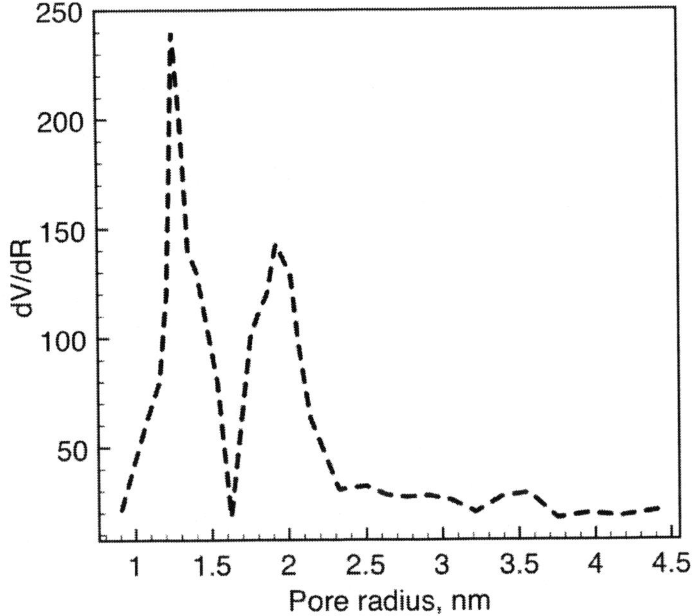

Figure 9. Pore size distribution for the microporous film after treatment in He plasma for 1 min, followed by etching to a depth of 100 nm.

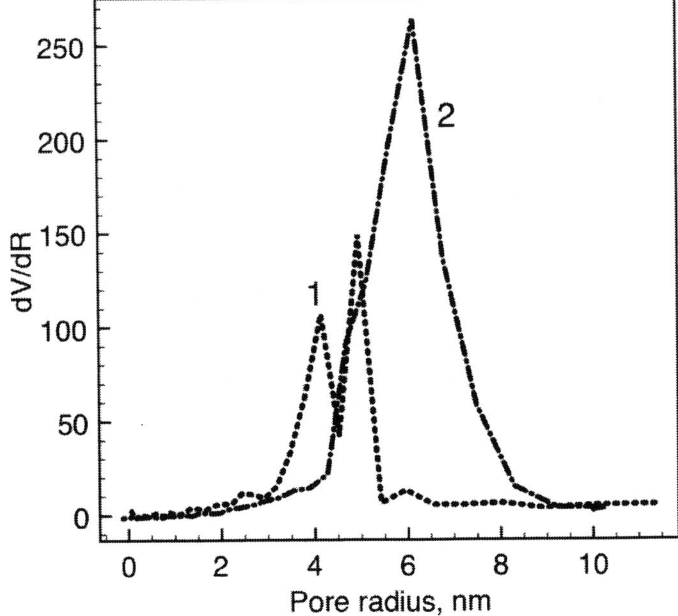

Figure 10. Pore size distribution for the mesoporous film: 1 – film after treatment in He plasma for 1 min, followed by etching to a depth of 100 nm, 2 – film after long (> 6 min) treatment in He plasma.

Conclusion

Thus, the rate of film etching in the plasma after UV irradiation depends on pore size. In particular, it is shown that the layers with the pores of different radii (1.1 and 5.5 nm) exhibit principally different kinds of behaviour under treatment in helium plasma. The main reason of the difference is that a decrease in pore size is accompanied by the changes in the electronic structure of the surface atoms. A decrease in pore radius causes an increase in the number of bonds with Si-O-Si angle less than 140 deg, which causes a decrease in dπ-pπ overlapping and an increase in the reactivity of porous films. The presence of methyl groups allows this structure to get stabilized; detachment of hydrogen atom under the action of UV irradiation causes an increase in dπ-pπ overlapping, which leads to a decrease in layer reactivity, that is, layer passivation. The effect of UV radiation may be considered mainly as the detachment of hydrogen atom, though the energy of UV radiation in plasma is sufficient not only for C-H or Si-C bond rupture but also for Si-O bond rupture. The experimental data and results of modeling allow us to state that the most probable process is detachment of hydrogen atom from the methyl group; rupture of Si-O bond should most probably be accompanied by recombination with the distribution of the excess energy over the lattice.

References

[1] Haase, J.; Ferretti, R.; Pille, J., *Mat. Res. Soc. Symp. Proc.* 1991, 190, 241-246.

[2] Dultsev, F.N.; Solowjev, A.P., *Thin Solid Films* 2002, 419, 27.

[3] Isai, G.I.; Kovalgin, A.Y.; Holleman, J.; Woerlee, P.H.; Wallinga, H., *Thin Film Transistor Technologies 2000*, Vol. 31, 169-175.

[4] Lim, S.; Kim, S.J.; Jung, J.H.; Ju, B.K.; Oh, M.H.; Wager, J.F., *Vacuum Microelectronics Conference*, 1996, 406 – 410.

[5] Isai, G.I.; Kovalgin, A.Y.; Holleman, J.; Woerlee, P.H.; Wallinga, H.; Cobianu, C. *Proceeding of the 30th European Solid-State Device Research Conference*, 2000, The Netherlands, 424- 427.

[6] Pakdee, U.; Suttisiri, N.; Hoonnivathana, E.; Chiangga, S., Kasetsart J. *(Nat. Sci.)*, 2007, 41, 173 – 177.

[7] Gao, J.; Gao, T.; Li, Y. Y.; Sailor, M. J., *Langmuir* 2002, 18, 2229.

[8] Worsley, M.A.; Bent, S.F.; Gates, S.; Fuller, N.C.M.; Volksen, W.; Steen, M.; Dalton, T., J., *Vac. Sci. Technol. B* 2005, 23, 395.

[9] Hua, X.; Kuo, M.; Oehrlein, G.S.; Lazzeri, P.; Iacob, E.; Anderle, M.; Inoki, C.K.; Kuan, T.S.; Jiang, P.; Wu, W., J. *Vac. Sci. Technol. B* 2006, 24(3), 1238.

[10] Grill, A.; Patel, V., J. Electrochem. Soc. 2006, 153(8), F169.

[11] Bao, J.J.; Shi, H. L.; Liu, J. J.; Huang, H.; Ho, P.S.; Goodner, M.D.; Moinpour, M.; Kloster, G.M.; *IEEE* 2007, 71.

[12] Silverstein, M.S.; Shach-Caplan, M.; Khristosov, M.; Harel, T., *Plasma Process. Polym.* 2007, 4, 789.

[13] Xu, S.; Qin, C.; Diao, L.; Gilbert, D.; Hou, L.; Wiesnoski, A.; Busch, E.; McGowan, R.; White, B.; Weber, F.; *J. Vac. Sci. Technol. B* 2007, 25(1), 156.

[14] Inoue, N.; Furutake, N.; Ito, F.; Yamamoto, H.; Takeuchi, T.; Hayashi Y., *Jpn. J. Appl. Phys.* 2008, 47, 2468.

[15] Urbanowicz, A.M.; Baklanov, M.R.; Heijlen, J.; Travaly, Y.; Cockburnb, A.; *Electrochem. Solid St. Lett.* 2007, 10(10), G76.

[16] Peng, H.G.; Chi, D.Z.; Wang, W.D.; Li, J.H.; Zeng, K.Y.; Vallery, R.S.; Frieze, W. E.; Skalsey, M.A.; Gidley, D. W.;Yeea, A.F., *J. Electrochem. Soc.* 2007, 154(4), G85.

[17] Tatsumi, T.; Fukuda, S.; Kadomura, S., Jpn. *J. Appl. Phys.* 1993, 32, 6114.

[18] Tatsumi, T.; Fukuda, S.; Kadomura S., *Jpn. J. Appl. Phys.* 1994, 33, 2175.

[19] Dultsev, F.N.; Baklanov, M.R., *Electrochem. Solid St. Lett.* 1999, 4(2), 92.

[20] Dultsev, F.N.; *Thin Solid Films* 2004, 458, 137.

[21] Gerber, Th.; Himmel, B., *J. Non-Cryst. Solids* 1986, 83, 324.

[22] Lee, H.J.; Soles, C.L.; Liu, D.W.; Bauer, B.J.; Lin, E.K.; Wu, W.L. *IEEE* 2003, 103.

[23] Baklanov, M.R.; Dultsev, F. N., *MRS* 2008, N7.3.

[24] Dultsev, F.N., *J. Struct. Chem.* 2007, 48(2), 236.

[25] Dultsev, F.N.; Dultseva, G.G., *Chem. Phys. Lett.* 2006, 429, 445.

[26] Kurihara, K.; Ono, T.; Kohmura, K.; Tanaka, H.; Fujii, N.; Hata, N.; Kikkawa, T., *J. Appl. Phys.* 2007, 101, 113301.

[27] *Comprehensive Handbook of Chemical Bond Energies*, Yu-R Luo, CRC Press, Taylor &Francis Group Boca Raton, 2007.

[28] Dultsev, F.N.; Mikhailovskii, I.P., *Appl. Surf. Sci.* 2007, 253, 3181.

In: Helium: Characteristics, Compounds…
Editors: Lucas A. Becker
ISBN: 978-1-61761-213-8

Chapter 6

LIGHTER THAN AIR VEHICLES: AUTONOMOUS AIRSHIPS FOR BRIDGE MONITORING

Yasmina Bestaoui *
Laboratoire IBISC, Université d'Evry, Evry, France

ABSTRACT

From its discovery, helium had a leading role in many scientific developments. In this chapter, we are interested by its use as a lifting agent for lighter than air vehicles, more specifically unmanned airships. An important application, bridge monitoring, is highlighted in this chapter. The disaster caused by the collapse of one of Minneapolis (Minnesota) highway bridges points to the need for better technologies to inspect bridges. A flight simulator software is first developed then kinematic and dynamic models of this lighter than air vehicle are developed. Finally, the trim trajectories, used in the flight simulator, are detailed and a robust control method introduced.

1. INTRODUCTION

Lighter than air vehicles suit a wide range of applications, ranging from advertising, aerial photography and survey work tasks. They are safe, cost-effective, durable, environmentally benign and simple to operate. Since their renaissance in the 1990's, airships have been increasingly considered for varied tasks such as transportation, surveillance, freight carrier, advertising, monitoring, research, spatial exploration and military roles [8, 10, 14, 19, 25].

An airship is a lighter-than-air aircraft having propulsion and steering systems. Unlike conventional heavier-than-air vehicles such as airplanes and helicopters whose lift is aerodynamically generated by moving an airfoil through the air, airships stay aloft using a light lifting gas. Unlike the lift force generated over a wing surface which is directly proportional to the square of the flight speed, aerostatic lift comes from Archimedes'

* Corresponding author: bestaoui@iup.univ-evry.fr

principle: it can be calculated by multiplying the volume of air displaced by the lifting gas, by the difference in density between such gas and air. Such force is known as lift. Therefore, only gases whose densities are lower than air at a given temperature and pressure, can be used. Helium is the most commonly used lifting gas nowadays. Under sea-level International Standard Atmosphere conditions (15°C, 1013.25 mbar, 1.225 kg/m^3 air density), its lifting capacity is 1.06 kg/m^3. This value is about 8% lower than hydrogen.

Non rigid airships or pressure airships are the most common form nowadays. They are basically large gas balloons. Their shape is maintained by their internal overpressure. The only solid parts are the gondola, the set of propellers and the tail fins. The envelope holds the helium. In addition to the lift provided by helium, airships derive aerodynamic lift from the shape of the envelope as it moves through the air. The most common form of a dirigible is an ellipsoid. It is a highly aerodynamically profile with good resistance to aerostatics pressures. When the airship is powered by fuel-burning engines, it will undergo a mass decrease over a given period of operation. If not properly accounted for, it can come to land with the buoyancy force far exceeding the weight of the complete vehicle. In that case, the landing operation can become quite a hazardous activity, to be avoided at all costs, since it could ultimately mean valving off expensive helium gas.

For manned aircrafts, the nominal path is specified by the air traffic controller by a sequence of times waypoints and is typically a piecewise linear function. Aircraft motions are subject to various random perturbations such as wind, air turbulence *etc*... thus may deviate from the nominal path. This cross-track deviation may be corrected by the onboard Flight Management System (FMS). In addition, aircraft dynamics may exhibit several distinct modes, *e.g.*, keeping constant heading, turning, ascending, descending and may switch modes at proper times when following the nominal paths. The aircraft is assigned some flight plan to follow that consists of an ordered sequence of waypoints $\{O_i, i=1,2,..,M+1\}, O_i=(x_i, y_i)\in\Re^2$ Ideally, the aircraft should fly at some constant speed along the reference path composed of the concatenation of the ordered sequence $\{I_i, i=1,2,..,M\}$ of line segments, I_i with starting point O_i and ending point $O_{i+1}, i=1,2,..,M$. Deviations from the reference path may be caused by the wind affecting the aircraft position and by limitations in the aircraft dynamics in performing sharp turns resulting in cross-track error. The onboard 3D FMS tries to reduce the cross-track error by issuing corrective actions based on the aircraft's current geometric deviation from the nominal path, however, without taking into account timing specifications. The obtained results are applied to a clearance changing the flight plan. The position of the aircraft may be measured via secondary radar or other localization systems such as Automatic Dependent Surveillance Broadcast (ADS-B). These measurements are precise enough to neglect measurement noise. Modern methods of position localization produce errors of order of terms of meters, which are negligible compared to the effect of other sources of nominal uncertainty such as winds. The flight plan possibly involves altitude changes that can be used as resolution maneuvers to avoid severe weather areas or other conflict situations with other aircraft. Forbidden airspace areas may have an arbitrary shape, which can also change in time, as for example, in the case of a storm that covers an area of irregular shape and evolves dynamically.

To provide autonomy to unmanned aerial vehicles has become a significant challenge, requiring the development of methods to conduct and decisions for implementing the various operations of a mission. With sufficient autonomy, an autonomous airship can be considered as a lighter than air robot. The autonomous airshiph must be able not only to follow its flight plan, but also generate a new plan in response to events occurring during the mission that may invalidate the flight plan in progress. The trajectory planning is to generate an optimal path between two points in configuration space. This trajectory minimizes a particular criterion such as time, energy or distance traveled.

Current research focuses on autonomous airships due to their important application fields: bridge monitoring. The disaster caused by the collapse of one of Minneapolis (Minnesota) highway bridges points to the need for better technologies to inspect bridges.

2. BRIDGE MONITORING FLIGHT SIMULATOR

The objective of this section is to present a software simulation of missions to demonstrate innovative concepts for airborne systems. This flight simulator software is used to accurately represent the system behavior and to generate multiple numerical data useful for analysis of results and algorithms improvement. Thus it is interesting to see a graphical representation of the displacement of the object simulated. Furthermore, the development of a standardized interface between software modeling and visualization environments can more easily use a variety of tools. The proposed simulator is based on Matlab/ Simulink® developed by MATHWORKS (www.mathworks.com). This flight simulator implements the numerical models: the model of airship, the simulated models of the environment, the atmosphere and its evolution, the databases needed for scenarios, and finally tools for recording simulation data. This flight simulator software enables to present an easy and understandable way of monitoring mission specialist. The environment of virtual reality can be connected directly with all possible commands and all types of translation. To demonstrate the possibilities of such a tool, a specific example of bridge was chosen: the bridge of Cheviré (Nantes, France) which an overview is shown in Figure 1. It is a typical example where the monitoring work of a bridge can be automated. This bridge has three distinct parts to watch for:

On deck,
The bearings that support the deck: abutments at both ends and a dozen piles,
The foundation that allows the transmission of forces of structure and land.

It will be difficult to monitor the foundations using an unmanned airship. However monitoring of the deck, abutments and piers lends itself well to automation by monitoring with an airborne platform.

Figure 1. Cheviré Brigde (Nantes, France)

Figure 2. Simulation in nominal conditions

Figure 3. Simulation with little wind in the x and z

In the simulated mission, the airship should fly from its point of rest, watch the first bridge pier in translation, monitor the span and continue monitoring the second cell, eventually returning to another point of rest. Several views are possible: on the road, on the airship, the whole trajectory and an observer far away. It is possible to move around the stage with arrows separated by 45 °: North, Northeast, East, South East, South, Southwest, West, North-West. Thus, it is possible to follow the airship in his motion.

To perform the simulation, the position in space (x, y, z) and orientation (Euler angles: roll ρ, pitch θ, yaw ψ) of the vehicle are sent to the interface located in a Simulink model with an "s-function" Subsequently, the interface converts the data into the coordinate system of the viewer selected. This is a model which is used to calculate the data needed to display graphics. This model consists of several blocks from library sources. These blocks represent the mission planning, trajectory generation, the airship model and 3D visualization. Figure 2 shows the path followed by the autonomous airship in nominal conditions.

Figure 3 shows a simulation where the breeze is supposed to have a constant speed in directions x and z. It is easy to see that the influence of wind increases with time, because uncorrected errors are added to each other in following the initial flight plan.

The feasibility of the trajectory generated depends on the technique used, the cost function chosen and the various constraints. In this case, constraints on the airship concerning its geometry, kinematics and dynamics have to be considered. The constraints from the environment are mainly non-collision with fixed obstacles cumbersome environment and taking into account proximity interactions with the airship. If the obstacle avoidance depends on the geometry of the environment and is common to all robotic tasks, the second point depends in general physical characteristics of the airship. The criteria to be met during the troubleshooting planning concern that a solution must optimize a cost function expressed in terms of distance traveled by the airship between two configurations ends, duration or energy necessary to the execution of his movement. Other criteria may also be considered such as the inclusion of safety distance to obstacles. Solving the problem of planning in an environment cluttered with obstacles, is given by a method of assimilating the airship to a particle constrained to move in a potential field shadow obtained by the composition of a first field attractive goal and to a set of fields repellents modeling the presence of obstacles in airship space. The displacement of the latter is then calculated iteratively by an algorithm of gradient descent of the potential obtained.

The objectives of the planning function are:

- Ordering the passage on the various mission areas
- Calculate a path between each element of the route;
- Require the implementation of the monitoring operation.

Flight planning involves creating a plan to guide an aerial vehicle from its initial position to a destination waypoint. A mission describes the operation of an airship in a given region, during a certain period of time while pursuing a specific objective. A flight plan is defined as the ordered set of movements executed by the airship during a mission. It can be decomposed in phases. Each phase is described by the coordinates of a pair of way-points and by the speed and acceleration at which the airship is to fly between these way-points. A phase is completed when the second way-point is reached by the aircraft. Along the way, there may be a set of regions to visit and a set of regions to avoid. In addition, the airship may have certain motion constraints. The mission planning strategy could be either static or dynamic depending on whether the mission planning problem is to create a path in static or in dynamic environment. Flight planning routines attempt to create paths that are fully consistent with the physical constraints of the airship, the obstacle avoidance, shortest and optimum flight path and weighed regions. Weighed regions are regions with abnormally low or high pressure, wind speeds or any other factor affecting flight. Mission planning in an autonomous vehicle provides the level of autonomy by having minimal ground control. Vehicle autonomy is a discipline fertilized by the robotics and computer science fields [4-6].

For the purpose of flight path generation and control, it is necessary to introduce modeling of this airship. This is presented in the next section.

3. MODELING

The lighter than air platform considered is the AS200 by Airspeed Airships (see figure 4). It is an airship designed for remote sensing. It is a non rigid airship equipped with two vectorable engines on the sides of the gondola, one tail rotor (option) and 4 control surfaces at the stern. The four stabilizers are externally braced on the full and rudder movement is provided by direct linkage to the servos. Envelope pressure is maintained. The engines are standard model aircraft type units.

3.1. Kinematic Modeling

Three reference frames are considered in the derivation of the kinematics and dynamics equations of motion. These are the Earth fixed frame R_f, the body fixed frame R_m and the wind frame R_w. The position and orientation of the vehicle should be described relative to the inertial reference frame while the linear and angular velocities $V=\begin{pmatrix}u & v & w\end{pmatrix}^T ; \Omega=\begin{pmatrix}p & q & r\end{pmatrix}^T$ of the vehicle should be expressed in the body-fixed coordinate system. The origin **N** of R_m coincides with the nose of the vehicle. Its axes $\begin{pmatrix}X_c & Y_c & Z_c\end{pmatrix}$ are the principal axes of symmetry when available. They must form a right handed orthonormal frame. The position of the vehicle **N** in R_f can be described by: $\eta_1=\begin{pmatrix}x & y & z\end{pmatrix}^T$ while the orientation is given by $\eta_2=\begin{pmatrix}\phi & \theta & \psi\end{pmatrix}^T$ with ϕ Roll, θ pitch and ψ Yaw angles. The orientation matrix R is given by

$$\mathbf{R}=\begin{pmatrix} c\psi c\theta & -s\psi c\phi+c\psi s\theta s\phi & s\psi s\phi+c\psi s\theta c\phi \\ s\psi c\theta & c\psi c\phi+s\psi s\theta s\phi & -c\psi s\phi+s\psi s\theta c\phi \\ -s\theta & c\theta s\phi & c\theta c\phi \end{pmatrix}$$ eq 1

with

$$\mathbf{R}^{-1}=\mathbf{R}^T \qquad \dot{\mathbf{R}}=\mathbf{R}Sk(\Omega)$$

where $c\theta = \cos(\theta)$ and $s\theta = \sin(\theta)$. This description is valid in the interval $-\frac{\pi}{2} < \theta < \frac{\pi}{2}$. A singularity of this transformation exists for $\theta = \frac{\pi}{2} \pm k\pi; k \in Z$. $Sk(\Omega)$ denotes the anti-symmetric cross product matrix

Figure 4. Body fixed frame

$$Sk(\Omega) = \begin{pmatrix} 0 & -r & q \\ r & 0 & -p \\ -q & p & 0 \end{pmatrix}$$

The kinematics of the airship can be expressed in the following way [11]:

$$\begin{pmatrix} \dot{\eta}_1 \\ \dot{\eta}_2 \end{pmatrix} = \begin{pmatrix} \mathbf{R} & 0_{3*3} \\ 0_{3*3} & \mathbf{J}(\eta_2) \end{pmatrix} \begin{pmatrix} V \\ \Omega \end{pmatrix} \qquad \text{eq 2}$$

Where

$$\mathbf{J}(\eta_2) = \begin{pmatrix} 1 & s\phi.\tan\theta & c\phi.\tan\theta \\ 0 & c\phi. & -s\phi \\ 0 & s\phi / c\theta & c\phi / c\theta \end{pmatrix} \qquad \text{eq 3}$$

3.2. Mass Characteristics

Significant difference of a buoyant like vehicle from a typical aircraft is that its mass characteristics strongly depends on the change of altitude z

$$m = m(z); \mathbf{I} = \mathbf{I}(z); r_{cg} = r_{cg}(z) \qquad \text{eq 4}$$

The pressure difference between the surrounding atmosphere and the inner gaz (Helium) should be kept as constant as possible at each altitude level. This permanent pressure difference is required for maintaining the aerodynamic shape of the envelope under most operational conditions. As the atmospheric pressure $P_A(z)$ changes with the altitude z, it should be compensated by the internal pressure. For this purpose, the envelope is equipped with two air filled ballonets, namely the fore and the aft ballonets located inside the hull. The volume occupied by the inner gas and the ballonets represents the inner volume of the airship's hull envelope and is nearly constant. By filling the ballonets with the air, they displace the volume of the inner gas, increasing the total pressure of the gas in the envelope [7, 16]. In general, the total mass of the airship can be expressed by

$$m(z,t) = m_g + m_{Bal}(z) + m_R + m_F(t) + m_P \qquad \text{eq 5}$$

m_g is the mass of the inner gas (Helium), $m_{Bal}(z)$ is the total mass of air ballonets, m_R represents the mass of all internal components (skin, structures..), $m_F(t)$ the time varying fuel mass and finally m_P the payload mass. The mass of the helium m_g can be considered as constant if leakage through the hull's skin is insignificant. The mass of all internal components can be derived by accounting all elements of the airship as a consolidation of point and distributed masses. Each ballonet is modeled as a fixed point with variable mass. The volume of ballonets depends on change of the atmospheric pressure gradient. At sea level, where the atmospheric level is high, the ballonet volume has its maximum level and reduces with the increased altitude.

3.3. Six Degrees of Freedom Modeling

In this section, analytic expressions for the forces and moments on the airship are derived [1-3]. There are in general two approaches in deriving equations of motion. One is to apply Newton-Euler's law which can give some physical insight through the derivation. The other one provides the linkage between the classical framework and the Lagrangian or Hamiltonian framework. Newton-Euler approach is used in this chapter. The dynamics model is defined as the set of equations relying the situation of the vehicle in its position, velocity and acceleration to the control vector. The forces and moments are referred to a system of body-fixed axes, centered at the airship nose **N**. Let the motion of the airship be described by its inertial velocity **V** a 6D vector including the inertial linear **V** and angular **Ω** velocities. Let the surrounding air be described by an inertial wind velocity $\mathbf{V}_W = \begin{pmatrix} \mathbf{V}_W \\ \mathbf{\Omega}_W \end{pmatrix}$. We start defining the wind coordinates vector as $\eta_W = \begin{pmatrix} x_W & y_W & z_W & \phi_W & \theta_W & \psi_W \end{pmatrix}^T$ whose time derivative is related to the wind velocity:

$$\dot{\eta}_W = \mathbf{R}\mathbf{V}_W \qquad \text{eq 6}$$

3.3.1. Rigid body mechanics

The airship has thus a relative air velocity $\mathbf{V_a} = \mathbf{V} - \mathbf{V}_W$. Some geometrical and kinematical relations are first given, G being the center of gravity: $\overrightarrow{\mathbf{OG}} = \overrightarrow{\mathbf{ON}} + \overrightarrow{\mathbf{NG}}$ where $\overrightarrow{\mathbf{NG}} = \begin{pmatrix} x_g & 0 & z_g \end{pmatrix}$. Both velocities are related via the cross product matrix $Sk\left(\overrightarrow{\mathbf{NG}}\right)$ skew matrix associated to the vector $\overrightarrow{\mathbf{NG}}$:

$$\vec{\mathbf{V}}_G = \vec{\mathbf{V}}_N - \overrightarrow{\mathbf{NG}} \times \mathbf{\Omega} = \vec{\mathbf{V}}_N - Sk\left(\overrightarrow{\mathbf{NG}}\right)\mathbf{\Omega} \qquad \text{eq 7}$$

The Euler-Lagrange equations of motion may be given by the following relations

Translational motion

$$\sum \overrightarrow{\mathbf{F}_{ext}} = \left.\frac{d\mathbf{P}_d}{dt}\right|_{R_0} = \left.\frac{d\mathbf{P}_d}{dt}\right|_{R} + \mathbf{\Omega} \times \mathbf{P}_d \qquad \text{eq 8}$$

F_{ext} is the generalized force vector, with the linear momentum

$$\mathbf{P}_d = m\vec{\mathbf{V}}_G = m\left(\vec{\mathbf{V}}_N - Sk\left(\overrightarrow{\mathbf{NG}}\right)\mathbf{\Omega}\right) \qquad \text{eq 9}$$

Rotational motion

$$\sum \overrightarrow{\mathbf{M}_{ext}} = \left.\frac{d\mathbf{\Pi}_N}{dt}\right|_{R_0} + \overrightarrow{\mathbf{V}_N} \times \mathbf{P}_d = \left.\frac{d\mathbf{\Pi}_N}{dt}\right|_{R} + \mathbf{\Omega} \times \mathbf{\Pi}_N + \overrightarrow{\mathbf{V}_N} \times \mathbf{P}_d \qquad \text{eq 10}$$

with the angular momentum

$$\mathbf{\Pi}_N = \mathbf{\Pi}_G + \overrightarrow{\mathbf{NG}} \times m\vec{\mathbf{V}}_G = \mathbf{I}_G\mathbf{\Omega} + \overrightarrow{\mathbf{NG}} \times m\vec{\mathbf{V}}_G = \mathbf{I}_N\mathbf{\Omega} + mSk\left(\overrightarrow{\mathbf{NG}}\right)\vec{\mathbf{V}}_N \qquad \text{eq 11}$$

Where $\mathbf{\Pi}_N, \mathbf{\Pi}_G$ are the angular moments computed respectively at the points N and G. From Koenig theorem, the inertia operator in N is calculated as:

$$\mathbf{I}_N = \mathbf{I}_G + m\overrightarrow{\mathbf{NG}} \times \left(\mathbf{\Omega} \times \overrightarrow{\mathbf{NG}}\right) \qquad \text{eq 12}$$

Thus, these forces and torques equations may be written as:

$$\sum \vec{\mathbf{F}}_{ext} = m\dot{\vec{\mathbf{V}}}_{N/R} - mSk\left(\overrightarrow{\mathbf{NG}}\right)\dot{\mathbf{\Omega}}_{/R} + \dot{m}\vec{\mathbf{V}}_{N/R} - \dot{m}Sk\left(\overrightarrow{\mathbf{NG}}\right)\mathbf{\Omega}_{/R} - m\mathbf{\Omega}\times\vec{\mathbf{V}}_{N/R} - m\mathbf{\Omega}\times Sk\left(\overrightarrow{\mathbf{NG}}\right)\mathbf{\Omega}_{/R} \quad \text{eq 13}$$

And

$$\sum \overrightarrow{\mathbf{M}_{N_{ext}}} = \mathbf{I}_N\dot{\mathbf{\Omega}}_{/R} + mSk\left(\overrightarrow{\mathbf{NG}}\right)\dot{\overrightarrow{\mathbf{V}_{N/R}}} + \dot{\mathbf{I}}_N\mathbf{\Omega}_{/R} + \dot{m}Sk\left(\overrightarrow{\mathbf{NG}}\right)\overrightarrow{\mathbf{V}_{N/R}} + \mathbf{\Omega}\times\mathbf{I}_N\mathbf{\Omega} +$$

$$+\mathbf{\Omega}\times mSk\left(\overrightarrow{\mathbf{NG}}\right)\overrightarrow{\mathbf{V}_N} + \overrightarrow{\mathbf{V}_N}\times m\overrightarrow{\mathbf{V}_N} - \overrightarrow{\mathbf{V}_N}\times mSk\left(\overrightarrow{\mathbf{NG}}\right)\mathbf{\Omega} \quad \text{eq 14}$$

The total kinetic energy $\mathbf{W}$ of the mechanical system is defined as:

$$\mathbf{W} = \frac{1}{2}\begin{pmatrix}\vec{\mathbf{V}}_G \\ \mathbf{\Omega}\end{pmatrix}\begin{pmatrix}\vec{\mathbf{P}}_D \\ \mathbf{\Pi}\end{pmatrix} = \frac{1}{2}\begin{pmatrix}\vec{\mathbf{V}}_N \\ \mathbf{\Omega}\end{pmatrix}\mathbf{M}_d\begin{pmatrix}\vec{\mathbf{V}}_N \\ \mathbf{\Omega}\end{pmatrix} \quad \text{eq 15}$$

With

$$\mathbf{M}_d = \begin{pmatrix} m\mathbf{I}_{3*3} & -mSk\left(\overrightarrow{\mathbf{NG}}\right) \\ mSk\left(\overrightarrow{\mathbf{NG}}\right) & \mathbf{I}_N \end{pmatrix}$$

or

$$\mathbf{M}_d = \begin{pmatrix} m & 0 & 0 & 0 & mz_g & 0 \\ 0 & m & 0 & -mz_g & 0 & mx_g \\ 0 & 0 & m & 0 & -mx_g & 0 \\ 0 & -mz_g & 0 & I_x & 0 & -I_{xz} \\ mz_g & 0 & -mx_g & 0 & I_y & 0 \\ 0 & mx_g & 0 & -I_{xz} & 0 & I_z \end{pmatrix}$$

The whole vehicle still has the symmetry about the XZ plane.
The equations of motion are presented in the general form of 6 dimensional vector [5-6]:

$$\mathbf{M}_d\begin{pmatrix}\dot{\vec{\mathbf{V}}} \\ \dot{\mathbf{\Omega}}\end{pmatrix} = -\mathbf{T}_d\begin{pmatrix}\vec{\mathbf{V}} \\ \mathbf{\Omega}\end{pmatrix} + \dot{\mathbf{M}}_d\begin{pmatrix}\vec{\mathbf{V}} \\ \mathbf{\Omega}\end{pmatrix} + \mathbf{T}_{ext} \quad \text{eq 16}$$

Where the Coriolis and centrifugal tensor are given by:

$$\mathbf{T}_d\begin{pmatrix}\mathbf{V} \\ \mathbf{\Omega}\end{pmatrix} = \begin{pmatrix} \mathbf{\Omega}\times m\mathbf{V} - \mathbf{\Omega}\times mSk\left(\overrightarrow{\mathbf{NG}}\right)\mathbf{\Omega} \\ \mathbf{\Omega}\times\mathbf{I}_N\mathbf{\Omega} + \mathbf{V}\times m\mathbf{V} + \mathbf{\Omega}\times mSk\left(\overrightarrow{\mathbf{NG}}\right)\mathbf{V} - \mathbf{V}\times mSk\left(\overrightarrow{\mathbf{NG}}\right)\mathbf{\Omega} \end{pmatrix} \quad \text{eq 17}$$

And the external torques

$$\mathbf{T}_{ext} = \begin{pmatrix} \mathbf{F}_{aero} \\ \mathbf{M}_{aero} \end{pmatrix} + \begin{pmatrix} \overrightarrow{\mathbf{F}_{gravity}} \\ \overrightarrow{\mathbf{M}_{gravity}} \end{pmatrix} + \begin{pmatrix} \overrightarrow{\mathbf{F}_{lift}} \\ \overrightarrow{\mathbf{M}_{lift}} \end{pmatrix} + \mathbf{T}_{propulsion} \qquad \text{eq 18}$$

3.3.2. Gravity and lift forces and moments

These forces and moments are given by:

$$\vec{F}_{gravity} = \mathbf{R}^T \vec{F}_G = \mathbf{R}^T \begin{pmatrix} 0 \\ 0 \\ mg \end{pmatrix} \qquad \vec{F}_{lift} = -\mathbf{R}^T \vec{F}_B = \mathbf{R}^T \begin{pmatrix} 0 \\ 0 \\ F_B \end{pmatrix} \qquad \text{eq 19}$$

The gravitational force vector is given by the difference between the airship weight (acting at the center of gravity) and the lift force (acting upwards on the center of lift):

$$\mathbf{R}^T e_3 (mg - B) = \begin{pmatrix} (mg - B) s\theta \\ -(mg - B) c\theta . s\phi \\ -(mg - B) c\theta . c\phi \end{pmatrix} \qquad \text{eq 20}$$

The gravitational and lift moments are given by:

$$M_G = \vec{NG} \times \vec{F}_G + \vec{NC} \times \vec{F}_B = \begin{pmatrix} z_g mg \cos\theta \sin\phi \\ z_g mg \sin\theta + (x_g mg - x_c B)\cos\theta \cos\phi \\ -(x_g mg - x_c B)\cos\theta \sin\phi \end{pmatrix} \qquad \text{eq 21}$$

where $\overline{NG} = (x_g \quad 0 \quad z_g)$ $\quad \overline{NC} = (x_c \quad 0 \quad 0)$ represents the position of the centre of lift with respect to the body fixed frame.

In aerostatics hovering (floating), the airship stability is mainly affected by its centre of lift in relation to the centre of gravity. The airship's centre of gravity can be adjusted to obtain either stable, neutral or unstable conditions. Putting all weight on the top would create a highly unstable airship with a tendency to roll over in a stable position. In aerodynamics flight, stability can be affected by fins and the general layout of the envelope. Control inertia can be affected by weight distribution, dynamic (static) stability and control power (leverage) available.

3.3.3. Propulsive forces

It is a non rigid airship equipped with two vectorable engines on the sides of the gondola. The propulsion torques can be written as:

$$T_p = \begin{pmatrix} F_m \cos\mu \\ F_T \\ -F_m \sin\mu \\ 0 \\ F_m O_z \cos\mu + F_m O_x \sin\mu \\ F_T O y \end{pmatrix} \qquad \text{eq 22}$$

Where μ is the angle of vectorization, F_m, F_T represent respectively the main and the tail torques.

3.3.4. Aerodynamic forces and moments

The total aerodynamic forces and moments are related to [11-13, 15, 17, 20, 21, 24, 26-28]:

Non stationary terms related to translational and rotational acceleration
Terms involving translation rotation products equivalent to Coriolis forces
Terms involving rotation-rotation products equivalent to centrifugal forces
Terms involving translation-translation products equivalent to stationary phenomena

The following relation can thus be proposed:

$$T_a(\nu_A) = A\dot{\nu}_A + D_1(\nu_2)\nu_A + T_{sta}\left(\nu_a^{\,2}\right) \qquad \text{eq 23}$$

$T_{sta}\left(\nu_a^{\,2}\right)$ represents the stationary aerodynamic coefficients, while the other terms represent the non-stationary coefficients.

a - Stationary coefficients

Some elements of the theory of slender bodies are given here. To introduce the effects of the vertical and horizontal control surfaces, the aerodynamic forces and moments are given by [12]:

$$T_{sta}\left(\nu_a^{\,2}\right) = \frac{1}{2}\rho V_a^2 S_{ref} \begin{pmatrix} C_T \\ C_L \\ C_N \\ L_{ref} C_{lN} \\ L_{ref} C_{mN} \\ L_{ref} C_{nN} \end{pmatrix} \qquad \text{eq 24}$$

where

$$C_N = \sqrt{\sin^2\alpha\cos^2\beta + \sin^2\beta}\ \sin\alpha\cos\beta\bar{C}_x\frac{S_{lat}}{S_{ref}} + \frac{S_q}{S_{ref}}\sin 2\alpha\cos^2\beta \qquad \text{eq 25}$$

$$C_L = \sqrt{\sin^2\alpha\cos^2\beta + \sin^2\beta}\ \sin\beta\bar{C}_x\frac{S_{lat}}{S_{ref}} + \frac{S_q}{S_{ref}}\cos\alpha\sin 2\beta \qquad \text{eq 26}$$

$$\begin{aligned} C_T &= \frac{S_q}{S_{ref}}\left(\bar{C}_{p_a} - \bar{C}_{p_q} - \sin^2\alpha - \sin^2\beta\right) \\ &= \cos^2\alpha\cos^2\beta\left(K_1\left(sign(C_L)\sqrt{C_N^2 + C_L^2} + K_2\right)^2 + K_3\right) \end{aligned} \qquad \text{eq 27}$$

$$\begin{aligned} C_{mN} &= C_{nN} = \\ &= -\sin\alpha\cos\beta\sqrt{\sin^2\alpha\cos^2\beta + \sin^2\beta}\bar{C}_{x_m}\frac{V_{lat}}{S_{ref}L_{ref}} - \sin 2\alpha\cos^2\beta\frac{S_q L - Vol}{S_{ref}L_{ref}} \end{aligned} \qquad \text{eq 28}$$

$$C_{lN} = \frac{\sin^2\beta}{\rho L_{ref}S_{ref}}\left(K_4\sin\alpha + K_5\rho\cos\alpha\right) \qquad \text{eq 29}$$

The coefficients $K_1 - K_5$ must be identified by wind tunnel experiments [9, 11-13, 17]:

Normal coefficient

$$\begin{aligned} C_N &= 0.024 + 0.937\sin 2\left(\alpha + 0.085\delta_e\right)\cos^2\beta + \\ &\quad 1.855\sin\left(\alpha + 0.085\delta_e\right)\cos\beta\sqrt{\sin^2\left(\alpha + 0.085\delta_e\right)\cos^2\beta + \sin^2\beta} \end{aligned} \qquad \text{eq 30}$$

Lateral coefficient

$$\begin{aligned} C_L &= 0.1226\cos\alpha\sin 2\beta + 0.372\sin\beta\sqrt{\sin^2\alpha\cos^2\beta + \sin^2\beta} + 0.937\cos\alpha\sin 2\left(\beta - 0.085\delta_g\right) \\ &\quad + 1.855\sin\left(\beta - 0.085\delta_g\right)\sqrt{\sin^2\alpha\cos^2\left(\beta - 0.085\delta_g\right) + \sin^2\left(\beta - 0.085\delta_g\right)} \end{aligned} \qquad \text{eq 31}$$

Tangential coefficient

$$C_T = \left(K_1 \left(\operatorname{sgn}(C_L)\sqrt{C_N^2 + C_L^2} + K_2 \right)^2 + K_3 \right) \cos^2\alpha \cos^2\beta$$

$$K_1 = -0.0553 + 0.0129\delta_g + 0.0488\delta_g^{\ 2}$$
$$K_2 = -0.061 + 0.4132\delta_g + 0.6899\delta_g^{\ 2} \qquad \text{eq 32}$$
$$K_3 = 0.1069 + 0.0087\delta_g + 0.0932\delta_g^{\ 2}$$

Roll, pitch and yaw moments coefficients

$$C_{lN} = \frac{1}{\rho S_{ref} L_{ref}} (0.548 \sin\alpha + 1.045 \rho \cos\alpha) \sin 2\beta$$
$$C_{mN} = -0.04 - 0.173 \sin 2\alpha - 1.234 \sin\alpha |\sin\alpha|$$
$$C_{nN} = 0.012 \cos\alpha \sin 2\beta + 0.069\sqrt{\sin^2\alpha \cos^2\beta + \sin^2\beta}\ \sin\beta + \qquad \text{eq 33}$$
$$+0.173 \cos\alpha \sin 2(\beta - 0.2\delta_g) +$$
$$+1.234 \sin(\beta - 0.2\delta_g)\sqrt{\cos^2\alpha \cos^2(\beta - 0.2\delta_g) + \sin(\beta - 0.2\delta_g)}$$

b - Nonstationary coefficients

When the airship moves, the air close to its body is moved. Contrary to the other aerial vehicles, the mass of displaced air is close to those of the airship and consequently cannot be neglected. The displaced air mass is known as"added mass" or"virtual mass". The added mass matrix A is, in general, an extra-diagonal matrix. The inertial effects of this added mass constitute the first component of the aerodynamic tensor. Another part of the aerodynamic forces is coming from the translation-rotation and rotation-rotation coupling motions and can be assimilated to Coriolis-centrifuge effects associated to the added mass and can also be represented as a damping effect representation. Due to the importance of the added mass, in the case of the airship, this tensor must be included. In addition, a pure translation depending aerodynamic tensor is considered. These phenomena come from the forces and moment coming from the distribution of the pressure around the airship body and also the friction forces due to the viscosity of the air. The added mass matrix of a rigid body airship includes the contributions of both the hull and the fins as,

$$M_{Arigid} = M_{AH} + M_{AF} \qquad \text{eq 34}$$

In practice, a simple approach to obtain the added mass and moments of inertia of the hull is to approximate the hull as an ellipsoid of revolution. All the off-diagonal terms in the added mass matrix of the ellipsoid are zero and the diagonal terms are given by:

$$X_{\dot{u}} = k_1 m' \qquad Y_{\dot{v}} = Z_{\dot{w}} = k_2 m' \quad L_{\dot{p}} = 0 \qquad M_{\dot{q}} = N_{\dot{r}} = k_3 I' \qquad \text{eq 35}$$

Where m' is the mass of air displaced by the hull, I' is the moment of inertia o the displaced air, the added mass factors are functions of the mass fineness ratio L/D where L is the length of the hull and F is its maximum diameter. The added mass and moment of inertia of the fins can be computed by integrating the 2D added mass of the cross section over the fin region. The contribution of the fins to these 2D added mass terms can be written as

$$\bar{m}_{F,22} = \bar{m}_{F,33} = \rho\pi\left(b - \frac{R^2}{b}\right)^2 \quad \bar{m}_{F,44} = \frac{2}{\pi}k_{44}\rho b^4$$

eq 36

Where R is the hull cross sectional radius and b is the fin semi span. The factor k_{44} is a function of R/b. The non-zero elements in the added mass matrix of the fins are obtained from the following integrals:

$$m_{F,22} = m_{F,33} = \eta_f \int_{x_{FS}}^{x_{FE}} \bar{m}_{F,22}dx \quad m_{F,35} = -\eta_f \int_{x_{FS}}^{x_{FE}} \bar{m}_{F,22}xdx = -m_{F,26}$$

$$m_{F,44} = \eta_f \int_{x_{FS}}^{x_{FE}} \bar{m}_{F,44}dx \qquad m_{F,55} = m_{F,66} = \eta_f \int_{x_{FS}}^{x_{FE}} \bar{m}_{F,44}x^2dx$$

eq 37

Where x_{FS}, x_{FE} are respectively the x coordinates of the start and end positions of the fins. An efficiency factor η_f is included to account for 3D effects.

Remark : For a system with added masses, the term $V \times MV$ is different from zero. The terms $V \times MV$ $\Omega \times MV$ and $\Omega \times M\Omega$ show the centrifugal and Coriolis components.

$$A\dot{v}_A = \begin{pmatrix} a_{11} & 0 & 0 & 0 & a_{15} & 0 \\ 0 & a_{22} & 0 & a_{24} & 0 & a_{26} \\ 0 & 0 & a_{33} & 0 & a_{35} & 0 \\ 0 & a_{42} & 0 & a_{44} & 0 & a_{46} \\ a_{51} & 0 & a_{53} & 0 & a_{55} & 0 \\ 0 & a_{62} & 0 & a_{64} & 0 & a_{66} \end{pmatrix}\begin{pmatrix} \dot{u}_1 \\ \dot{u}_2 \\ \dot{u}_3 \\ \dot{p} \\ \dot{q} \\ \dot{r} \end{pmatrix}$$

eq 38

Using the Complementarity between the coefficients of Kirchoff and Bryson theories, the following damping coefficient is obtained.

$$D_1(\nu_2) = -\begin{pmatrix} 0 & a_{22}r & -a_{33}q & a_{24}r & -a_{35}q & a_{26}r \\ pm_{13}+r(xm_{11}-a_{11}) & 0 & a_{33}p & a_{35}q & -a_{15}q & 0 \\ (a_{11}-xm_{22})q & -a_{22}p & 0 & -a_{24}p-a_{26}p & a_{15}q & 0 \\ pm_{33}+r(xm_{13}+a_{15}) & -(a_{62}+a_{35})q & (a_{62}+a_{35})r+a_{24}p & -a_{64}q & (a_{55}-a_{66})r & 0 \\ (a_{35}+x^2m_{22})q & -a_{42}r+a_{62}p & -a_{15}q & (a_{65}-a_{44})r+a_{64}p & 0 & -a_{64}r \\ -(a_{51}+a_{24}-xm_{13})p-(a_{26}-x^2m_{11})r & (a_{15}+a_{42})q & -a_{33}p & (a_{44}-a_{55})q & a_{46}r & 0 \end{pmatrix}$$

eq 39

with

$$\begin{pmatrix} \nu_a \\ \nu_2 \end{pmatrix} = \begin{pmatrix} u_1 & u_2 & u_3 & p & q & r \end{pmatrix}^T \qquad \text{eq 40}$$

This 6 degrees of freedom model is necessary for control purpose however, for guidance purpose, the 3 degrees of freedom gives enough informations about the airship motion, while keeping some simplicity.

3.4. Three Degrees of Freedom Dynamic Modeling

The translational equations of an aerospace vehicle through the atmosphere are directly derived from Newton's law. The equations of motion are expressed in a velocity coordinate frame attached to the airship, considering the velocity of the wind $W = \begin{pmatrix} W_x & W_y & W_z \end{pmatrix}^T$ (components of the wind velocity in the inertial frame). It is assumed that the airship thrust vector is aligned along its velocity vector. The kinematic equations are given by:

$$\begin{aligned} \dot{x} &= V\cos\chi\cos\gamma + W_x \\ \dot{y} &= V\sin\chi\cos\gamma + W_y \\ \dot{z} &= -V\sin\gamma + W_z \end{aligned} \qquad \text{eq 41}$$

Where x (downrange), y (cross range) and z (altitude) are the vehicle's position, γ is the flight path angle, χ is the azimuth (heading) angle, σ is the bank angle and V is the velocity magnitude.

The approach followed in [22,23] is proposed in this section. The contribution of the added mass phenomenon is described as:

$$F_{AM} = -\left(C_w^b\right)^T M_a C_w^b \begin{pmatrix} \dot{V} \\ \dot{\chi} V \cos\gamma \\ \dot{\gamma} V \end{pmatrix} - \omega_B \times \left(C_w^b\right)^T M_a C_w^b \begin{pmatrix} V \\ 0 \\ 0 \end{pmatrix} =$$

$$= -\begin{pmatrix} m_{11} & m_{12} & m_{13} \\ m_{12} & m_{22} & m_{23} \\ m_{13} & m_{23} & m_{33} \end{pmatrix} \begin{pmatrix} \dot{V} \\ \dot{\chi} V \cos\gamma \\ \dot{\gamma} V \end{pmatrix} - \omega_B \times \begin{pmatrix} m_{11} V \\ m_{12} V \\ m_{13} V \end{pmatrix}$$

eq 42

where the added mass matrix is approximated by $M_a = diag\begin{pmatrix} m_{ax} & m_{ay} & m_{az} \end{pmatrix}$ and the matrix C_w^b can be written as:

$$C_w^b = \begin{pmatrix} \cos\alpha\cos\beta & \sin\beta & \sin\alpha\cos\beta \\ -\cos\alpha\sin\beta & \cos\beta & -\sin\alpha\sin\beta \\ -\sin\alpha & 0 & \cos\alpha \end{pmatrix}$$

with

$$m_{11} = m_{ax}\cos^2\alpha\cos^2\beta + m_{ay}\cos^2\alpha\sin^2\beta + m_{az}\sin^2\alpha$$

$$m_{12} = \cos\alpha\cos\beta\sin\beta\left(m_{ax} - m_{ay}\right)$$

$$m_{13} = \cos\alpha\sin\alpha\cos^2\beta\left(m_{ax} + m_{ay}\right) - m_{az}\cos\alpha\sin\alpha$$

$$m_{22} = m_{ax}\sin^2\beta + m_{ay}\cos^2\beta$$

$$m_{23} = \sin\alpha\cos\beta\sin\beta\left(m_{ax} - m_{ay}\right)$$

$$m_{33} = m_{ax}\sin^2\alpha\cos^2\beta + m_{ay}\sin^2\alpha\sin^2\beta + m_{az}\cos^2\alpha$$

while the angular velocity is given by

$$\omega_B = \begin{pmatrix} \omega_1 \\ \omega_2 \\ \omega_3 \end{pmatrix} = \begin{pmatrix} \dot{\gamma}\cos\alpha \\ \dot{\alpha} \\ \dot{\gamma}\cos\alpha \end{pmatrix} + \begin{pmatrix} 1 & 0 & -\sin\gamma \\ 0 & \cos\sigma & \sin\sigma\cos\gamma \\ 0 & -\sin\sigma & \cos\sigma\cos\gamma \end{pmatrix} \begin{pmatrix} \dot{\sigma} \\ \dot{\gamma} \\ \dot{\chi} \end{pmatrix}$$

After some calculations, the powered dynamic model used for flight over a flat Earth is the following:

$$\dot{V} = \frac{T\cos\alpha - D + (B - mg)\sin\gamma}{m + m_{11}} - \frac{m}{m + m_{11}}\left(\dot{W}_x\cos\gamma\cos\chi - \dot{W}_y\cos\gamma\sin\chi + \dot{W}_z\sin\gamma\right)$$ eq 43

$$\dot{\chi} = \frac{(L + T\sin\alpha)\sin\sigma}{(m+m_{12})V\cos\gamma} + m\frac{\dot{W}_x \sin\chi - \dot{W}_y \cos\chi}{(m+m_{12})V\cos\gamma}$$

$$\dot{\gamma} = \frac{L\cos\sigma + T\sin\alpha\cos\sigma + (B - mg)\cos\gamma}{(m+m_{13})V} + m\frac{\dot{W}_x \sin\gamma\cos\chi + \dot{W}_y \sin\gamma\sin\chi + \cos\gamma}{(m+m_{13})V}$$

eq 44

The forces D and L being respectively the drag and lift are given by:

$$D = \frac{C_D(M,\alpha)A_{ref}\rho V^2}{2} \qquad L = \frac{C_L(M,\alpha)A_{ref}\rho V^2}{2}$$ eq 45

ρ is the free stream mass density, m is the airship mass, A_{ref} is a characteristic area for the body, C_L, C_D are respectively the lift and drag coefficient functions that depend upon the Mach number M and angle of attack α. The dynamic pressure is $\overline{q} = 0.5\rho V^2$ where the air density ρ at altitude h is approximated using an exponential model $\rho = \rho_0 e^{-\beta h}$ where ρ_0 is the air density at sea level and β is the atmospheric density scale.

If one assumes that the control inputs are (T, σ, α) then the airship motion is differentially flat. By determining a suitable airship trajectory in Cartesian coordinates x, y, and z, the required airship controls can be calculated.

These equations have an important place in aerospace vehicle study because they can be assembled from trimmed aerodynamic data and simple autopilot designs. Nevertheless, they give a realistic picture of the translational and rotational dynamics unless large angles and cross coupling effects dominate the simulations. Trajectory studies, performance investigations, navigation, guidance evaluations can be successfully executed with simulations of these equations.

4. Trim Trajectories

In Aeronautics, trim trajectories have a significant place. Under the trim condition the vehicle motion is uniform in the body fixed frame. The trim trajectories have the advantage of facilitating the planning and control problems. A linear control technique could be sufficient to stabilize the vehicle in the neighbourhood of trim conditions. Another advantage is that the aerodynamic coefficients which are variable in time and space become stationary under this condition and their identification becomes easier.

4.1. Trim Conditions Determination Algorithm

The vehicle configuration in space is defined by its linear and angular velocities and the pitch and roll angles. If we refer to some simplifying assumptions, neglecting the ground curvature and the air density variation with altitude, a vector of eight state variables $(V,\alpha,\beta,p,q,r,\theta,\phi)^T$ would be enough to parameterize any trim trajectory in space.

Performance index

In the case of a trim trajectory, the external forces and moments are constant or equal to zero. The aim is to determine the trim conditions i.e. all accelerations vanish ($\dot{V},\dot{\alpha},\dot{\beta},\dot{p},\dot{q},\dot{r}=0$). The state and control vectors are determined by the resolution of a nonlinear equations system. Thus, we can formulate a numerical optimization problem seeking to minimize all accelerations. The performance index chosen in the algorithm is the sum of the squares of accelerations:

Constraints

The algorithm must take account of some constraints: under-actuation, flight envelope, trajectory geometry, actuators limitations and environment [17]-[20].

Under-actuation constraints

The autonomous airship has three control inputs: main and torque torques and main angle of vectorization, the number of degrees of freedom being six. Thus we can formulate three equality dynamical differential equations as constraints due to the under-actuation.

Flight envelope constraints

In order to guarantee some flight performances and for safety reasons we define a flight envelope. The maximum available power provided by the engines imposes a maximum limit speed. Limitations on the load factor are imposed to limiting the effort exerted on the structure during the turn and pull up.

Trajectory geometry

The choice of the trajectory geometry (line, circle, helix) imposes kinematics constraints. Figure 5 shows the trim helix:

In [4], we can find two examples of these constraints. The first one is an algebraic relation expressing the pitch angle as function of desired rate of climb (rate of climb constraint). The second example is an algebraic constraint allowing a coordinated turn by expressing the roll angle as function of desired heading rate. This ensures that the vehicle turning without skidding (coordinated turn constraint). These two constraints can be used in any optimization algorithm.

Constraints dues to the actuators limitations

Throttle position is included between zero and maximum power available. The deflection angles of the control surfaces vary between two limits. These limitations can be introduced into any algorithm as algebraic inequalities:

$$0 \leq F_m \leq F_{\max}, \mu_{\min} \leq \mu \leq \mu_{\max} \qquad \text{eq 46}$$

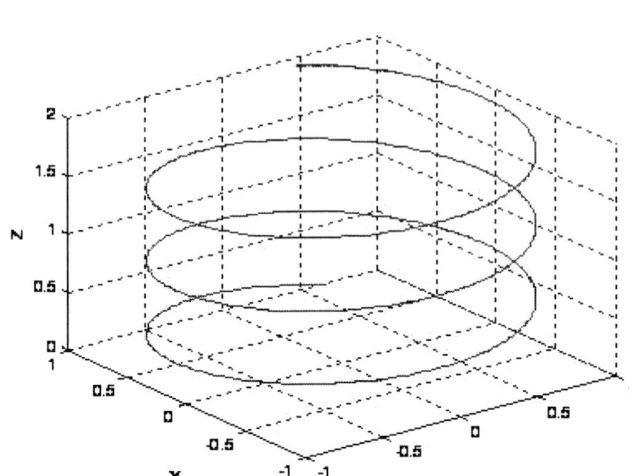

Figure 5. General trim trajectory

4.2 Trajectory Planner

A. Problem formulation

We consider the vehicle in a trim flight with constant airspeed. The more general trim trajectory is a vertical helix with constant curvature and rate of climb. This trajectory can be described by the following equations:

$$\begin{aligned} \dot{x} &= A\cos(\dot{\chi}t) + B.\sin(\dot{\chi}t) \\ \dot{y} &= -B\cos(\dot{\chi}t) + A.\sin(\dot{\chi}t) \\ \dot{z} &= C \end{aligned} \qquad \text{eq 47}$$

With

$$\begin{aligned} A &= V\cos\alpha\cos\beta\cos\theta + V\sin\beta\sin\theta\sin\phi + V\sin\alpha\cos\beta\sin\theta\cos\phi \\ B &= -V\sin\beta\cos\phi + V\sin\alpha\cos\beta\sin\phi \\ C &= -V\cos\alpha\cos\beta\sin\theta + V\sin\beta\cos\theta\sin\phi + V\sin\alpha\cos\beta\cos\theta\cos\phi \end{aligned}$$

We assume that the vehicle is equipped with a control system ensuring the reference trajectory tracking. The role of this system is to maintain the vehicle airspeed V attitude (Euler angles) and aerodynamic angles within their respective desired values. The state constraints defining the flight envelope are:

$$-\dot{\psi}_{\max} \leq \dot{\psi} \leq \dot{\psi}_{\max}, \; 0 < V_{\min} \leq V \leq V_{\max} \; \alpha_{\min} \leq \alpha \leq \alpha_{\max}$$

$$|\beta| \leq \beta_{\max} \; |\phi| \leq \phi_{\max}$$

The constant $\dot{\psi}_{\max}$ and the maximal value of angle of attack $\alpha_{\max}$ are determined by the maximal value of the load factor. The available engine power allows determining maximal vehicle airspeed $V_{\max}$. A last constraint on the state allows maintaining the sideslip angle β in the neighbourhood of zero. To ensure the regularity of the coordinate transformation matrix, we suppose: $\theta \neq \pm\frac{\pi}{2}$.

A trajectory $t\hat{r}(t) = (\hat{x}(t), \hat{y}(t), \hat{z}(t))^T$ is dynamically feasible if there exist inputs ψ_c, ϕ_c, $\theta_c, V_c, \alpha_c, \beta_c$ such as $tr(t) = t\hat{r}(t)$ for all $t \geq 0$, the dynamics and the constraints being satisfied. The input of the trajectory planner algorithm are the desired airspeed $\hat{V}$ and the waypoints coordinates $(x, y, z)_i \in R^3$ expressed in the Earth fixed frame: $\left\{\hat{V}, (x, y, z)_1, (x, y, z)_2, ..., (x, y, z)_n\right\}$.

Thus, the trajectory planner equations are given by:

$$
\begin{aligned}
&\dot{\hat{x}} = \hat{A}\cos(\dot{\hat{\psi}}t) + \hat{B}\sin(\dot{\hat{\psi}}t) \\
&\dot{\hat{y}} = -\hat{B}\cos(\dot{\hat{\psi}}t) + \hat{A}\sin(\dot{\hat{\psi}}t) \\
&\dot{\hat{z}} = \hat{C}\sin u_2 \\
&u_1 = \dot{\hat{\psi}} \qquad |u_1| \leq \dot{\psi}_{\max} \\
&u_2 = \hat{\gamma} \qquad |u_2| \leq \gamma_{\max}
\end{aligned}
\qquad \text{eq 48}
$$

With

$$
\begin{aligned}
\hat{A} &= \hat{V}\cos\alpha\cos\beta\cos\theta + \hat{V}\sin\beta\sin\theta\sin\phi + \hat{V}\sin\alpha\cos\beta\sin\theta\cos\phi \\
\hat{B} &= -\hat{V}\sin\beta\cos\phi + \hat{V}\sin\alpha\cos\beta\sin\phi \\
\hat{C} &= -\hat{V}\cos\alpha\cos\beta\sin\theta + \hat{V}\sin\beta\cos\theta\sin\phi + \hat{V}\sin\alpha\cos\beta\cos\theta\cos\phi
\end{aligned}
$$

If the trajectories are traversed with a constant airspeed $\hat{V}$ included between $V_{\min}$ and $V_{\max}$, the initial condition $t\hat{r}(0) = tr(0)$ guarantee a priori the dynamical feasibility of the trajectory.

If $u_1 = \dot{\psi}_{\max}$, the trajectory generated by (eq. 48) is a circle on the right side of the plane where the centre is given by : $(x, y, z)^T_{cent} = (\hat{x}, \hat{y}, \hat{z})^T + R_{\min}(-\sin(\dot{\psi}t), \cos(\dot{\psi}t), 1)^T$

If $u_1 = -\dot{\psi}_{\max}$, the trajectory generated by (eq. 48) is a circle on the left side of the plane where the centre is given by : $(x, y, z)^T_{cent} = (\hat{x}, \hat{y}, \hat{z})^T + R_{\min}(\sin(\dot{\psi}t), -\cos(\dot{\psi}t), 1)^T$

If $u_2 = +\gamma_{\max}$, the flight path angle is positive (up) and given by : $\dot{z} = V\sin\gamma_{\max}$

If $u_2 = -\gamma_{max}$ the flight path angle is negative (down) and given by: $\dot{z} = -V \sin \gamma_{max}$.

B. Extremal curves

In this section we present briefly the Dubins principle [4] in the optimal trajectory generation with a limited curvature for a vehicle moving in a plan [6], [8]. A direct application of this principle is possible in the case a flight at constant altitude with a constant airspeed. Let's suppose that the yaw angle is controlled by a stability augmentation system, the kinematics equations can be reduced as :

$$\begin{aligned} \dot{x} &= V.\cos\psi \\ \dot{y} &= V.\sin\psi \\ \dot{\psi} &= u = K_\psi.(\psi_c - \psi) \end{aligned}$$

With V the airspeed, ψ the yaw angle, x and y are respectively the north and east UAV positions. These parameters are represented in Fig. 5.

We assume also a trajectory curvature constraint corresponding to a turn minimal radius R_{min}. The approach minimizes the trajectory length to go from an initial point p_i to a final point p_f :

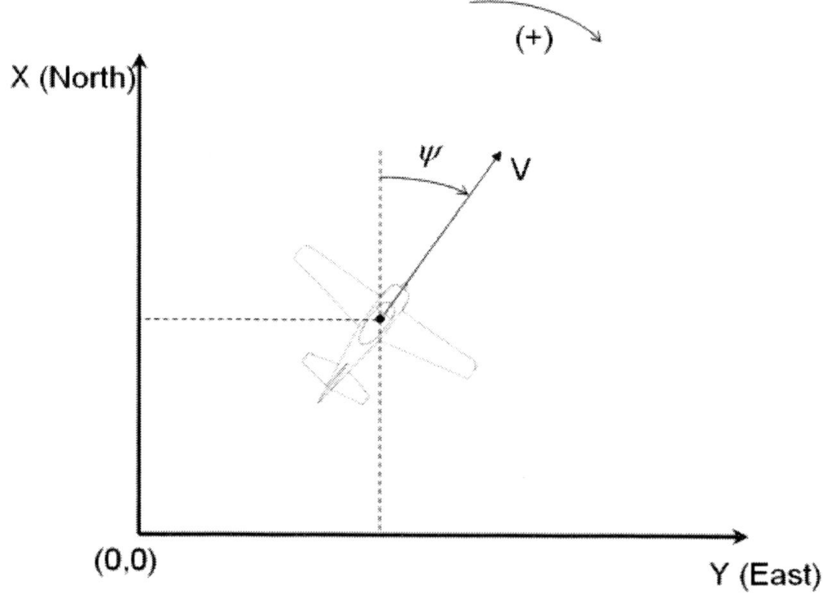

Figure 6. The directions and orientation used in the case of flight in the horizontal plan

$$J(p_i, p_f, u) = \int_0^{t_f} \sqrt{\dot{x}^2(t) + \dot{y}^2(t)}dt \qquad \text{eq 49}$$

Dubins [4] shows that between any two configurations, it is possible to construct the shortest path with a combination of a maximum of three primitives. In each primitive we apply a constant action during a time interval. Any extremal trajectory can be indicated by a sequence of three symbols. This sequence corresponds to the order of which the primitives are applied. It is not necessary to have two symbols of a same type consecutively, because they can be reduced to one symbol. Dubins proved that among all the possible sequences only six can be optimal: $\{RLR\,;LRL\,;RSR\,;RSL\,;LSL\,;LSR\}$ . . S indicates a straight line, R and L indicate respectively, the right and left level (sharpest) turn. The shortest path between two any configurations can be characterized by one of these sequences. They are called the Dubins curves. We can allot to each primitive an index which indicates the execution time. In the case of a level turn the index (μ,ν or o) indicates the rotation angle carried out during the application of the primitive. In the case of a straight line the index d indicates the total covered distance during the application of the primitive. With these indexes the sequences will be presented as follows:

$$\left\{R_\mu L_\nu R_o\,;L_\mu R_\nu L_o\,;R_\mu S_d R_o\,;R_\mu S_d L_o\,;L_\mu S_d L_o\,;L_\mu S_d R_o\right\}$$

With these definitions the problem is formulated as follows: being given the points p_i and p_f, which is the shortest sequence in eq. (49) and which are the values of indices of this sequence?

For each couple of points, a simple algorithm can answer these two questions, by evaluation of the six sequences, than by choosing the shortest among them.

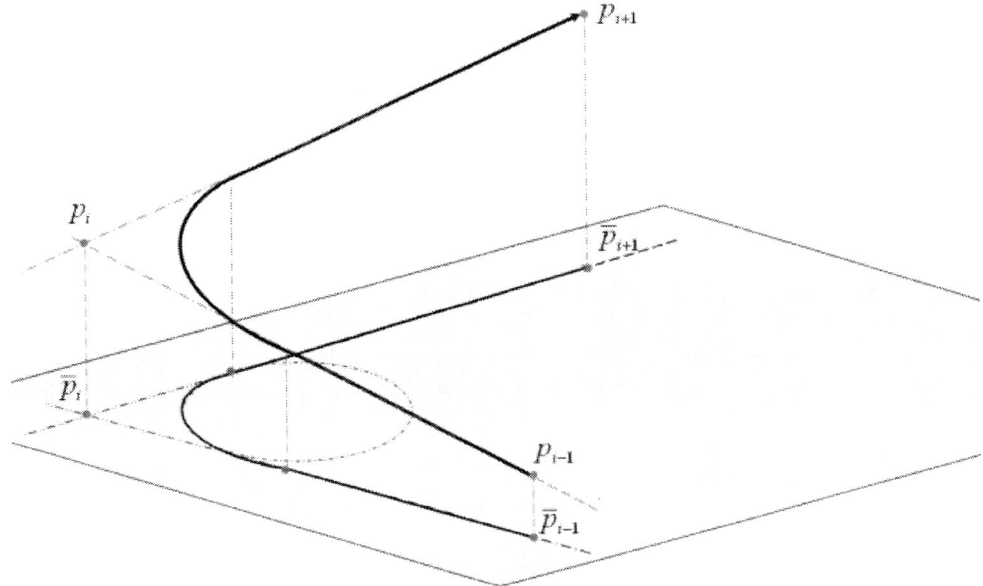

Figure 7. Trajectory projection in the horizontal plan (x,y)

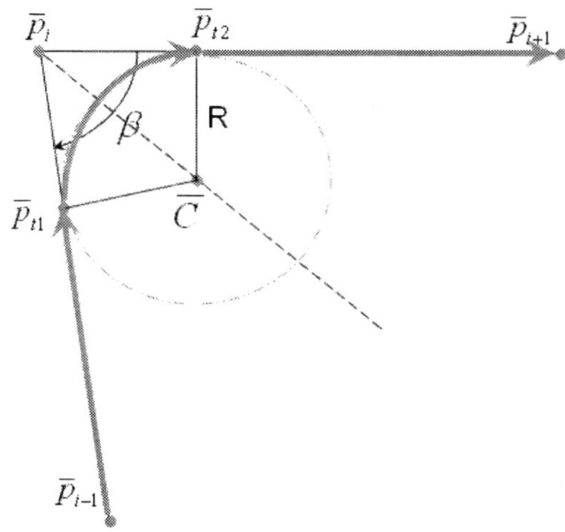

Figure 8. Time optimal control problem in the new coordinates

Extremal trajectoires

Let's define the way path by three points p_{i-1}, p_i and p_{i+1}, see Fig. 3. The projection of segments $\left[p_{i-1}p_i\right]$ and $\left[p_i p_{i+1}\right]$ in the horizontal plan gives respectively the segments $\left[\overline{p}_{i-1}\overline{p}_i\right]and\left[\overline{p}_i\overline{p}_{i+1}\right]$. Let's define $\overline{C}$ the circle tangent with the two segments$\left[\overline{p}_{i-1}\overline{p}_i\right]$and $\left[\overline{p}_i\overline{p}_{i+1}\right]$of radius $R_{\min} = V / \dot{\psi}_{\max}$ of which the centre belongs to the bisectrix of the angle formed by the three waypoints.

The trajectory between the points $\overline{p}_{i-1}$ and $\overline{p}_{i+1}$ is constructed as follows:

1. The segment $\left[\overline{p}_{i-1}\overline{p}_i\right]$ is followed until $\overline{p}_{t1}$ the intersection point with the circle $\overline{C}$.
2. Then the arc of circle $\overline{C}$ from $\overline{p}_{t1}$ to $\overline{p}_{t2}$ the intersection point with the segment $\left[\overline{p}_i\overline{p}_{i+1}\right]$.
3. Finally, the segment $\left[\overline{p}_{t2}\overline{p}_{i+1}\right]$ is followed until the point p_{i+1}.

The circle $\overline{C}$ is in fact the projection of the vertical helix with constant rate of climb $\dot{z}$.

Algorithm description

The proposed algorithm allows the characterization of the time parameterized optimal trajectory between two arbitrary points at the same altitude. The algorithm presented here is intended to generate a trajectory for the specified type of missions: bridge monitoring, the area being represented by many way-points coordinates. The presence of possible obstacles prohibited areas between the start point (the runway) the target point (monitoring area), requires that the airship passes by a number of waypoints before reaching the target.

The algorithm comprises five steps (see figure 9):

1. **Initialisation**: The airship cruise speed and altitude (constants), the runway data (coordinates, direction and altitude), the target and waypoints coordinates are introduced as well as the maximal trajectory curvature and rate of climb (descent).
2. **Initial climb**: the horizontal distance which the airship must traverse before starting the manoeuvres is computed, than the initial climb trajectory is generated.

3.a. **To move towards the next waypoint**: For each waypoint, the direction of turn and the transition times straight/circle/straight are determined, this operation is repeated as many times as the number of waypoints.

3.b. **To move towards the target point:** in the case when there is no waypoint, the airship makes one turn to direct towards the target point.

4. **Flying over the target area:** from the target point coordinates, the transition times are determined, depending on the trim: straight line/right turn/left turn/right turn. The circle of which belongs the left turn will be traversed several times to allow the sensors in board to carry out the necessary measurements.
5. **Return to the landing/take-off area:** while leaving the target area, the airship traverses the same generated trajectory in the opposite direction.

During step 3 the algorithm determines the turn centre position and direction (left or right) from the position of the waypoint. In Figure 10, the coordinate space is decomposed into four subspaces separated by the direction line of actual waypoint and the perpendicular line to this direction in the horizontal plan.

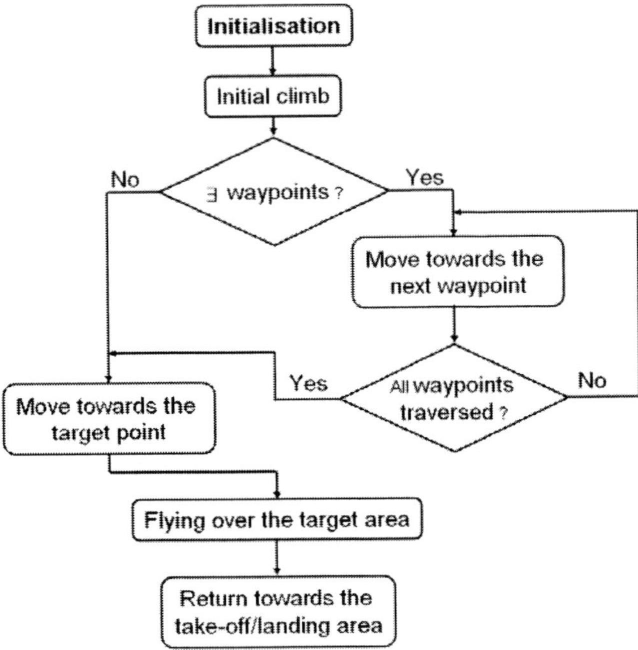

Figure 9. Diagram of the trajectory planner

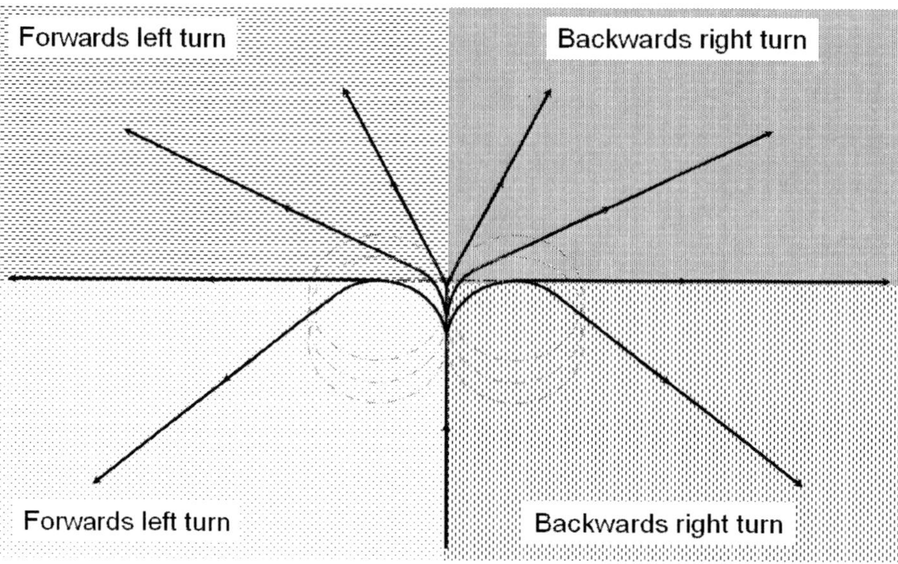

Figure 10. The four (subspaces) types of turn

Case of backward right turn

If the next waypoint is rather behind the actual waypoint, the turn (circle) centre and the start point are determined by the actual waypoint coordinates, the end point only depends from the next waypoint coordinates. The airship must approach enough the actual waypoint before starting a new turn. Indeed, the role of a waypoint is to allow the airshp to avoid an obstacle or a prohibited area. Without this restriction the airship is likely to start a turn towards the next waypoint before reaching the neighbourhood of actual waypoint as we can see in Figure 11.

Case of a forward right turn

If the next waypoint is rather ahead the actual point, the algorithm computes the turn (circle) centre from the next waypoint coordinates, than it determines the start and end points of the arc of circle.

5. Flight Control Methodology

In a variety of missions, inertial trajectory control of aircraft in three-dimensional space is essential. Considerable research has been done for the control systems design for maneuvering aircraft. Dynamic inversion approach has been widely used for the control of a variety of output variables of interest including the angle of attack α, sideslip β, roll ϕ and velocity roll p angles.

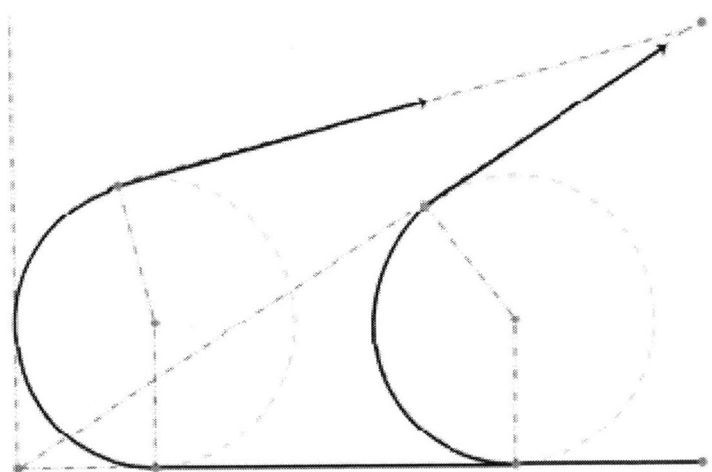

Figure 11. The two possible cases of turn construction of which the next waypoint is behind the actual one.

In the presence of uncertainties, robust design using variable structure and adaptive control theories has been performed. Unlike control of α, β, ϕ, p variables, the design of flight path controllers is relatively complicated. The reason lies in the higher relative degrees (4, 4, 4) (ignoring small forces contributed by the control surfaces) of the position coordinates (x, y, z) with second derivative of thrust and control surface deflections as control inputs. Each of the variables α, β, ϕ, p has relative degree only 2. Relative degree of an output of a system is defined to be the order of the derivative of the output in which control input appears for the first time. Thus, dynamic inversion approach requires derivative of fourth order of x, y, and z. The presence of uncertainties in the model creates additional difficulties. One can use an adaptive back stepping design for the compensation of uncertainties. However, computational difficulties arise because the virtual control inputs appear nonlinearly and depend on uncertain aerodynamic parameters.

The approach followed in this paper is robust control design using the Variable Structure Control theory (VSC), one first selects an appropriate sliding surface on which the trajectory has desirable property. The control input is taken as $U = \begin{pmatrix} \dot{T} & \dot{\sigma} & \dot{\alpha} \end{pmatrix}^T$, the output $Y = \begin{pmatrix} x & y & z \end{pmatrix}^T$ and the state vector $X = \begin{pmatrix} X_1 \\ X_2 \end{pmatrix}; X_1 = \begin{pmatrix} x & y & z \end{pmatrix}^T, X_2 = \begin{pmatrix} V & \chi & \gamma \end{pmatrix}^T$. The following sliding surface is chosen

$$S = \ddot{e} + K_v \dot{e} + K_p e + K_I \int_0^t e \, d\tau$$

$$e = \begin{pmatrix} x \\ y \\ z \end{pmatrix} - \begin{pmatrix} x_r \\ y_r \\ z_r \end{pmatrix} = X_1 - X_{1r}$$

eq 50

These gains are chosen so that S=0 yields exponentially stable response for e; Integral feedback provides additional flexibility for a robust design.

The motion of the closed loop system including the VSC law evolves in two phases. First, the trajectory beginning from arbitrary initial state is attracted towards S=0. In the second phase, which is termed a sliding phase, e converges to zero because S=0. Once the choice of sliding surface has been made, one must design a controller such that S=0 becomes an attractive surface;

Differentiating X gives (s is for sin and c for cos)

$$\ddot{X} = \begin{pmatrix} c\chi c\gamma & Vs\chi c\gamma & -Vc\chi s\gamma \\ s\chi c\gamma & Vc\chi c\gamma & -Vs\chi s\gamma \\ -s\gamma & 0 & -Vc\gamma \end{pmatrix} \begin{pmatrix} \dot{V} \\ \dot{\chi} \\ \dot{\gamma} \end{pmatrix} = B_1(V,\chi,\gamma)(f_0 + f_1) \qquad \text{eq 51}$$

Where the functions f are given by

$$f_0(V,\gamma) = \begin{pmatrix} -gs\gamma \\ 0 \\ -\dfrac{gc\gamma}{V} \end{pmatrix} \qquad f_1(T,V,\gamma) = \begin{pmatrix} \dfrac{T-D}{m} \\ Ls\sigma/mV \\ -\dfrac{Lc\sigma}{Vc\gamma} \end{pmatrix} \qquad \text{eq 52}$$

Differentiating one more time, the following relation is obtained:

$$\dddot{X} = \dot{B}_1(f_0 + f_1) + B_1(\dot{f}_0 + \dot{f}_1) = f_3 + B_1 B_2 U \qquad \text{eq 53}$$

Where

$$f_3 = \dot{B}_1(f_0 + f_1) + B_1\dot{f}_0 + B_1\left(\frac{\partial f_1}{\partial V}\dot{V} + \frac{\partial f_1}{\partial \chi}\dot{\chi} + \frac{\partial f_1}{\partial \gamma}\dot{\gamma}\right)$$

$$B_4 = B_1 B_2 \qquad \text{eq 54}$$

$$B_2 = \begin{pmatrix} \dfrac{1}{m} & 0 & \dfrac{-1}{m}\dfrac{\partial L}{\partial \alpha} \\ 0 & \dfrac{Lc\sigma}{mV} & 0 \\ 0 & \dfrac{Ls\sigma}{Vc\gamma} & \dfrac{c\sigma}{Vc\gamma}\dfrac{\partial L}{\partial \alpha} \end{pmatrix} \qquad \text{eq 55}$$

Differentiating S, the following relation is obtained :

$$\dot{S} = f_3 + B_4 U + K_v B_1 (f_0 + f_1) + K_p \dot{e} + K_I e - \dddot{X}_r - K_v \ddot{X}_r =$$
$$= f_5 + B_4 U = (f_5^* + \Delta f_5) + (B_4^* + \Delta B_4) U \qquad \text{eq 56}$$

Here starred functions denote nominal values of functions and $\Delta B, \Delta f$ denote uncertain functions.

For the existence of a VSC (or even a feedback linearizing control), invertibility of matrix B_4 is required. Of course, B_4 is function of its arguments and singularities of this matrix belong to certain hypersurfaces. In view of (eq 56), if B is singular, U cannot directly affect S, computing the determinant gives:

$$\det(B_4) = -\frac{\partial L}{\partial \alpha} \frac{Lc^2 \sigma}{m} \neq 0 \qquad if \sigma \neq \pm \frac{\pi}{2} \qquad \text{eq 57}$$

Thus in a neighborhood of the trim value, B is invertible and a VSC law can be designed. Of course computation of exact region in which it is invertible is interesting, but it is quite involved because B_4 is a complicated nonlinear function. For trajectory control, maneuver through the region in which singularity lies must be avoided by a proper trajectory planner. For the derivation of the control law, the Lyapunov approach is used:

$$W = \frac{1}{2} S^T S \qquad \text{eq 58}$$

The derivative of W is given by:

$$\dot{W} = S^T \left(f_5 + \Delta f_5 + (B_4 + \Delta B_4) U \right) \qquad \text{eq 59}$$

In view of (eq. 59) for making $\dot{W}$ negative, the control law is chosen of the form

$$U = (B_4)^{-1} (-f_5 - KS - g \operatorname{sgn}(S)) \qquad \text{eq 60}$$

Where K is a diagonal matrix, and g >0 is yet to be chosen. Now the following assumption is made

$$\left\| \Delta B_4 (B_4)^{-1} \right\|_\infty \leq a_0$$
$$\left\| \Delta f_5 - \Delta B_4 (B_4)^{-1} (f_5 + KS) \right\|_\infty \leq a_1 \qquad \text{eq 61}$$

It is pointed out that a_0 and a_1 can be taken as positive constants. Note that the first inequality restricts the uncertainty in the input matrix which requires that any uncertain U dependent term appearing in (eq 59) should be small. A good estimate of this term can be used to satisfy assumption (eq 61). In view of (eq 60), for compensating the uncertainty, one chooses the gain g as

$$g \geq (1-a_0)^{-1}(\eta + a_1) \qquad \eta > 0 \qquad \text{eq 62}$$

Then using the uncertainty bounds of assumption 1, one can show that

$$\dot{W} \leq -\eta \|S\|_1 - S^T KS \qquad \text{eq 63}$$

$\forall S \neq 0$ almost everywhere on $t \in [0, +\infty)$ where $\|S\|_1 = \sum_1^3 S_i$ Thus the surface S=0 is reached in a finite time and by the definition of S, the trajectory tracking error e tends to zero. It is noted that the VSC law, can be synthesized using the computed value of $\ddot{X}$ using the nominal values of the parameters.

The control system includes discontinuous functions this can cause chattering phenomenon, here it is avoided by smoothing the discontinuous functions by replacing sign function by saturation function. This modifies the control law in small boundary layers surrounding S=0. Outside the boundary layers, (eq 59) is valid and as ε tends to zero, the control laws with the sign functions are recovered.

CONCLUSIONS

Airships are a highly interesting study object due to their stability properties. In this chapter, an important application, bridge monitoring, is highlighted. In the first part, a flight simulator was presented . In the second part, kinematics and dynamics of an airship are discussed using Newton-Euler approach. Here, motion is referenced to a system of orthogonal body axes fixed in the airship, with the origin assumed to coincide with the nose. The equations of motion are derived from the Newton-Euler approach. Trim trajectories have been generated for different flight operating modes. The fundamentals of flight are in general: straight and level flight (maintenance of selected altitude), ascents and descents, level turns, wind drift correction and ground reference maneuvers. Trim is concerned with the ability to maintain flight equilibrium with controls fixed. A trimmed flight condition is defined as one in which the rate of change (of magnitude) of the airship state vector is zero (in the body-fixed frame) and the resultant of the applied forces and moments is zero. In a trimmed maneuver, the airship will be accelerated under the action of non-zero resultant aerodynamic and gravitational forces and moments, these effects will be balanced by effects such as centrifugal and gyroscopic inertial forces and moments.

After establishing the equations of motion of airships, some questions arise:

What are the handling qualities of this airship?
What are their controllability and stabilizability properties?
How can closed loop control systems be solved?

A partial answer is given in the fifth section where a robust control method is presented.

The design of advanced control system must take into account the strong non linearity of the dynamic model. The control characteristics of the vehicle have to be evaluated by considering specific tasks such as ability to maneuver from hover, ability to accelerate into a heavy wind or cross wind, and ability to hover a point on the ground in a variable, shifting wind. Subsequently, the control power characteristics of the vehicle with or without payload are to be determined by considering the proposed control concepts.

The next step is the implementation of this guidance and control methods to the autonomous airship and test its effectiveness when the Venturi effects are present.

REFERENCES

[1] Azinheira, JR; de Paiva, EC; Bueno, SS. 'Influence of wind speed on airship dynamics' *J. of Guidance, Control & Dynamics*, 2002, vol 25, #6, 1116-1124.

[2] Azouz, N; Bestaoui, Y; Lemaître, O. 'Dynamic analysis of airships with small deformations' proc. *IEEE Workshop on Robot Motion & Control*, Bukowy-Dworek, Poland, Nov., 2002, 209-215.

[3] Bateman, H. 'The inertia coefficients of an airship in a frictionless fluid' NACA Technical report, #164, 1924.

[4] Bestaoui, Y. 'Nominal Trajectories of an autonomous under-actuated airship' Inter. *J. of Control, Automation and Systems,* Aug., 2006, 395-404.

[5] Bestaoui, Y; Hima, S. *'Modelling and Trajectory Generation of Lighter Than air Aerial robot'* Robot Motion and Control 2007, K. Kozlowski (editor), 2007, Springer , LNCIS 360, 3-28

[6] Bestaoui, Y; Kuhlmann, H. *'Modeling of a quad-rotor airship with wind and freight effect : a Newton Euler approach'* proc. AIAA Aerospace Sciences Meeting, Orlando, Fl., Jan. 2010, paper AIAA-2010-39.

[7] Cai, Z; Qi, W; Xi, Y. 'Dynamic modelling for airship equipped with ballonets and ballasts' *Applied Mathematics and Mechanics*, 2005, vol. 26, 1072-1082.

[8] Elfes, A; Siqueira Bueno, S; Bergerman, M; Guimaraes Ramos, JJ. 'A semi autonomous robotic airship for environmental monitoring mission' *IEEE international Conference on Robotics and Automation,* Detroit, MI, May, 1999, 3449-3455.

[9] Evans, JR; Laurier, JD; Scholaert, DH. 'A study of airship six degrees of freedom flight dynamics' Research report #79, july 1981, University of Toronto, Canada.

[10] Elfes, A; Bueno, S; Bergerman, M; de Paiva, E; Ramos, J. 'Robotic airships for exploration of planetary bodies with an atmosphere: autonomy challenges' *Autonomous Robots*, 2003, vol. 14, 147-164.

[11] Fossen, T. *'Guidance and control of ocean vehicles'*, J. Wiley press, Chichester, 1996.

[12] Hygounenc, E. 'modélisation et commande d'un dirigeable pour le vol autonome' *Phd thesis*, univ. de toulouse (france), 2003.

[13] Jones, SP; DeLaurier, JD. 'Aerodynamics estimation techniques for aerostats and airships' *J. of aircraft*, 1982, vol. 20,#2, 120-126.

[14] Kample, T; Elfes, A. 'Optimal wind assisted flight planning for planetary aerobots' *IEEE Conf on Robotics and Automation*, New Orleans, Louisiana, 2004.

[15] Katz, J; Plotkins, A. 'Low-speed Aerodynamics, from Wings Theory to Panel Methods', McGraw-Hill Book Co, New York 1991.

[16] Kayuk, Y; Denisenko, V. 'Motion of a mechanical system with variable mass inertia characteristics' *Int. J. of Applied Mechanics*, 2004, vol. 40, 814-820.

[17] Khoury, GA; Gillet, JD. *'Airship Technology'*, Cambridge Univ. Press, 1999.

[18] Kim, J; Keller, J; Kumar, V. 'Design and verification of controllers for airships' *IEEE/RSJ, Int. Conf. On Intelligent Robots and Systems*, Las Vegas, Nevada, Oct. 2003, 54-60.

[19] Kungl, P; Schlenker, M; Winner, DA; Kroplin, BH. 'Instrumentation of remote controlled airship 'Lotte' for in-flight measurements' *Aerospace Science & Technology*, 2004, vol. 8, 599-610.

[20] Kornienko, A; Well, K. 'Estimation of longitudinal motion of a remotely controlled airship' *AIAA Atmospheric flight mechanics* conf., Aug. 2003, Austin, TX.

[21] Lamb, H. *'On the motion of solids through a liquid, Hydrodynamics'* Dover, New York, 6th edition, 1945, 120-201.

[22] Lee, S; Bang, H. '3D ascent trajectory optimization for stratospheric airship platforms in the Jet Stream' *J. of Guidance, Control and Dynamics*, 2007, vol. 30, #5, 1341-1352.

[23] Mueller, JB; Zhao, YJ; Garrard, WL. 'Optimal ascent trajectories for stratospheric airships using wind energy' *J. of Guidance, Control and Dynamics*, 2009, vol. 32, #4, 12321245.

[24] Munk, M. '*The aerodynamic forces on Airship Hulls'*, NACA, report # 184, 1924.

[25] Prentice, B; Thompson, J. editors '*Symposium Airships to the Arctic 'Sustainable Northern Transportation'* Winnipeg, Manitoba, Canada, 2005.

[26] Thomasson, PG. 'Equations of motion of a vehicle in a moving fluid' *Journal of Aircraft*, 2000, vol. 37, 630-639.

[27] Watt, GD. *'Estimates for the added mass of a multi-component deeply submerged vehicle'* Technical mem 88/213, DREA, Canada, 1988.

[28] Zipfel, PH. 'Modeling and simulation of aerospace vehicle dynamics' *AIAA Education series*, 2007.

In: Helium: Characteristics, Compounds...
Editors: Lucas A. Becker

ISBN: 978-1-61761-213-8

Chapter 7

FIRST-PRINCIPLES STUDY OF HELIUM BEHAVIOR IN NUCLEAR FUEL MATERIALS

Younsuk Yun[1], Olle Eriksson[2] and Peter M. Oppeneer[2]
[1]Laboratory for Reactor Physics and Systems Behaviour,
Paul Scherrer Institut, CH-5232 Villigen PSI, Switzerland
[2]Department of Physics and Astronomy, Uppsala University,Uppsala, Sweden

ABSTRACT

The α-decay product helium (He) is an important safety-limiting factor of nuclear fuel materials due to its low solubility in the fuel matrix, especially in the actinide dioxide lattice with fluorite structure. When the concentration of He exceeds approximately 1%, He contributes to diminishing the mechanical strength of the fuel by initiating the precipitation. We present here the results of a first-principles study to understand the diffusion mechanism as well as clustering behavior of He in primarily UO_2. Performing energetic calculations, the favored locations of a He atom are calculated as an octahedral interstitial site (OIS) in a defect-free UO_2 lattice, and as a uranium vacancy when radiation induced vacancy defects are present. Helium has a strong agglomerating tendency, resulting in the creation of point defects in the fuel lattice. The He behavior in other oxide fuels, ThO_2 and PuO_2, is investigated and compared with that obtained for UO_2.

Keywords: helium, solubility, nuclear fuel materials, first-principles modeling, diffusion mechanism, He-clustering.

1. INTRODUCTION

The presence of He is an important issue for nuclear fuel materials, especially for the long-term stability in the nuclear waste management. In spent nuclear fuel, He is continuously generated by α-decay of actinides (^{238}Pu, ^{242}Cm, ^{244}Cm, ^{241}Am, etc), which are produced during irradiation in nuclear reactors. However, the solubility of He is very low in the nuclear fuel matrix, especially in ceramic materials like UO_2, ThO_2, and mixed oxide fuel of $(U,Pu)O_2$ [1-6]. For example, in UO_2, which is the most widely used nuclear fuel material in light-water reactors, He starts to precipitate if its atomic content exceeds approximately 1% [5]. Due to the very limited solubility, a sufficiently large amount of He could modify the micro and macroscopic structure of the fuel materials, leading to swelling and cracking of the materials. This can lead to a safety hazard under nuclear operation conditions, because a high burn-up of the fuel creates more He defects. Consequently, the mechanical properties of the fuel are significantly influenced by the presence of He.

For these reasons, the behavior of He has been experimentally [1-6] as well as theoretically [7-14] studied to understand their location and diffusion mechanism in actinide oxide fuels. Guilbert *et al.* [1,2] reported He migration in single- and poly-crystalline UO_2 disks using $^3He(d, \alpha)H$ Nuclear Reaction Analysis (NRS) method. They showed that He moves for relatively low temperatures as 600 °C and the diffusion coefficient was calculated as $\sim 6\times10^{-17}$ m^2s^{-1} at 1100 °C. The octahedral interstitial site was suggested as the preferential location of He in UO_2 single crystal by Garrido *et al.*'s experiment [5] using a combination of RBS (Rutherford backscattering) and NRS techniques. Sattoney *et al.* [6] performed a transmission electron microscopy (TEM) analysis to investigate the formation of He bubbles as a function of temperature and implantation conditions in UO_2. Roudil *et al.* [3,4] reported that He atoms trapped in vacancy defects in UO_2 are rapidly released when heated up to 1000 °C and coalesce into small He bubbles. Furthermore, the diffusion coefficient of He between 1123 and 1273 K was determined to have a common activation energy of 2 eV. Roudil *et al.* suggested that, assuming a lower He mobility in the fuel under disposal conditions, yet in the time-temperature domain of intrinsic evolutions, large quantities of He bubbles will be formed.

Also theoretical studies have contributed to understanding the behavior of He trapped at various defects and at the variation of the lattice parameter of UO_2 induced by He [7-14]. Petit *et al.* [9] and Crocombette [10] used a first-principles method based on density functional theory (DFT) within the local density approximation (LDA). From the calculated results, they suggested that the most stable trap site for He is a uranium vacancy (V_U) while two other theoretical groups predicted that an octahedral interstitial site (OIS) would be more stable [7-8,11]. Using the classical Mott-Littleton and embedded quantum cluster methodology, Grimes *et al.* [7-8] investigated the calculated energy barrier of He migration between two interstitial sites in UO_2. Compared to the diffusion within a perfect lattice, the energy barrier of He was calculated to be significantly lower in a defective lattice where He migrates via oxygen vacancies (V_O) and V_U, from 3.8 eV to ~ 0.3 eV. Despite the amount of work already done, there are still many unclarified issues regarding He behavior, both from the viewpoint of microscopic and macroscopic length scales. This is the case because He diffusion and its interaction within nuclear fuel materials are very complicated phenomena that are influenced by many parameters such as temperature, irradiation, chemical composition, and so on.

The purpose of this paper is to review theory results for the atomic diffusion behavior of He in the UO_2 lattice. We summarize the important calculated results of the trapping, diffusion, and clustering behavior of He in UO_2, which are reported in our previous papers [12-14]. Moreover, additional studies have been performed on other actinide oxides, the results of which we compared and discussed together with those for UO_2. All the results have been obtained by performing first-principles calculations based on DFT. First, the site stability of He is determined by calculating the incorporation energy. In order to understand the diffusion mechanism of He, nudged elastic band (NEB) calculations have been performed to determine the energy pathways of He migration between two trap sites. In addition to the diffusion behavior of a single atom, the He nucleation has been investigated by increasing the number of He atoms in the used supercell. More details on the computational methodology are presented in the following section.

2. Computational Methodology

In the here-reported first-principles calculations, the DFT framework has been adopted. In all the energetic calculations, we have used the projector-augmented wave (PAW) [15] method as implemented in the Vienna Ab Initio Simulation Package (VASP) [16-18]. For the exchange-correlation functional the conventional generalized gradient approximation (GGA) [19] is utilized. Furthermore, spin-polarization (SP) and spin-orbit coupling (SOC) are taken into account. In our fist-principles calculations, it has been shown that the GGA method without Hubbard *U* correction give reasonable results in energetic calculations, and formation and migration energies of defects in UO_2 computed using the spin-polarized GGA method agree well with experimental data [14], regardless of the fact that a wrong electronic band structure was predicted [20-24]. This is understandable, because for He diffusion and trapping the potential of the *f*-electrons, which are almost inner-shell electrons, is not essential. In addition, the use of the GGA+*U* method to large supercells as we employ here is non-trivial, because non-global energy minima can occur in the GGA+*U* calculations with increased meta-stable states. Amadon *et al.* [25] reported the difficulty to determine the ground states of a system within the GGA+*U* calculations due to the increase of metastable states. Recently, Dorado *et al.* [26] presented that the presence of metastable states in the GGA+*U* calculation strongly depends on the initial occupation matrix of electrons in UO_2 and suggested that the ground state can be reached unequivocally by imposing non-diagonal occupation matrices. However, the application of the density-matrix controlling scheme to supercells of hundreds of atoms seems to be still quite tricky. In this study, large supercells containing up to 96-atoms have been employed to reduce any artificial error due to the use of a small supercell. Figure 1(a) and (b) show the conventional unit cell of UO_2 in the CaF_2 structure where there exists an antiferromagnetic ordering along the (001) direction, and a 2×2×2 supercell containing 96 atoms, respectively. In figure 1(a), the center of the unit cell is a so-called octahedral interstitial site (OIS) in the FCC structure.

The incorporation energy is calculated as the energy difference between two systems, where He is trapped at an OIS and at a vacancy, respectively, and can be written as follows:

$$E^{In}_{He_X} = \left(E^{N}_{perfect} + E^{N}_{He_X}\right) - \left(E^{N+1}_{He_{OIS}} + E^{N-1}_{V_X}\right)$$

where V_X is a vacancy of element X, which is either a uranium or an oxygen atom, and He_{OIS} and He_X indicate that He is trapped at an OIS and at a vacancy of an X-element, respectively. N is the number of atoms in the supercell, which is 96 in this study, and $E_{perfect}$ is the total energy of a defect-free supercell. In this study, we assume that vacancy defects pre-exist in the fuel lattice, thus we do not explicitly consider the vacancy formation energy and focus on the relative incorporation energy between a vacancy and an OIS. Other possible locations including off-diagonal positions are also considered to understand the site stability of He in the UO_2 lattice. The diffusion behavior of He has been investigated by calculating the energy barrier at the saddle point in its movement pathways between two adjacent trap sites. The lattice parameter that gives the lowest total energy of UO_2 has been determined as a function of the number of He atoms in the supercell; subsequently, we have calculated the lattice expansion coefficient of UO_2 induced by He atoms.

The cut-off energy of the plane-wave expansion was used up to 400 eV, and the electron charge density in the used supercell was computed using a 2×2×2 k-points grid in the Brillouin zone. All the calculations have been done at constant volume while fully relaxing the atomic positions, and the force acting on each atom was relaxed until less than 0.01 eV/A.

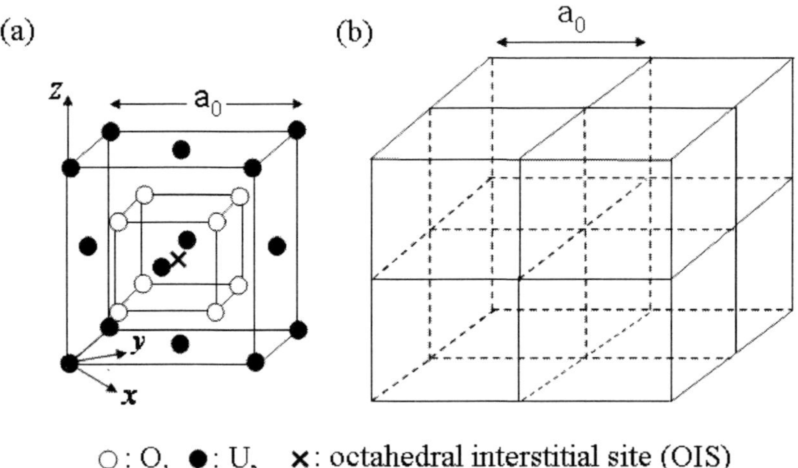

Figure 12.
(a) Conventional unit cell of UO_2, (b) 2×2×2 supercell containing 94 atoms

3. Results and Discussion

3.1. Trapping and Diffusion Behavior of a Single He Atom in the UO_2 Lattice

Our DFT calculations predict that an OIS is the most preferred position of noble gas atoms in a defect-free UO_2 lattice [12-14]. If we locate a He atom at various interstitial positions, and let the computed forces act on the He atom, He spontaneously moves up to the OIS during atomic relaxation. In a next step, we assume that point defects intrinsically pre-exist in the UO_2 lattice. Especially in nuclear fuel materials, defects are continuously provided through nuclear fission. In a defect structure, the incorporation energy is calculated relative to the value for an OIS as follows

$$E^{In}(V_X - OIS) = E^{In}_{V_X} - E^{In}_{OIS}$$

where V_X is a vacancy of a uranium or an oxygen atom. A negative value for $E^{In}(V_X - OIS)$ indicates an energy decrease of the system when He moves from an OIS to a V_X. Conversely, a positive value for $E^{In}(V_X - OIS)$ means that energy has to be provided to incorporate He at a V_X. Our collected results in Table 1 indicate that the V_U is the most favorable incorporation site of He in a UO_2 lattice that contains a V_U and V_O, and a V_O was obtained as most unstable.

Table 1. Relative incorporation energies (in eV) of He between an OIS and a V_U or V_O in the UO_2 lattice, as reported recently [8-12]. A negative energy indicates that He is more stable at the corresponding site than at an OIS, and positive value its opposite

	$E^{In}(V_U - OIS)$	$E^{In}(V_O - OIS)$
Grimes [8]	0.08	0.01
Freyss et al.[11]	0.50	2.50
Petit et al.[9]	-8.70	0.09
Crocombette [10]	-1.20	1.60
Yun [12]	-0.70	0.67

As recognizable in Table 1, there is a controversy over the He stability at an OIS and a V_U in the UO_2 lattice. According to Grimes [8] and Freyss [11], an OIS would be more stable for He than V_U while we [12] and two other groups [9,10] suggested that a V_U is energetically more stable than an OIS. In order to clarify this further, we have performed a NEB [27,28] calculation, which is an efficient method for finding the energy saddle point between a given initial and final state of a transition. The energy pathways of He between an OIS and a V_U have been calculated using the NEB method, and it was found that there is no energy barrier in the energy path of He between an OIS and a V_U, and, hence, He spontaneously moves from an OIS to a V_U during atomic relaxations. An interpretation of these findings can be that He is most likely to be found in OIS's when a very low concentration of vacancies is present in the UO_2 lattice. Although a V_U may not necessarily constitute the majority site available to He, our calculations nevertheless suggest that a V_U is an energetically more stable site for He than an OIS and a V_O. In order to support further these results, we have calculated the electron charge density of He when it occupies an OIS, V_U, or V_O, respectively. The lowest charge density of He was obtained at a V_U as 2.03 /atom, which is lower than that at a OIS and a V_O having 2.06 and 2.12, respectively. It is well known that He prefers to occupy the site with the lowest electron density due to its filled-shell electronic configuration [29]. From all the results, we determine that V_U is the most stable site of He among an OIS and the two vacancies of host atoms in UO_2, V_U, and V_O. In the present study, we do not consider bigger traps like a vacancy complex, such as a U-O divacancy, or a U-2O trivacancy, and so on. Those bigger traps are known to be very important to understand the fission atom's behavior [30-33]. In several articles, a trivacancy has been suggested as the most favorable trap for a Xe atom [30-33], which is one of highest fractional released fission gases in UO_2. The atomic radius of Xe is 2.15 Å and thereby remarkably larger than the ionic radii of uranium and oxygen of 1.01 and 1.40 Å, respectively. Hence, Xe mobility is inevitably affected by the presence of defect complexes. In contrast, He has a very small atomic radius of 0.3 Å, and a single V_U and V_O are considered to be large enough to accommodate a He atom. Therefore, the He behavior at the single vacancies is required to be investigated with priority.

Subsequent to the established site stability of He in the UO_2 lattice, we have calculated the migration energy of He between two adjacent incorporation sites to understand the diffusion mechanism of He. The migration energy was calculated as the energy difference between the saddle point along the migration pathway and the highest energy of initial and final states, as shown in Figure 2.

The calculated energy barriers of He are summarized in Table 2; the results indicate that interstitial site hopping is the rate-determining mechanism of the He diffusion process. The value that is closest to the experimental one [3] was calculated using the experimentally determined lattice parameter at 1200 K, i.e. 5.53 Å. Nevertheless, the energy barrier between two OISs without a vacancy is still relatively high and implies that the vacancy-assisted hopping is a dominant diffusion mechanism of He in UO_2.

From the calculated energy barriers, we suggest a schematic diagram for He diffusion by hopping through V_O and V_U, shown in Figure 3.

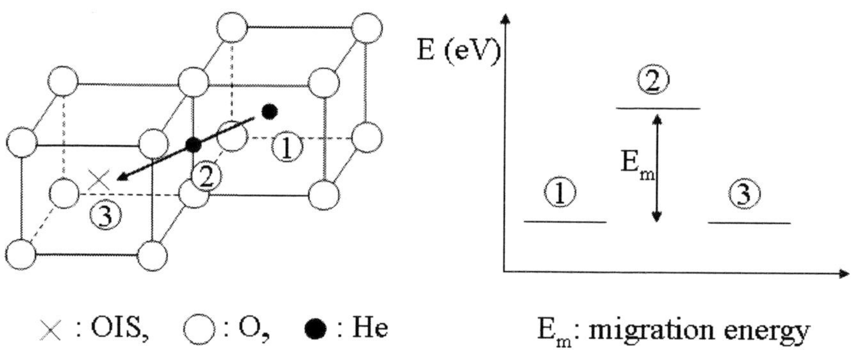

Figure 13 Left: The migration path of He between two OISs. Right: The migration energy of He is defined as the energy difference between the saddle-point energy and that of initial and final sites

Table 2. Calculated energy barriers (in eV) for migration of He between two OIS's

Initial and final position	This study	Exp(3)
OIS-OIS (direct)	2.79	-
OIS-V_U-OIS	0.79	-
OIS-V_O-OIS	0.41	-
OIS-OIS (1200 K, direct)	2.09*	-
-	-	2.00

* Calculated by using the temperature-expanded lattice parameter (i.e. larger) experimentally determined at 1200K, 5.53 Å [38].

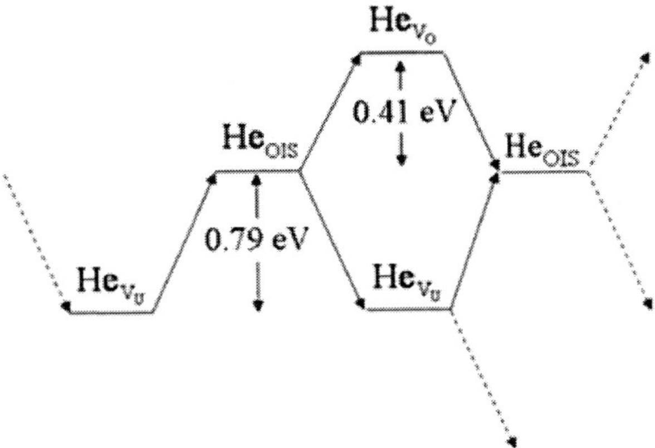

Figure 14. Schematic energy pathways of He diffusion in the UO_2 lattice

These results provoke the importance of understanding the vacancy self-diffusion process in UO_2. The defect characteristics including the formation and migration energies of V_U and V_O were reported in our study [30],where it was reported that V_O is highly mobile, moving around owing to its low migration energy. Also, V_O hopping contributes to stimulating the V_U diffusion by decreasing the energy barrier for V_U by up to about 1 eV. Apart from the thermal-activated process, the vacancy self-diffusion is caused by temperature independent processes of recoil and knock-out which follow the nuclear fission. Therefore, it is expected that the defects created and mobilized by irradiation also assist He atoms to diffuse.

3.2. Atomistic Modeling of He Nucleation

Experimental measurements have reported that He is released above 800 °C from UO_2 pellets and, as a consequence, the thermal diffusion of He scarcely occurs in the fuel under waste disposal conditions [3,4]. It is therefore anticipated that the He concentration increases in the fuel matrix and can lead to the formation of a He bubble. To understand the nucleation process of He in the UO_2 lattice, we performed atomic relaxations in which we increased the number of He atoms in the given supercell. If we start by placing them far apart from each other, then the He atoms migrate each to the their nearest OIS. However, if they get closer to each other around an OIS, then He atoms are likely to cluster together. Figure 4 (a) shows the initial and obtained final configurations of two He atoms in a process of self-consistent atomic relaxation. The simulations show that two He atoms form a dumbbell in the vicinity of an OIS.

Increasing the number of He atoms further in the supercell, we have computed that their collective action is sufficiently strong to create point defects in the UO_2 lattice. If the number of He atoms exceeds five, a uranium Frenkel pair which is a defect pair of a uranium vacancy and an interstitial is created around the He agglomeration. Figure 4(b) shows the initial and final configurations of the simulation of a He cluster of six atoms. As the He atoms cluster, they push the nearest uranium and oxygen atoms from their normal equilibrium lattice sites, to create sufficient space to form a He bubble. The nearest uranium atoms are displaced by about 2.27 Å and reach the nearest OIS. The nearest four oxygen atoms are concurrently displaced up to about 0.65 Å. The collective behavior of the He atoms implies that an OIS can play the role of a nucleation center for He cluster formation when other defects are not present in the UO_2 lattice The agglomerating tendency of He can cause degradation of the local mechanical properties of UO_2 by creating point defects which are known to contribute to the swelling of the material and reduction of its mechanical strength.

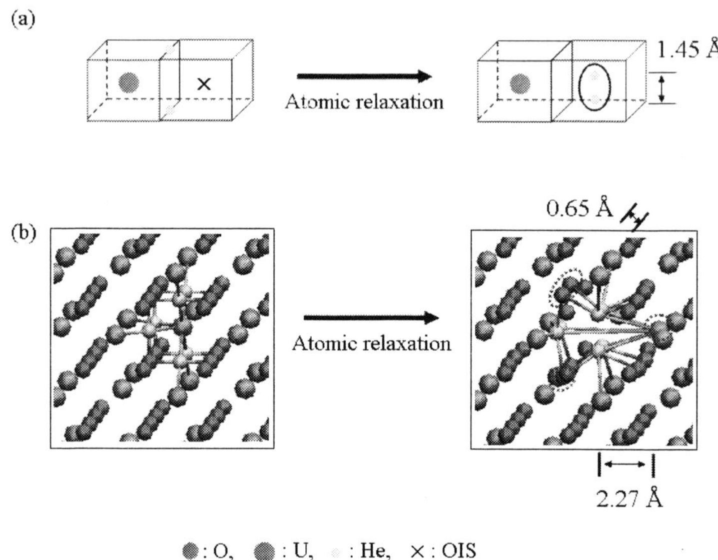

Figure 15. Atomic relaxation results showing that (a) two He atoms cluster and form a He dumbbell, and (b) the collective action of six He atoms creates a U Frenkel pair defect and displacements of the surrounding four oxygen atoms

3.3. He Behavior in Other Actinide Dioxides

The actinide dioxides ThO_2 and PuO_2 are also important nuclear fuel materials. To compare the He behavior in these respective nuclear fuel materials, it is foremost necessary to understand the defect characteristics in each material. The reason for this is, as mentioned above, that defect characteristics are a critical factor governing the trapping and diffusion behavior of He in the material. Recently, we reported the defect formation energy and diffusion behavior in ThO_2 [34]. The migration energy for single vacancies was calculated to be significantly higher for ThO_2 than for UO_2 as is summarized in Table 3. In addition, if the uranium vacancies diffuse through the effective movement of V_{UO}, namely V_O-assisted diffusion of V_U, the migration energy of V_U became decreased by about 1 eV.

However, the effect of V_O-assisted diffusion was not obtained for ThO_2. Table 4 shows that formation energy of an O Frenkel pair and a Schottky trio–which are known as dominant intrinsic defects in dioxides–are more than twice higher in ThO_2 as well. This implies that the concentration of vacancies that are thermally created is very low in ThO_2 as compared to UO_2. Even if vacancies are created in ThO_2, they are much less mobile than those in UO_2 owing to the high required migration energy. The inactivity of vacancies leads to a lower release rate of He atoms from the ThO_2 lattice. Furthermore, the higher formation energy of defects in ThO_2 suggests that the creation of defects by He nucleation is significantly less probable in ThO_2 than in UO_2. As a consequence, it is anticipated that the mechanical stability of ThO_2 is less affected by the increased He concentration.

Meanwhile, for PuO_2 it has been reported that the formation energies of intrinsic defects as the Frenkel pair and Schottky trio are within a similar range of those in UO_2 [33,35]. Our calculated results in Table 4 agree well with other theoretical and experimental results, which give that the formation energies of Frenkel pair and Schottky trio defects are in the range of 4–5 eV and 8–10 eV, respectivly [33,34]. We suggest that the significant difference obtained for the defect characteristic in ThO_2 compared with those in UO_2 and PuO_2 is associated with the absence of 5*f* electrons in ThO_2. In UO_2 and PuO_2, the presence of a partially filled 5*f* shell can be related with their stoichiometry deviations, as the various possible valences of the U and Pu ions easily permit the formation and migration of vacancies [36]. In contrast, ThO_2 is known as a highly stable stoichiometric oxide where the valency of the Th ion unlikely is changed from 4+ [37].

Table 3. Calculated migration energies (in eV) of single vacancy defects in ThO_2 and UO_2

Migration energy	ThO_2 [34]	UO_2 [30,34]
V_O	1.27	0.63
V_U	4.47	3.09 (2.19*)

* Migration energy of V_U in the effective movement of V_{UO}

Table 4. Calculated formation energy (in eV) of intrinsic defects in ThO_2, UO_2, and PuO_2

Formation energy	ThO_2 [34]	UO_2 [34]	PuO_2
O Frenkel pair	9.8	4.5	4.4
Schottky trio	20.6	7.2	7.1

CONCLUSION

We have presented a theoretical investigation of He behavior in UO_2 using first-principles atomistic modeling. In a defect-free UO_2 lattice, an OIS is determined as the most stable trap site for He. In the UO_2 lattice containing single vacancy defects, He is energetically more stable in a V_U than in an OIS or a V_O. NEB calculations showed the energy barrier-less movement of He from an OIS to a V_U. In addition, the lowest electron charge density of He at a V_U confirms that a V_U is more stable for He than an OIS or V_O. In the UO_2 lattice, He is likely to diffuse by hopping through a V_U or V_O with relatively low energy barriers of less than 1 eV. When He atoms get close to each other, they tend to agglomerate at the vicinity of an OIS and form He bubbles. If the number of He atoms exceed five, their collective action becomes strong enough to create a uranium Frenkel pair which could affect to the local mechanical properties of UO_2. From a comparison of the defect characteristics of UO_2, ThO_2, and PuO_2, we derive that the He mobility is much lower in

ThO_2, and also other defects are relatively inactive in ThO_2. The defect characteristics of UO_2 and PuO_2 are obtained to be rarther similar.

References

[1] Guilbert, S; Sauvage, T; Erramli, E; Barthe, MF; Desgardin, P; Blondiaux, G; Corbel, C; Piron, JP; *J. Nucl. Mater.*, 2003, 321, 121-128.

[2] Guilbert, S; Sauvage, T; Garcia, P; Carlot, G; Barthe, MF; Desgardin, P; Blondiaux, G; Corbel, C; Piron, JP; Gras, JM. *J. Nucl. Mater.*, 2004, 327, 88-96.

[3] Roudil, D; Deschanels, X; Trocellier, P; Jégou, C; Peuget, S; Bart JM. *J. Nucl. Mater.*, 2004, 325, 148-158.

[4] Roudil, D; Bonhoure, J; Pik, R; Cuney, M; Jégou, C; Gauthier-Lafaye, F. *J. Nucl. Mater.*, 2008, 378, 70-78.

[5] Garrido, F; Nowicki, L; Sattonnay, G; Sauvage, T; Thomé, L; Nucl. Instr. Meth. *In Phys. Res.*, 2004, 219-220, 196-199.

[6] Sattonnay, G; Vincent, L; Garrido, F; Thomé, L. *J. Nucl. Mater.*, 2006, 355, 131-135.

[7] Grimes, RW; Miller, RH; Catlow, CRA. *J. Nucl. Mater.*, 1990, 172, 123-125.

[8] Grimes, RW. *Fundamental Aspects of Inert Gases* In: SE. Solids; Donnelly, & J. H. Evans; Eds; NATO ASI Series B; Plenum Press; New York; N.Y;, 1991, Vol. 279, 415-429.

[9] Petit, T; Freyss, M; Garcia, P; Martin, P; Ripert, M; Crocombette, JP; Jollet, F. *J. Nucl. Mater.*, 2003, 320, 133-137.

[10] Crocombette, JM. *J. Nucl. Mater.*, 2002, 305, 29-36.

[11] Freyss, M; Vergnet, N; Petit, T. *J. Nucl. Mater.*, 2006, 352, 144-150.

[12] Yun, Y; Eriksson, O; Oppeneer, PM. *J. Nucl. Mater.*, 2009, 385, 510-516.

[13] Yun, Y; Eriksson, O; Oppeneer, PM. *J. Nucl. Mater.*, 2009, 385, 72-74.

[14] Yun, Y; Eriksson, O; Oppeneer, PM; Kim, H; Park, K. *J. Nucl. Mater.*, 2009, 385, 364-367.

[15] Perdew, JP; Chevary, JA; Vosko, SH; Jackson, KA; Pederson, MR; Singh, DJ; Fiolhais, D. *Phys. Rev. B,* 1992, 46, 6671-6687; 1993, 48, 4978(E).

[16] Kresse, G; Furthmüller, J. *Comput. Mat. Sci.*, 1996, 6, 15-50.

[17] Kresse, G; Furthmüller, J. *Phys. Rev.*, B 1996, 54, 11169-86.

[18] Kresse, G; Joubert, D. *Phys. Rev.*, B 1999, 59, 1758-1775.

[19] García, A; Elsässer, C; Zhu, J; Louie, SG; Cohen, ML. *Phys. Rev.*, B. 1992, 46, 9829-9832.

[20] Geng, HY; Chen, Y; Kaneta, Y; Kinoshita, M. *Phys. Rev.*, B 2007, 75, 054111 (1-8).

[21] Kudin, KN; Scuseria, GE; Martin, RL. *Phys. Rev.*, Lett. 2002, 89, 266402 (1-4).

[22] Pickard, CJ; Winkler, B; Chen, RK; Payne, MC; Lee, MH; Lin, JS; White, JA; Milman, V; Vanderbilt, D. *Phys. Rev.*, Lett. 2000, 85, 5122-5125.

[23] Freyss, M; Petit, T; Crocombette, JP. *J. Nucl. Mater.*, 2005, 347, 44-51.

[24] Yun, Y; Kim, H; Kim, H; Park, K. *Nucl. Eng. Tech.*, 2005, 37, 293-298.

[25] Amodon, B; Jollet, F; Torrent, M. *Phys. Rev.* B 2008, 77, 155104 (1-9).

[26] Dorado, B; Amadon, B; Freyss, M; Bertolus, M. *Phys. Rev.*, B. 2009, 79, 235125 (1-8).

[27] Mills, G; Jónsson, H. *Phys. Rev. Lett.*, 1994, 72, 1124-1127.

[28] Mills, G; Jónsson, H; Schenter, G. *Surf. Sci.*, 1995, 324, 305-337.

[29] Puska, MJ; Nieminen, RM. *Phys. Rev.*, B 1991, 43, 12221-33.

[30] Yun, Y; Eriksson, O; Oppeneer, PM; Kim, H; Park, K. *J. Nucl. Mater.*, 2008, 378, 40-44.

[31] Grimes, RW; Catlow, CRA; Phil. Trans. R. *Soc. Lond.*, A 1991, 335, 609-634.

[32] Matzke, HJ. *Radiat. Eff.*, 1980, 53, 219-242.

[33] Matzke, HJ. *J. Chem. Soc. Faraday Trans.*, II, 1987, 83, 1121-42.

[34] Yun, Y; Oppeneer, PM; Kim, H; Park, K. *Acta Mat.*, 2009, 57, 1655-1659.

[35] Jackson, RA. Catlow, CRA. 1986, 53, 27-50.

[36] Olander, DR; *Fundamental Aspect of Nuclear Reactor Fuel Elements*; Energy Research and Development Administration; Oak Ridge, TN, 1976, 145-171.

[37] Skomurski, FN; Shuller, LC; Ewing, RC; Becker, U. *J. Nucl. Mater.*, 2008, 375, 290-310.

[38] Kang, KH; Ryu, HJ; Song, KC; Yang, MS. *J. Nucl. Mater.*, 2002, 301, 242-244.

In: Helium: Characteristics, Compounds...
Editor: Lucas A. Becker

ISBN: 978-1-61761-213-8

Chapter 8

HELIUM THERMODYNAMICS, ANALYTICAL MODEL

Yasser Safa *
Zurich University of Applied Sciences, Switzerland

Abstract

The thermodynamic properties of helium at wide range of temperature and pressure are analytically derived from the knowledge of the pair spherical potentials to which quantum corrections are superposed. The double Yukawa potential model is considered to describe the intermolecular attraction and repulsion energies. Low temperature quantum effects are incorporated by using the first order quantum correction of the Wigner-Kirkwood expansion. A fundamental equation of state is formulated including Helmholtz energy as an explicit function of temperature and density. The thermodynamic properties are expressed as an explicit combination of the Helmholtz energy and its derivatives. The obtained values are compared to the thermodynamic data and the Molecular Dynamic calculations, a satisfactory correspondence with simulation results is realized. The feature of helium thermodynamics in the critical region is discussed. Contrary to most previous similar works, the present theory retrieves the main features of the helium at wide temperature and pressure from analytical formulation.

PACS 64.10.+h, 64.75.Gh, 64.75.Ef.

Keywords: Equations of state, Thermodynamic properties, Hard spheres.

1. Introduction

Helium is the second most ubiquitous element in the universe, it is detected in great abundance in stars and considered as an important constituent of giant planets and interstellar gas. Helium thermodynamics is obviously an attractive research topic to garner special attention in astrophysics; it serves for dynamic models describing birth of the planets and their evolution.

*E-mail address: yasser.safa@zhaw.ch

Though much less abundant on earth, helium is of scientific interest both experimentally and theoretically. Indeed, the thermodynamic properties of helium are a rewarding topic for study especially because it is exploited in a number of high-technology applications. On the other hand, helium is one of the lightest elements that are ideally used for testing the quantum statistical theories of matter.

One of the special properties of the most abundant form of helium (^{4}He) is its extremely low boiling point (4.2 K), which is the lowest of common cryogenic fluids. As reported in Fig. 1, helium has no triple point (the intersection on a phase diagram where three phases: solid, liquid and gas coexist in equilibrium), this because helium will not solidify even at absolute zero unless an external pressure of greater than 2.5 MPa is applied. These properties make helium ideal for a variety of industrial and research applications.

Actually, helium cryogenics is a subfield of cryogenics which cuts across many disciplines, including refrigeration, fluid mechanics and heat transfer, material science, and instrumentation [1]. Its transparency to neutrons, and its stable heat transfer regime are some of the important features of helium-coolant in power plant applications [2]. Further, helium is a monatomic molecule and has the smallest atomic size of any element. This extremely small molecular size gives rise to an important use: helium leak detection systems which are applied in many industrial fields [3].

Clearly, an advanced understanding of helium thermodynamics at different exploitation conditions is highly relevant to such a helium based technology.

Liquid helium ^{4}He can be classified in either of two phases (Fig. 1): He I which is the normal liquid, existing between the critical point (T_c = 5.195 K; P_c = 0.227 MPa) and the lambda transition (T_λ = 2.176 K). He II, denoted sometimes as super-fluid helium, exists between T_λ and absolute zero and has exceptional transport properties. However, helium also has a rare isotope, ^{3}He, which is of particular importance to very low-temperature refrigeration, the present chapter is focused on ^{4}He feature.

The thermodynamic properties of fluid are calculated with some sort of model since it is not possible to measure each property of interest at each combination of temperature, pressure and (in the case of mixture) composition. Often, these thermodynamic values are represented by a variety of equations. Multiple equations may be used to span wide ranges of density, where one equation is often used for the liquid, one for the vapor, and possibly a third for solid (e.g. [4]). The application of multiple equations may result in discontinuities at the borders between thermodynamic sub-domains and inconsistencies between related properties.

In order to avoid as possible these shortcomings we use a single Equation of State (EOS) of wide range applicability to represent all of the thermodynamic properties for helium. We use a fundamental equation where the Helmholtz energy is formulated explicitly as a function of density and temperature. The thermodynamic properties can thus be obtained through combinations of derivatives of the Helmholtz energy with respect to temperature or/and density.

Moreover, the strong point of method presented in this chapter is that the fundamental equation is derived analytically from the sole knowledge of the intermolecular potentials.

In an effort to limit the scope and the length of this chapter, the focus here is on a Helmholtz fundamental equation of state for helium and the related thermodynamic properties. We introduce also in detail an analytical derivation of the compressibility factor and

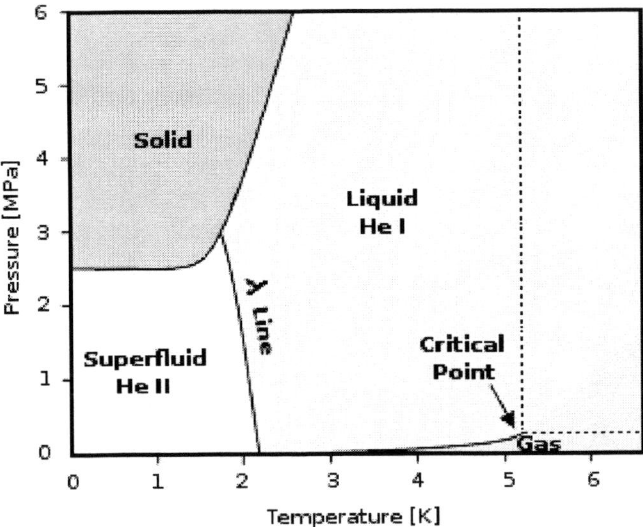

Figure 1. Phase diagram of helium ^{4}He

the pressure.

2. Development of fluid EOS concept and formulation

The exploitation of a fluid in an advanced technology requires the improvement of both the representation accuracy and the thermodynamic consistency in EOS formulation. This is necessary in order to insure stability during process simulation. The development of EOS formulation for the fluid arose from the need for a reasonable mathematical expression which would be usable, not only for the calculation of pressure, density and temperature values, but also for the derivation of important extensive and intensive thermodynamic properties over a wide range of values.

Originally, an equation of state is a mathematical formulation which relates the thermodynamic properties of the fluid. The quest for a simple EOS form has been persisting for more than a century; several practical equations of state have been formulated empirically and have been fitted to experimental data. Different approaches have been proposed in varied forms of modeling, notably: cubic, virial and Helmholtz energy EOS.

2.1. Cubic EOS

The original equation in this class was proposed in 1873 by van der Waals [5] for fluids relating the pressure P, the molar volume V and the temperature T:

$$P = \frac{RT}{V-b} - \frac{a}{V^2},$$

where R is the universal gas constant. This equation introduces two corrections to the ideal-gas law: a is the parameter characterizing the long range attractive force between the molecules and b is an excluded volume correction depending on the hard core volume of the molecules. The volume that the molecules have to move around in is not just the molar volume V, but is reduced to $(V-b)$. For helium we have $a = 0.00341$ Joule m^3/mole^2 and $b = 2.37\ \ 10^{-5}$ m^3/mole.

The shortcoming of this cubic formulation is the limited range of applicability. Therefore several developments were carried out to improve this cubic EOS, (e.g. Dieterici in 1899 and Guggenheim in 1965 and others [6]). Some improved formulations of both repulsive and attractive terms have been contributed by several researchers between 1949 and 1990 and are summarized in [7].

2.2. Virial EOS

The virial equation of state can be derived from statistical mechanics, it expresses the deviations from the ideal gas law as a power series in molar density $1/V$:

$$\frac{PV}{RT} = 1 + \frac{B}{V} + \frac{C}{V^2} + \frac{D}{V^3} + \ldots$$

Here the terms B, C, D,... are the virial coefficients depending on the temperature. At very low density $(1/V \to 0)$ this equation reduces to the ideal gas law. Helium virial EOS was early described in [8] and [9] where values for virial coefficients are provided for a limited range of temperature $100 < T < 400$ K. Although it is possible to extract values for the low order virial coefficients from the fundamental relations, the high order terms in the virial equation are empirical and depend on the experimental data for the fitted fluid. Virial EOS is not expected to be applied to the liquid phase properties. In the case of helium, the limited region of applicability of this virial equation is discussed by [10] and [11].

2.3. Helmholtz energy (fundamental) EOS

Fundamental equations of state garned special importance in recent years since they allow a consistent derivation of the thermodynamic properties from a single mathematical formulation.

The molecular Helmholtz energy F [Joules per molecule] is introduced as a fundamental property with independent variables: number density n and temperature T, it can be split into: $F(n,T) = F^{id}(n,T) + F^{r}(n,T)$. Here the term F^{id} denotes the ideal gas contribution to the Helmholtz energy where the term F^{r} represents the residual Helmholtz energy which corresponds to the influence of both intermolecular forces and quantum effects. An equation in such a format is a fundamental EOS: all the thermodynamic properties can be obtained through combinations of derivatives of the explicit Helmholtz energy function with respect to temperature or/and density. For example, from the first order derivative $n^2\left(\frac{\partial F}{\partial n}\right)_T$ we extract the classical equation of state explicit in pressure $P = P(n,T)$. Similarly, extraction of other properties such as the speed of sound still requires only differentiations of the Helmholtz energy.

3. Main Features of the Presented Helium Model

The thermodynamic model for helium as presented in this chapter can be specified by the following features:

- The fundamental Helmholtz EOS formulation is adopted to describe the thermodynamic properties of helium.
- The inter-molecular potential model used to describe the pairwise interaction of helium molecules is the spherical pair potential containing a short range repulsion and a long range attraction components. To describe the intermolecular repulsive and attractive interaction, we use the double Yukawa potential (DY) which provides an accurate analytical expression for the Helmholtz free energy.
- The model of hard sphere system is used to represent the geometry of helium molecules. A system of hard spheres is the simplest realistic prototype for modeling the vapor-fluid phase separation in such a fluid. The idea of representing a fluid by a system of hard spheres was originally proposed by van der Waals, his classical equation of state was derived using essentially such a simple representation.
- Until recently, most Equations of State (EOS) have resulted from mathematical approximations of experimental data and any functional connection to theory is not completely justified. The strong point of the method presented in this chapter is that the relation between pressure, temperature and density is derived analytically from the knowledge of the intermolecular potentials.
- When dealing with light species such as He and H_2 at low temperature, we need to take into account the quantum mechanical effects. We use the Wigner-Kirkwood expansion (see [12]) to introduce first order quantum effects of such a system.

4. Intermolecular Potential Energy

The estimation of the intermolecular potential energy involves some assumptions concerning the nature of attraction and repulsion between molecules. Intermolecular interaction is resulting from both short-ranged hard sphere repulsion u^{HS} and long-ranged attraction (or "traction") u^{t}. At a radial distance r from molecule's center we have

$$u(r) = u^{\mathrm{HS}}(r) + u^{\mathrm{t}}(r), \tag{1}$$

while the long-ranged attraction is treated as a perturbation and the short-ranged repulsion acts as an unperturbed reference.

The Lennard-Jones formulation is the most widely used expression to describe the intermolecular potential u^{LJ} for molecular simulation. It is a simple continuous potential that provides an adequate representation of intermolecular interactions for many applications at low pressure:

$$u^{LJ}(r) = 4\varepsilon\left(\left(\frac{\sigma^0}{r}\right)^{12} - \left(\frac{\sigma^0}{r}\right)^{6}\right), \tag{2}$$

Table 1. DY potential parameters for helium.

$\sigma^0\,[\text{Å}]$	$\frac{\varepsilon}{k}\,[\text{K}]$	A	λ	ν
2.634	10.57	2.548	12.204	3.336

where σ^0 is the position at which the potential is zero (see Fig. 2) and ε represents the depth of the potential minimum which is located at $r = r^M$. But the inverse-power repulsion in LJ potential is inconsistent with quantum mechanical calculations and experimental data, which show that the intermolecular repulsion has an exponential character. Therefore, the exponential-6 (or α-exp-6) potential can be a reasonable choice instead of the LJ potential (see [13]):

$$u_{\text{exp6}}(r) \quad = \quad \frac{\varepsilon}{\underline{\alpha}-6}\left(6\,e^{\underline{\alpha}\left(1-r/r^M\right)} - \underline{\alpha}\left(\frac{r^M}{r}\right)^6\right), \tag{3}$$

the term $\underline{\alpha}$ determines the softness of the repulsive part of u_{exp6} with respect to the "standard" LJ potential i.e. $u_{\text{exp6}}(r) < u^{LJ}(r)$ for $r < r^M$, [14]. In the case of pure helium molecules we have $r^M = 2.97$ Å and $\underline{\alpha} = 13.1$.

However, an anomalous property of the α-exp-6 potential is observed in the region of high temperature ($T > 2000\,\text{K}$). Indeed, at a small distance r_c, the potential reaches a maximum value and, in the limit $r \to 0$, it diverges to $-\infty$ [15].

4.1. Double Yukawa potential

The double Yukawa potential u^{DY} may be considered as advantageous since it can fit many other forms of empirical potentials (see [12] and [16]), and, in addition, the related integral equation of the Helmholtz free energy and compressibility factor can be solved analytically:

$$u^{\text{DY}}(r) \quad = \quad \varepsilon A\frac{\sigma^0}{r}\left(\mathrm{e}^{\lambda\left(1-r/\sigma^0\right)} - \mathrm{e}^{\nu\left(1-r/\sigma^0\right)}\right). \tag{4}$$

The terms A, λ and ν control the magnitude of the repulsive ant attractive contributions of the double Yukawa potential. The parameters in Table 1 are suitably chosen to provides a close fit to the exp-6 potential proposed in [13].

In Fig. 2 the double Yukawa potential for helium molecules is introduced. At $r = r^M = 2.97$ Å, the absolute value of the potential u^{DY} is equal to the potential depth ε. We observe a consistent representation of the repulsive and the attractive subdomains located in $r < r^M$ and $r > r^M$ respectively. At $r = \sigma^0 = 2.634$ Å, the potential is zero.

5. Fundamental Helmholtz free energy equation

In this section we introduce an analytical derivation of the different components of the Helmholtz free energy. We proceed with an explicit formulation of the molecular energy (with unit [Joules per molecule]) which is a function of the temperature T and the number density n (with unit [inverse of a cubic Angström]). Here n takes the value of the previously mentioned molar density $1/V$ multiplied by a factor of $Na \times 10^{-30}$ where Na is the Avogadro number.

5.1. Hard sphere free energy

The total Helmholtz energy F for a molecule of helium can be written as

$$F = F^{\mathrm{id}} + F^{\mathrm{HS}} + F^{\mathrm{t}} + F^{\mathrm{Q}}, \tag{5}$$

where F^{id} is the Helmholtz energy per molecule arising from the ideal gas. It is defined with

$$\beta F^{\mathrm{id}} = \ln\left(n\left(\frac{h^2}{2\pi kTm_a}\right)^{\frac{3}{2}}\right) - 1, \tag{6}$$

here h is the Planck's constant, m_a is the atomic mass, β is the inverse temperature $\beta = 1/kT$and k is Boltzmann's constant.

The term F^{HS} is the Helmholtz energy of hard sphere. The expression of F^{HS} reads (see for example [16])

$$\beta F^{\mathrm{HS}} = \frac{3\eta_3}{1-\eta_3} + \frac{\eta_3}{(1-\eta_3)^2}, \tag{7}$$

where the parameter η_3 in Eq. (7) is the packing factor related to the hard sphere diameter σ by

$$\eta_3 = \frac{\pi}{6} n\sigma^3. \tag{8}$$

The distance σ is calculated via the integration of the correlation function

$$\sigma = \int_0^{\sigma^0} \left(1 - e^{-u(r)/kT}\right) dr. \tag{9}$$

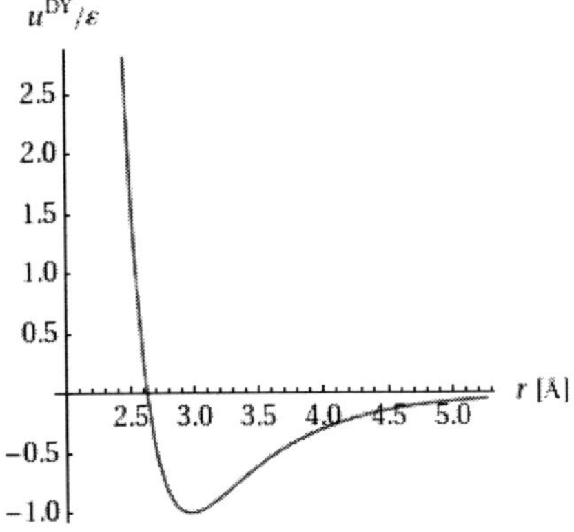

Figure 2. The double Yukawa potential for helium molecules (plotted with Mathematica 7)

In [17] we found Eq. (9) interpreted as a result of the minimization of the free energy difference between the reference fluid (a purely short range repulsive model) and the effective hard sphere model (including the long range attraction).

The use of Eq. (9) makes σ temperature dependent and enables an in-deep investigation of the temperature effects on the thermodynamic behavior of the fluid.

5.2. Attraction free energy

At extreme values of temperature and pressure, the stiffness and the range of the intermolecular repulsion play dominant roles. In contrast, at low temperature and pressure, for predicting properly the vapor-liquid transition both the repulsive and attractive effects must be considered.

In Eq. (5), the term F^{t} is the first order perturbation contribution due to long-ranged attraction. Statistical mechanics provides an evaluation of F^{t} via the integral equation including the radial distribution functions $g(r)$ and the potentials $u^{\mathrm{DY}}(r)$. We have

$$\beta F^{\mathrm{t}} = \beta \frac{n}{2} \int_{\sigma^0}^{\infty} u^{\mathrm{DY}}(r)\, g^{\mathrm{HS}}(r)\, 4\pi r^2\, dr. \tag{10}$$

Using the respective Laplace transform of the function $r g^{\mathrm{HS}}(r)$

$$G(s) = \int_0^{\infty} r g^{\mathrm{HS}}(r)\, e^{-sr}\, dr, \qquad \forall s \in \mathbb{R}, \tag{11}$$

Eq. (10) can be brought to

$$\beta F^{\mathrm{t}} = \frac{2\pi n}{kT} \varepsilon \sigma^0 A \left(e^{\lambda} G\left(\frac{\lambda}{\sigma^0}\right) - e^{\nu} G\left(\frac{\nu}{\sigma^0}\right)\right) - \delta F^{\mathrm{t}}, \tag{12}$$

where δF^{t} is the value of the integral on the interval $[\sigma, \sigma^0]$

$$\delta F^{\mathrm{t}} = \beta \frac{n}{2} \int_{\sigma}^{\sigma^0} u^{\mathrm{DY}}(r)\, g^{\mathrm{HS}}(r)\, 4\pi r^2\, dr. \tag{13}$$

The details of the analytical expressions of the functions $G(s)$ are given in [18].

The substraction of δF^{t} is important due to the fact that the interval $[\sigma, \sigma^0]$ is covered by Eq. (11) and does not belong to the attractive range $[\sigma^0, \infty]$.

Regarding the intersection of the functions $u^{\mathrm{DY}}(r)\, g^{\mathrm{HS}}(r)$ and $u^{\mathrm{DY}}(r)\, g^{\mathrm{HS}}(\sigma)$ at $r = \sigma$ and $r = \sigma^0$ respectively, by considering the close variations of these functions in the interval $[\sigma, \sigma^0]$, we can approach the value of δF^{t} by numerical integration of the expression

$$\delta F^{\mathrm{t}} \approx \beta \frac{n}{2} \int_{\sigma}^{\sigma^0} u^{\mathrm{DY}}(r)\, g^{\mathrm{HS}}(\sigma)\, 4\pi r^2\, dr, \tag{14}$$

here the term $g^{HS}(\sigma)$ refers to (RDF) the radial distribution function $g(r)$ for a hard sphere model at the contact point $r = \sigma$ (conventionally the term $g^{HS}(r)$ is noted RDF and it measures the extent to which the positions of particle center deviate from those of uncorrelated ideal gas).

The contact value of RDF $g^{HS}(\sigma)$ is the improved version of [19] and [20], denoted by $g^{BMCSL}(\sigma)$:

$$g^{HS}(\sigma) = g^{BMCSL}(\sigma) + g^{BS}(\sigma), \tag{15}$$

the values of the terms $g^{BMCSL}(\sigma)$ are introduced in [18] as

$$g^{BMCSL}(\sigma) = g^{(0)}(\sigma) + g^{(1)}(\sigma), \tag{16}$$

where $g^{(0)}(\sigma)$ is the contact value of the Precus-Yevick radial distribution function (PY RDF)

$$g^{(0)}(\sigma) = \frac{1}{1-\eta_3} + \frac{3\eta_2}{(1-\eta_3)^2}\frac{\sigma}{2} \tag{17}$$

and $g^{(1)}(\sigma)$ is the first-order RDF at contact point, it is given by

$$g^{(1)}(\sigma) = \frac{2\eta_2^2}{(1-\eta_3)^3}\left(\frac{\sigma}{2}\right)^2, \tag{18}$$

where η_i is defined as $\eta_i = \frac{\pi}{6} n \sigma^i, \quad i = 0, 1, 2, 3, ...$

5.3. Quantum free energy

The term F^{Q} in Eq. (5) corresponds to the first order quantum correction of the Wigner-Kirkwood expansion [21], [22]

$$\beta F^{\mathrm{Q}} = \frac{h^2 \beta^2 N_A n}{96\pi^2} \frac{1}{m_a} \int_\sigma^\infty \nabla^2 u^{\mathrm{DY}}(r)\, g^{\mathrm{HS}}(r)\, 4\pi r^2\, dr. \tag{19}$$

Using Eq. (11) we can obtain F^{Q} in term of Laplace transform

$$\beta F^{\mathrm{Q}} = \frac{h^2 \beta^2 N_A n}{24\pi} \frac{\varepsilon A}{m_a \sigma^0} \left(\lambda^2 e^{\lambda} G\left(\frac{\lambda}{\sigma^0}\right) - \nu^2 e^{\nu} G\left(\frac{\nu}{\sigma^0}\right)\right). \tag{20}$$

After checks we find that the first order quantum correction exhibits a poor convergence for temperature $T < 40\,\mathrm{K}$. Although the second order correction of Wigner-Kirkwood expansion extends somewhat the convergence over colder systems, it is still by far insufficient at cryogenic temperatures, e.g., in the case of pure He with temperature T less than 40 K, [23].

We recall that one may suggest the application of quantum correction to the hard sphere diameter σ, [24], via the relation

$$\sigma_{\mathrm{cor}} = \sigma + \frac{\lambda}{8}, \tag{21}$$

where λ is the Broglie wavelength $\lambda = \sqrt{\beta \hbar^2 / 2m_a}$ and $\hbar = \frac{h}{2\pi}$.

This correction is usable at high temperature, but insufficient for obtaining reasonable description of quantum effects at temperature $T < 40\,\mathrm{K}$.

6. The thermodynamic properties

The intensive or extensive properties such as temperature, density, pressure, internal energy, enthalpy, entropy, Gibbs energies, heat capacities, speed of sound, chemical potential, and the Joule-Thompson coefficient as well as phase-equilibrium properties are all considered as thermodynamic properties. On the other hand, the transport properties include viscosity, thermal conductivity, diffusion and dispersion coefficients are considered as transport properties. The term "thermophysical properties" denotes all of the thermodynamic, transport, and interfacial (surface tension) properties.

The calculation of the transport properties of helium as function of temperature and density and the dependency of helium surface tension on temperature are described in [25]. Our focus in this chapter is on the thermodynamic properties.

The properties common to engineering applications are summarized in Table 2. They are obtained by derivation of the fundamental equation with respect to the independent variables: density and temperature.

We write the dimensionless Helmholtz energy, $\alpha = \beta F = \frac{F}{kT}$, as an explicit function of the independent variables: the dimensionless number density $\delta = \frac{n}{n_r}$ and the inverse dimensionless temperature $\tau = \frac{T_r}{T}$. The terms n_r and T_r can be chosen as the values of the critical number density $n_r = n_c = 0.0104\ \mathring{A}^{-3}$ and critical temperature $T_r = T_c = 5.1953$ K. The reduced Helmholtz energy, α can be split into the reduced ideal gas value α^{id} and the reduced residual value α^r:

$$\alpha(\delta,\tau) = \alpha^{id}(\delta,\tau) + \alpha^r(\delta,\tau), \tag{22}$$

where

$$\alpha^{id} = \frac{F^{id}}{kT}, \tag{23}$$

and

$$\alpha^r = \frac{F^r}{kT} = \frac{F^{HS}}{kT} + \frac{F^t}{kT} + \frac{F^Q}{kT}. \tag{24}$$

In Table 2 the thermodynamic properties are introduced as function of the reduced Helmholtz energy α and its derivatives with respect to reduced variables δ and τ.

This analytical model for Helmholtz energy formulation (and subsequently the corresponding formulations of the thermodynamic properties introduced in Table 2) is valid with high accuracy for a wide range of temperature [26].

Due to the current advances in computing facilities the calculation of the thermodynamic properties can be carried out by mean of numerical discretization (finite difference) of the derivatives presented in Table 2.

However the important feature of Laplace transforms Eq. (11) in this model that it allows an analytical representation of the compressibility factors as an explicit function of density and temperature. Subsequently, an explicit representation of the pressure and the Gibbs energy can be obtained analytically as a function of density and temperature.

Table 2. The thermodynamic properties

The property	Relation to $\alpha = \beta F$ and its derivatives
Compressibility factor $Z(T,n) = n\left(\frac{\partial \beta F}{\partial n}\right)_T$	$Z(T,n) = 1+\delta\left(\frac{\partial \alpha^r}{\partial \delta}\right)_\tau$
[a] Pressure $P(T,n) = nkTZ(T,n)$	$P(T,n) = nkT\left[1+\delta\left(\frac{\partial \alpha^r}{\partial \delta}\right)_\tau\right]$
1st derivative at constant T	$\left(\frac{\partial P}{\partial n}\right)_T = kT\left[1+2\delta\left(\frac{\partial \alpha^r}{\partial \delta}\right)_\tau+\delta^2\left(\frac{\partial^2 \alpha^r}{\partial \delta^2}\right)_\tau\right]$
2nd derivative at constant T	$\left(\frac{\partial^2 P}{\partial n^2}\right)_T = \frac{kT}{n}\left[2\delta\left(\frac{\partial \alpha^r}{\partial \delta}\right)_\tau+4\delta^2\left(\frac{\partial^2 \alpha^r}{\partial \delta^2}\right)_\tau+\delta^3\left(\frac{\partial^3 \alpha^r}{\partial \delta^3}\right)_\tau\right]$
1st derivative at constant n	$\left(\frac{\partial P}{\partial T}\right)_n = kn\left[1+\delta\left(\frac{\partial \alpha^r}{\partial \delta}\right)_\tau-\delta\tau\left(\frac{\partial^2 \alpha^r}{\partial \delta \partial \tau}\right)\right]$
Entropy $S(T,n) = -\left(\frac{\partial F}{\partial T}\right)_V$	$\frac{S}{k} = \tau\left[\left(\frac{\partial \alpha^{id}}{\partial \tau}\right)_\delta+\left(\frac{\partial \alpha^r}{\partial \tau}\right)_\delta\right]-\alpha^{id}-\alpha^r$
Internal energy $U(T,n) = F+TS$	$\frac{U}{kT} = \tau\left[\left(\frac{\partial \alpha^{id}}{\partial \tau}\right)_\delta+\left(\frac{\partial \alpha^r}{\partial \tau}\right)_\delta\right]$
Isochoric heat capacity $C_v(T,n) = \left(\frac{\partial U}{\partial T}\right)_V$	$\frac{C_v}{k} = -\tau^2\left[\left(\frac{\partial^2 \alpha^{id}}{\partial \tau^2}\right)_\delta+\left(\frac{\partial^2 \alpha^r}{\partial \tau^2}\right)_\delta\right]$
Enthalpy $H(T,P) = U+kTZ$	$\frac{H}{kT} = 1+\tau\left[\left(\frac{\partial \alpha^{id}}{\partial \tau}\right)_\delta+\left(\frac{\partial \alpha^r}{\partial \tau}\right)_\delta\right]+\delta\frac{\partial \alpha^r}{\partial \delta}$
Isobaric heat capacity $C_p(T,P) = \left(\frac{\partial H}{\partial T}\right)_P$	$\frac{C_p}{k} = -\tau^2\left[\left(\frac{\partial^2 \alpha^{id}}{\partial \tau^2}\right)_\delta+\left(\frac{\partial^2 \alpha^r}{\partial \tau^2}\right)_\delta\right]+\frac{\left[1+\delta\left(\frac{\partial \alpha^r}{\partial \delta}\right)_\tau-\delta\tau\left(\frac{\partial^2 \alpha^r}{\partial \delta \partial \tau}\right)\right]^2}{1+2\delta\left(\frac{\partial \alpha^r}{\partial \delta}\right)_\tau+\delta^2\left(\frac{\partial^2 \alpha^r}{\partial \delta^2}\right)_\tau}$
Gibbs energy $G(T,P) = F+kTZ$	$\frac{G}{kT} = 1+\alpha^{id}+\alpha^r+\delta\left(\frac{\partial \alpha^r}{\partial \delta}\right)_\tau$
[b] Speed of sound $W = \left(\frac{1}{m_a}\left(\frac{\partial P}{\partial n}\right)_s\right)^{1/2}$	$\frac{w^2 m_a}{kT} = 1+2\delta\left(\frac{\partial \alpha^r}{\partial \delta}\right)_\tau+\delta^2\left(\frac{\partial^2 \alpha^r}{\partial \delta^2}\right)_\tau-\frac{\left[1+\delta\left(\frac{\partial \alpha^r}{\partial \delta}\right)_\tau-\delta\tau\left(\frac{\partial^2 \alpha^r}{\partial \delta \partial \tau}\right)\right]^2}{\tau^2\left[\left(\frac{\partial^2 \alpha^{id}}{\partial \tau^2}\right)_\delta+\left(\frac{\partial^2 \alpha^r}{\partial \tau^2}\right)_\delta\right]}$

[a] The pressure in this table is obtained in [Joules per cubic Angstr ̈om] = [10^{-30} Pa].
[b] The term m_a denotes the atomic mass of Helium given in [Kg].

6.1. Analytical compressibility factors

In this section the compressibility factors corresponding to the repulsive, attractive and quantum effects are derived analytically. Therefore, the equation of state where the pressure is an explicit function of temperature and density can be readily obtained, and subsequently, the Gibbs free energy.

Let us introduce the superscripts: $\chi \in \{\mathrm{id}, \mathrm{HS}, \mathrm{t}, \mathrm{Q}\}$ which carry the same meaning as in Eqs. (5) and (7). The compressibility factor Z^χ is expressed via the thermodynamic relation

$$Z^\chi = n\frac{\partial}{\partial n}\left(\frac{F^\chi}{kT}\right). \quad (25)$$

It is easy to see that from Eqs. (25) and (6) we have the behavior of the ideal gas:

$$Z^{\text{id}} = n\frac{\partial}{\partial n}\left(\frac{F^{\text{id}}}{kT}\right) = 1. \tag{26}$$

With Eq. (25) and by taking into account Eq. (7), the compressibility factor Z^{HS} can be expressed as

$$Z^{\text{HS}} = n\frac{\partial}{\partial n}\left(\frac{F^{\text{HS}}}{kT}\right) = \frac{1}{1-\eta_3} + \frac{3\eta_1\eta_2}{\eta_0(1-\eta_3)^2} + \frac{\eta_2^3(3-\eta_3)}{\eta_0(1-\eta_3)^3}. \tag{27}$$

By applying the partial derivative with respect to n given in (25) in the relation (12), we obtain the compressibility factor corresponding to the attractive effects:

$$\begin{aligned} Z^{\text{t}} &= \frac{2\pi n}{kT}\varepsilon\sigma^0 A\left(e^{\lambda}\left(G\left(\frac{\lambda}{\sigma^0}\right) + n\frac{\partial}{\partial n}G\left(\frac{\lambda}{\sigma^0}\right)\right) - \right. \\ &\qquad \left. e^{\nu}\left(G\left(\frac{\nu}{\sigma^0}\right) + n\frac{\partial}{\partial n}G\left(\frac{\nu}{\sigma^0}\right)\right)\right) - \delta Z^{\text{t}}, \end{aligned} \tag{28}$$

where δZ^{t} corresponds to the integral in the interval $[\sigma, \sigma^0]$, which is evaluated by numerical integration of the expression

$$\delta Z^{\text{t}} \approx \beta\frac{n}{2}\left(g^{\text{HS}}(\sigma) + n\frac{\partial g^{\text{HS}}(\sigma)}{\partial n}\right)\int_{\sigma}^{\sigma^0} u^{\text{DY}}(r)\,4\pi r^2\,dr, \tag{29}$$

the term $g^{\text{HS}}(\sigma)$ and the derivative $\frac{\partial g^{\text{HS}}(\sigma)}{\partial n}$ can be readily obtained from Eq. (15)-(18).

The expression of the compressibility factor Z^{Q} corresponding to the first order quantum correction of Wigner-Kirkwood expansion is obtained from Eq. (20)

$$\begin{aligned} Z^{\text{Q}} &= \frac{h^2\beta^2 N_A n}{24\pi}\frac{\varepsilon A}{m\sigma^0}\left(\lambda^2 e^{\lambda}\left(G\left(\frac{\lambda}{\sigma^0}\right) + n\frac{\partial}{\partial n}G\left(\frac{\lambda}{\sigma^0}\right)\right) - \right. \\ &\qquad \left. \nu^2 e^{\nu}\left(G\left(\frac{\nu}{\sigma^0}\right) + n\frac{\partial}{\partial n}G\left(\frac{\nu}{\sigma^0}\right)\right)\right). \end{aligned} \tag{30}$$

Finally, the compressibility factor is $Z = 1 + Z^r$ where $Z^r = Z^{HS} + Z^{\text{t}} + Z^Q$.

6.2. Analytical explicit pressure EOS

The pressure P can be directly obtained by summing the respective compressibility factors

$$P = nkT\left(1 + Z^{\text{HS}} + Z^{\text{t}} + Z^{\text{Q}}\right). \tag{31}$$

This analytical model has been validated for the case of Helium and Hydrogen (where shape factor effect was introduced). In [26] a comparison is presented between the computed values of the pressure and the results of Monte Carlo (MC) simulation presented by

Ree [13] where the intermolecular potential is the exp-6 potential. The range of the domain of comparison is $50 < T < 4000$ K and $0.043 < P < 16$ GPa. A reasonable agreement is observed between results. The quantum effect was not included.

Nevertheless, in the calculation based on this model, for a temperature $T = 100\,\mathrm{K}$, the quantum corrections as introduced raise P by about 15%, which corresponds to the first order correction of the Wigner-Kirkwood expansion as estimated in [13]. For higher temperature the obtained values of P are not affected significantly by the quantum contribution.

Further, an other comparison is established with the results from the work of Koči et al. [27]. In this latter work the Buckingham potential is used to perform molecular dynamics (MD) simulations of He for studying the phase transitions and the melting points. In Fig. 3-6 the variation of the pressure with respect to the density is shown for some given values of the temperature. The squares represent the variation resulting from our computations.

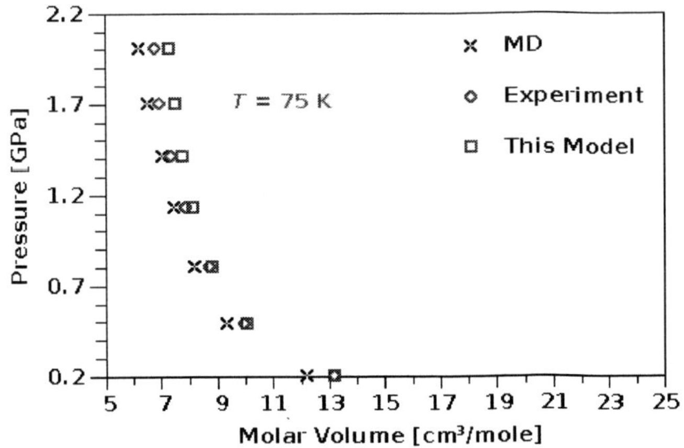

Figure 3. Pressure at $T = 75\,\mathrm{K}$ compared to values from experiments and MD simulations

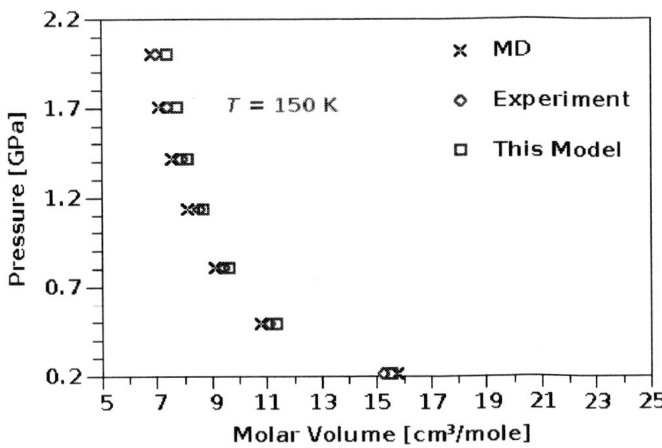

Figure 4. Pressure at $T = 150\,\mathrm{K}$ compared to values from experiments and MD simulations

The cross symbols correspond to the MD results, whereas the other symbols are related to the experimental data as reported in [27]. The match of the values resulting from the model with the experimental data is almost perfect especially for the region of low pressure.

In Fig. 5 and Fig. 6 the MD results exhibit a good agreement with the high pressure reference data, since the model in [27] is expected to be valid for high pressures, but not for very low pressures and low temperature (Fig. 3), where quantum effects are more important.

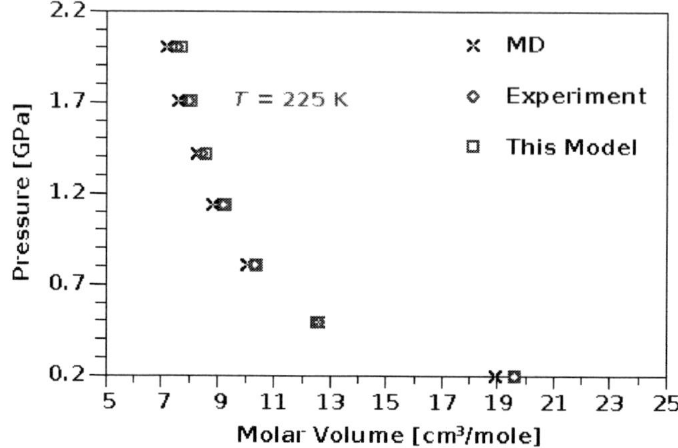

Figure 5. Pressure at $T = 225\,\mathrm{K}$ compared to values from experiments and MD simulations

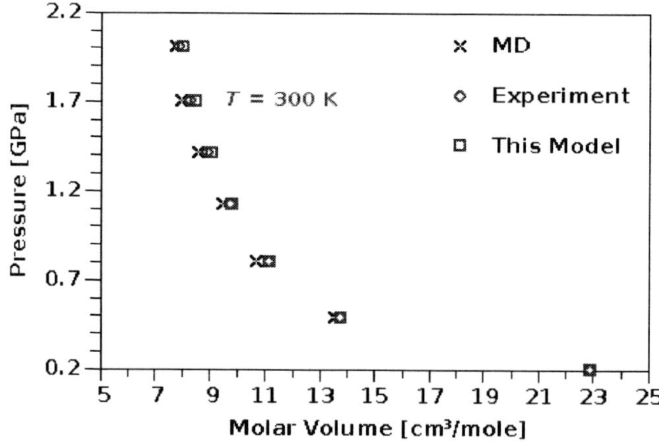

Figure 6. Pressure at $T = 300\,\mathrm{K}$ compared to values from experiments and MD simulations

7. Helium at temperature below 40 K

To our knowledge, no physically founded equation is formulated yet to describe accurately the thermodynamic behavior of helium at temperature $0 < T < 40$ K. More efforts still needed for an analytical formulation of the quantum energy contribution especially in critical and sub-critical regions. Since the methodology in this chapter is to derive the thermodynamic properties from Helmholtz energy, it is instructive to mention the proposed formulation of Helmholtz free energy in this region of high quantum effects.

In [26] the low temperature quantum effects are represented in the formulation of F^Q via cumulant approximations of the Wigner-Kirkwood expansion proposed in [28]. This is a reasonable choice since it is usable down to zero temperature. The quantum correction was introduced with an adjustable factor. The critical cryogenic region was calculated with a relatively "enough accuracy" for astrophysical study of cold interstellar medium (see Fig. 7-8). In sub-critical temperature region the resulting pressure is treated by a Maxwell construction when density $n(P,T)$ at a fixed T becomes multi-valued.

The development of a new equation of state for helium had to fulfil the highest demands on the accuracy in technical applications. Thus, it is practical in this chapter to mention some advanced fitted parameters formulations for Helmholtz energy in the region of low temperature.

The temperature range $2.2 < T < 40$ K belongs to the domain of validity of the fitted formulation of Helmholtz energy proposed in [29] for helium. More precisely, for a pressure up to 100 Mpa and temperature ranging from 2.2 to 400 K (the lower temperature limit $T =$ 2.2 K is the lambda transition point between helium I and helium II) the residual part of the Helmholtz free energy of the equation of state for helium is determined with the structure-optimization method (see section 4 in [29]). Further, a fitted Helmholtz energy model for helium has been established in [30] especially for temperature bellow 0.8 K.

However, the detailed descriptions of the adjustable parameters of those formulations are beyond the scope of this chapter which is devoted to present an analytical "non fitted" model.

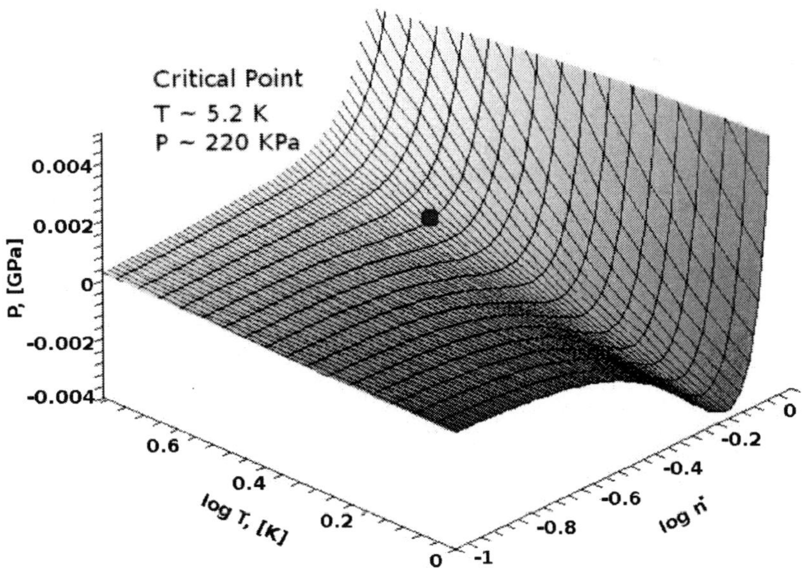

Figure 7. The pressure of helium resulting from this analytical model without Maxwell Construction. It is plotted against log T and log $n^\star$ where $n^\star$ is a reduced density = $(\sigma^0)^3 n$. Quantum effects are qualitatively represented via cumulant approximations of the Wigner-Kirkwood expansion. The reported critical point is located at $P_c \approx 220\,\text{kPa}$ and $T_c \approx 5.2\,\text{K}$.

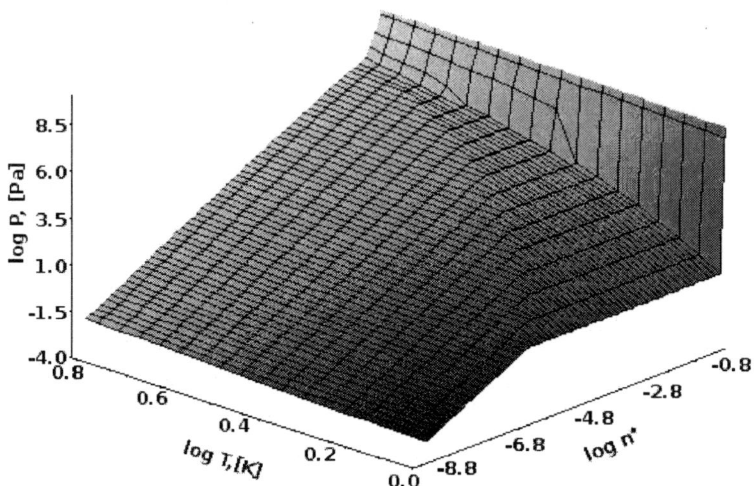

Figure 8. The pressure of helium resulting from this analytical model with Maxwell Construction. It is plotted as log P against log T and log $n^{\star}$ with $n^{\star} = (\sigma^0)^3 n$.

8. Conclusion

In this chapter an analytical model for helium thermodynamics was presented with a wide range of applicability. The feature of helium in a mixture is not treated in the present chapter. However an analytical model for the excess Gibbs energy of mixing still is feasible by applying a suitable mixing rule and by introducing the shape factor in the case of non spheric elements such as hydrogen, nitrogen or oxygen molecules. In the region of low temperature (critical and sub critical region) the quantum effects have a strong influence on the thermodynamic behavior of helium. More efforts still are needed to formulate a physically founded model for an accurate representation of the thermodynamic data of cryogenic helium.

9. Acknowledgments

I thank D. Pfenniger for several discussions on related topics during a research work achieved at Geneva University that was supported by the Swiss National Science Foundation, Grant No 200020-107766.

References

[1] Van Sciver, S. W. Advances in Cryogenic Engineering 161-178 (2007)

[2] Wang, C. P. C. Helium-Cooling in Fusion Power Plant, International Conference Plasma Physics and Controlled Nuclear Fusion Research. Seville, Spain, (1994)

[3] Smith, D. M.; Goodwin, T. W. & Schillinger, J. A. Advances in Cryogenic Engineering: Transactions of the Cryogenic Engineering. 119-138 (2004)

[4] Kerley, G. I. A Kerley Technical Services Research Report, December (2004)

[5] Van der Waals, Johannes, D. The Equation of State for Gases and Liquids. Nobel Lecture, December 12. (1910)

[6] Dharmadurai, G. Journal De Physique III France. Vol 6, 505-509 (1996)

[7] Wei, Y. S.; Sadus, R. J. AIChE Journal. Vol. 46, No. 1, 169-196 (2000)

[8] Kirkwood, J. G.; Keyes, F. G. Physical Review, Vol. 37, 832-840 (1931)

[9] Buckingham, R. A. Proc. R. Soc. Lond. A, Vol. 168, 264-238 (1938)

[10] McLinden, M. O.; Lemmon, E. W. & Jacobsen, R. T. International Journal of Refrigerant, 322-338, (1998)

[11] McCarty, R. D.; Stewart, R. B. The second Symposium on Thermophysical Properties, Princeton University, (1962)

[12] Ali, I.; Osman, S. M.; Sulaiman, N. & Singh, R. N. Mol. Phys., 101, 22, 3239-3247 (2003)

[13] Ree, F. H.; Mol. Phys., 96, 87-99 (1983)

[14] Vörtler, H. L.; Nezbeda, I. & Sal, M. L. Molecular Physics, Vol. 92, No. 5, 813-824 (1997)

[15] Schouten, I. A.; Kuijper, A. & Michels, J. P. J., Phys. Rev. B, 44, 13 (1991)

[16] Ali, I.; Osman, S. M.; Sulaiman, N. & Singh, R. N. Phys. Rev. E, 59, 1028-1033 (2004)

[17] Tang, Y.; Jianzhong, W. J. Chem. Phys., 119, 7388-7397 (2003)

[18] Tang, Y.; Benjamin, C-Y. L. J. Chem. Phys., 103, 7463-7470 (1995)

[19] Boublik, T. J. Chem. Phys. 53, 471-472 (1970)

[20] Mansoori, G. A. et al. J. Chem. Phys. 54, 1523-1525 (1971)

[21] Wigner, E. P. Phys. Rev. 40, 749 - 759 (1932)

[22] Kirkwood, J. G. Phys. Rev., 44, 31-37 (1933)

[23] Boushehri, A. et al. J. of Phy. Soc. Japan, 69, 6 (2000)

[24] Churakov S. V.; Gottschalk M. Geochimica et Cosmochimica Acta, 67, 13, 2397-414 (2003)

[25] Arp V. D.; McCarty, E. D. Natl. Inst. Stand. Technol. (NIST), Technical Note 1334, (1989)

[26] Safa, Y.; Pfenniger, D. Eur. Phys. J. B - Cond. Mat. and Comp. Sys. 66, 337-352, (2008)

[27] Koĕi, L.; Ahuja, R.; Belonoshko, A. B. & Johansson, B. J. Phys. Condens. Matter, 19, 016206 (9pp) (2007)

[28] Royer, A. Phys. Rev. A 32, 1729-1743 (1985)

[29] Kunz, O.; Klimeck, R.; Wagner, W. & Jaeschke, M. The GERG-2004 Wide-Range Equation of State for Natural Gases and Other Mixtures. GERG Technical Monographe 15 (2007)

[30] Arp, V. International Journal of Thermophysics, Vol. 26, No. 5, September (2005)

In: Helium: Characteristics, Compounds...
Editor: Lucas A. Becker
ISBN 978-1-61761-213-8

Chapter 9

COMBINING HELIUM WITH ANTIMATTER: ANTIPROTONIC HELIUM AND ITS APPLICATIONS

*Nicola Zurlo**
Department of Chemistry and Physics for Engineering and Materials,
University of Brescia, Brescia, Italy
National Institute of Nuclear Physics Group of Brescia, Brescia, Italy

PACS 36.10-k, 32.70.Jz, 25.43.+t.

Keywords: Antiproton, Exotic Atoms

1. Introduction

Charged particles such as pions, kaons or antiprotons (produced either by nature or by man in High Energy Physics laboratories under ultrahigh vacuum conditions) are soon decelerated when brought into a medium made of ordinary matter, where their fate essentially depends on their charge. Positive particles (as π^+ or K^+) just hang around in the space between different atoms, until they spontaneously decay with their characteristic lifetime (see e.g. [1]). Negative particles (as π^-, K^- or $\overline{p}$), are instead quickly captured by the atomic nuclei in the medium, leading to their annihilation on nanosecond or picosecond timescale [2].

This is actually what happens most of the times yet with one exception: helium. In fact, when a negative particle such as π^-, K^- or $\overline{p}$is captured by an helium atom, ejecting one of the two electrons, an exotic atom π^-He^+, K^-He^+, or $\overline{p}He^+$respectively is formed. A small yet measurable fraction of these atoms, about 2-3 percent, shows a much longer lifetime, which is anyway limited by the intrinsic decay of the particle in the first two cases ($\tau_\pi \simeq 26$ ns, $\tau_k \simeq 12.4$ ns), though it can be as long as a few microseconds in the case of antiprotonic helium.

Although quite puzzling at the beginning, this behaviour was realized to stem from the fact that a small fraction of negative particles were trapped in long-lived states, giving rise to the formation of "metastable" atoms with lifetimes of the order of 1 μs. These very

*E-mail address: zurlo@bs.infn.it

special atoms must be very resilient to the effects usually shortening their lifetime, mainly Stark decay caused by the collisions with the surrounding helium atoms and Auger decay transforming $\overline{p}He^{+}$ into $\overline{p}He^{++}$. This fact leaves them affected by only the radiative decay, which is usually much slower and has timescales comparable to what is observed.

Subsequent studies have actually revealed that only a few states of the antiprotonic atom have these features, enabling one to identify them and eventually to measure, with very high accuracy, the energy involved in the transitions between one of these states and one of the normal, short-lived ones.

2. Historical Perspective

2.1. Early Observations

Evidence of the existence of these "surprisingly" long-lived exotic helium atoms was collected already in the 1960's with bubble chambers [3]. Experiments performed with π^- and K^- showed that far too many of these negatively charged mesons were seen to decay "at rest", still in atomic orbits, instead of being captured by the nucleus. This happened only when the bubble chamber was filled with liquid helium instead of liquid hydrogen.

To explain these data, Condo [4] and Russel [5] hypotesed that a small amount of the negative mesons could be somehow trapped in "special" atomic states of the exotic helium atom, with a lifetime much longer than the usual strong interaction time delay ($\sim 10^{-12}$ s). These special states were associated with (nearly) circular Bohr orbits having large principal quantum number n, i.e. $l \sim n-1 \gg 1$. The reason of the metastability lied in the relatively large energy required to eject the remaining electron in the exotic helium atom, much larger than the spacing between the different meson energy levels. As noticed by Russel, this effect should also be effective for the capture of the newly discovered antiprotons [6], where it had to be even most evident, due to the fact that antiproton itself is a stable particle. According to his calculations, n had to be around $\sqrt{M^*/m_e}$, where M^* is the reduced mass of the system and m_e the mass of the electron, that is $\sim$15 for the π^-He^+ case, $\sim$23 for the K^-He^+ case, and $\sim$ 38 in the $\overline{p}He^+$case.

But no more investigations were performed until the end of the 80's.

2.2. Metastability Rediscovered

Almost unaware of the background just reviewed, in 1988 the authors of an experiment performed (with a quite different purpose actually) at the National Laboratory for High Energy Physics (KEK) in Japan reported that about 2% of the negative mesons (K^-) stopped in liquid helium had a lifetime of 9.5 ns which, combined with its natural decay lifetime of 12.4 ns, implied a trapping lifetime of about 40 ns [7]. In a subsequent experiment at TRIUMF, in which negative pions (π^-) were stopped in liquid helium, the trapping fraction was 2.3%, and the measured lifetime was 7.26 ns, corresponding to a trapping lifetime of $\sim$10 ns (given the π^- decay time of 26 ns) [8].

2.3. Discovery of Antiprotonic Helium

A new experiment was then quickly arranged at KEK to see whether the same effect was present also with antiprotons ($\overline{p}$) in liquid helium: although the beam was including much more intruding pions than antiprotons, a tight selection of the events allowed to measure that the fraction of long-lived antiprotons was about 3.6% [9]. Their trapping lifetime was almost 3 orders of magnitude longer than in the two previous cases, and precisely of about 3 μs, due to the fact that $\overline{p}$ itself is a stable particle (i.e. a particle having infinite lifetime). The same experiment, a little later, reported that the same effect was present also using other phases of helium, namely gaseous helium, with very similar behaviour [10].

2.4. More Recent Results

Since then, a great effort has been done to investigate in great detail antiprotonic helium, especially at CERN: first using the 100–200 MeV/c beam supplied by the Low Energy Antiproton Ring (LEAR), from 1991 to 1996 (see e.g. [11]), and then using the antiprotons at the Antiproton Decelerator (AD), straight as they are delivered at 100 MeV/c or subsequently further decelerated down to ~100 keV through a Radio-Frequency Quadrupole Decelerator (RFQD), from 2000 up to the time being [12]. And the last experiments, more and more eleborated, are still under way [13].

3. Delayed Annihilation Time Spectra

3.1. Experimental Method

After the first observation of the antiprotonic helium metastability at KEK [10], it became soon clear that a much purer and higher quality antiproton beam was needed to perform more detailed investigations. Luckily enough, at the same time a new machine, dedicated to antiprotons, was just starting to be put into operation at CERN, the already mentioned LEAR. The low energy ($p \lesssim 200$ MeV/c) and the narrow energy spread ($\Delta p/p \sim 10^{-3}$) of the delivered beam made it possible to stop the antiprotons in a relatively small volume of helium (a few cubic centimeters) surrounded by plastic scintillators and Lucite Cerenkov counters aimed at detecting their annihilation [14].

The $\overline{p}$ beam was supplied in the form of a "continuous" flux of antiprotons, and precisely 10^4 $\overline{p}$/s for about 1 hour, or in the form of a single bunch of about 10^7 $\overline{p}$s lasting ~200 ns. For the subsequent laser spectroscopy experiments (presented in section 6.) the latter was preferable, but for the first experiments and the almost-complete removal of the background (see the analysis presented herein), the first kind of extraction was compulsory.

In this case, what was actually measured was the time between the arrival of the antiproton and the detection of its annihilation products; the resulting distribution of these time intervals was called "delayed annihilation time spectrum" (DATS), which in turn is:

$$DATS \equiv \frac{dN}{dt} \tag{1}$$

where N is the number of stopped antiproton still surviving after a time t.

Because of the high multiplicity of the annihilation products (more than three charged pions produced in one annihilation on average), the detection efficiency was quite close to 100%, but the main problem came up from the background associated with the muons/electrons produced by the positive pions produced in $\overline{p}$ annihilations and stopped in the material surrounding the target via:

$$\pi^+ \rightarrow \mu^+ \rightarrow e^+ \qquad (\tau_\mu = 2.2\ \mu s) \tag{2}$$

Anyway, this issue was overcome by selecting only the events with higher multiplicity (i.e. with at least two tracks in a 3.3 ns time window, identified by means of a detector with good granularity and time resolution properties).

An example of DATS is presented in figure 1. In this distribution, all the sources of possible noise (like pile-up or π-μ-e signal) were preempted or removed, but the contribution from the "prompt peak" hasn't been removed yet: on this timescale it actually appears like a huge spike at $t = 0$, and most ($\sim$ 97 %) of the $\overline{p}$He$^+$annihilations occur there. The delayed annihilation distribution itself is enhanced by the logarithmic scale, but doesn't contain more than 3% of the annihilations (still it was very surprising to see some antiprotons living as long as 30 μs). Its shape in not precisely exponential, but it is clearly bent downward: this is a signature of the antiprotons cascade along a series of metastable states (each of them having a different lifetime) until they reach a short-lived state and thus annihilate.

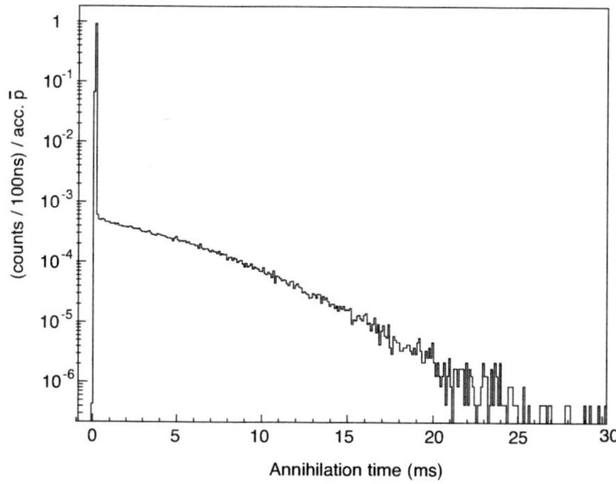

Figure 1. Example of DATS obtained with pure ^{4}He at 5.8 K and 405 mbar.

3.2. Results

The experiments started in 1991 and were performed with helium in different phases (gaseous, liquid and solid) and different physical conditions, i.e. with different temperature (spanning from $\sim$1 K, for experiments with solid and liquid helium, to room temperature) and pressure (spanning from a fraction of a bar up to some bar, and reaching tens of bar, for the liquid and solid case). This needed the usage of different helium containers, some of them being cryogenic to keep the very low temperature required.

As a result of all these experiments, it was quite interesting to see that the DATS shape and features were quite insensitive to the helium physical conditions: trapping fraction was alway about 2.5%-3.5%, and the average lifetime, which may be defined in different ways, was always comprised between $\sim 2\ \mu$s and $\sim 4\ \mu$s, slightly decreasing at very high densities. A summary of all the results obtained in all these experiments is reported in [15].

Furthermore, the effect of small quantities of impurities was investigated in detail almost from the very beginning, because it was found that some contaminants (even in very little quantity) had a dramatic effect on the delayed $\overline{p}He^+$ lifetime, shortening it by far. We will be back on this in section 5.2. For this reason, especial care was dedicated to the target gas preparation, with preliminary baking and continuous monitoring of possible contaminants.

As a final point, it was confirmed that the "delayed annihilations" effect is peculiar to helium: the same delayed annihilations were searched for also in other substances, like liquid neon or nitrogen, but were never found (the upper limit for a possible "long lived trapped antiproton" fraction in these cases was set to be about 10^{-4} at 90% confidence level).

4. Properties of Antiprotonic Helium

Here we will report some very simple-minded thoughts which are enough to infer some basic properties of this system.

4.1. Antiprotonic Helium Formation: Energy Levels

The antiprotonic helium is simply an helium atom in which one of the two electrons is replaced by an antiproton. So, let's start from his father, the helium atom; as discussed in great detail elsewhere in this book, its ionization energy from the ground state is $I_0 = 24.6$ eV, giving a total binding energy of the two electrons of $B_0 = 79.0$ eV. If we assume that each of the two electrons has a binding energy B_e that is one half of this value, B_0, we can attribute an effective charge $Z_e \simeq 1.70$ to the helium nucleus, being:

$$B_e = Z_e^2 R_\infty \quad \text{with} \quad R_\infty = \frac{e^2}{2a_0} = 13.6 \text{ eV} \quad \text{(Rydberg energy)} \tag{3}$$

$$\text{and} \quad a_0 = \frac{\hbar^2}{m_e e^2} = 0.53\ \mathring{A} \quad \text{(Bohr radius)} \tag{4}$$

and the mean orbit radius of each 1s-electron being:

$$\langle r_e \rangle = \frac{1}{Z_e} a_0 \tag{5}$$

When the antiprotonic helium is formed, one of the two electrons is simultaneously ejected, and it is rather intuitive that the capture probability is highest when the state (n, l) better matches the following conditions (usually called "Massey conditions" [16]):

- the binding energy of the antiproton equals the one had by the ejected electron;
- the mean radius of the antiproton equals the old mean radius of the ejected electron

Now, recalling that:

$$B_{\bar{p}} = Z_{\bar{p}} \frac{M^*}{m_e} \frac{1}{n^2} R_\infty \tag{6}$$

$$\langle r_{\bar{p}} \rangle = n^2 \frac{1}{Z_{\bar{p}}^2} \frac{m_e}{M^*} a_0 \tag{7}$$

where M^* is the reduced mass of the antiprotonic helium atom (about $\frac{4}{5}$ of the antiproton mass), from the Massey conditions we can infer that the two effective charges must be equal ($Z_{\bar{p}} = Z_e$) and that n^2 must be equal to $\frac{M^*}{m_e}$.

As a consequence, the newly formed state (n, l) is most likely when

$$n \simeq n_0 = \sqrt{\frac{M^*}{m_e}} \simeq 38 \tag{8}$$

By the way, this means that the antiproton is well localized in space (having a kind of semiclassical trajectory), while the electron is completely distributed around the nucleus an agreement with the quantum mechanics laws.

Pushing a bit further these simple-minded calculations, we expect that the energy level spacing around this value is:

$$\Delta E \simeq \frac{2}{n_0} B_{\bar{p}} \simeq 2 \text{ eV} \tag{9}$$

which does not take into account the effect of the angular quantum number l but anyway it is not far from the true value when states with the same l are considered (see figure 2).

Concerning the angular momentum distribution of the formed $\bar{p}He^+$, rough arguments indicate that the $l-$distribution is approximately proportional to $(2l+1)\log(\frac{l_{\max}}{l})$, where $l_{\max} = n - 1 \simeq n_0 - 1$ is the maximum angular momentum. This means that antiprotonic atoms are formed with all the allowed values for the angular momentum, once fixed the $n-$value. No particular values are forbidden.

Eventually, some information can easily be inferred about the recoil during $\bar{p}He^+$formation. In fact, energy conservation demands that the formed atom recoils with an energy $E_{\text{recoil}} \simeq I_0 \frac{M_{\bar{p}}}{M_{\bar{p}He^+}} \approx 5$ eV, which in turn imply that, for typical experimental conditions (gaseous target at 5 K and 1 bar), a mean free path of the order of nanometers and a mean collision time of the order of picoseconds. This also means that, upon the usual metastable atom timescales, we can assume the atom is already thermalized.

4.2. Antiprotonic Helium Evolution: Decay Processes and Rates

Once formed the antiprotonic helium decays until the antiproton comes close to the nucleus ($n \sim 1$) and annihilates with it. Every single level decays according to one of these three mechanisms:

- Stark mixing induced by collisions with other atoms. This effect can dramatically reduce the principal quantum number n of the state in a few collisions, so the related lifetime is very short, of the order of picoseconds. For $\overline{p}\mathrm{He}^+$ states with $n \simeq n_0$, it is anyway strongly suppressed, because the presence of the electron removes the l degeneracy for the same n. But according to Korenman [17] the states initially formed with $n > 40$ are quickly destroyed by this effect, even before reaching thermal equilibrium.

- Auger decay, which proceeds by the ejection of the remaining electron:

  $\overline{p}\mathrm{He}^+(n,l) \rightarrow \overline{p}\mathrm{He}^{++}(n',l') + e^-$

  leading to a system[1] that is very short lived because of the Stark mixing compelled by collisions with other atoms , which in turn cause annihilation of the antiproton within a few picoseconds.

  Simple-minded thoughts based on the energy conservation (see also figure 2) show that the principal quantum number n' of the resulting $\overline{p}\mathrm{He}^{++}$ cannot exceed the value of about $0.83n_0 = 0.83\frac{M^*}{m_e} \simeq 32$; this also imply a limit on its angular momentum, $l' < n' - 1 \lesssim 31$. As a consequence, if the original metastable atom had a nearly-circular orbit, i.e. $l \simeq n-1 \simeq n_0 - 1 \simeq 37$, this decay implies quite a large momentum change.

  On the other hand, the Auger lifetime depends on the multipolarity Δl: for the levels we are interested in, it generally spans from picoseconds from $\Delta l = 2$ to nanoseconds for $\Delta l = 3$ to microseconds for $\Delta l = 4$ (and so on).

- radiative decay, that essentially occurs via the dipole transition: $(n,l) \rightarrow (n-1,l-1)$. This usually is by far the slowest process amongst the three, with microsecond-scale lifetime, so when the other two are somehow suppressed this process results the only one to be effective, making the atom metastable.

So, only atoms which decay radiatively have a longer lifetime and are therefore metastable. This is usually true for near-circular states ($l \sim n-1$) whose Auger decay is associated with a large angular momentum change ($\Delta l \gtrsim 4$), and so strongly suppressed. In figure 2 the populated energy levels for $\overline{p}^4\mathrm{He}^+$ obtained from the calculations discussed in section 7. are depicted, distinguishing short-lived, Auger-dominated states (wavy lines) from long-lived radiative-dominated states (continuous lines). On the same figure also the subsequent $\overline{p}^4\mathrm{He}^{++}$ levels are drawn, destined to undergo Stark decay (and then annihilation) in picosecond-scale time.

4.3. Molecular Approach

From the results presented in 4.1., it is quite clear that the electron velocity is much higher than the antiproton velocity itself. This makes it possible to consider the $\overline{p}\mathrm{He}^+$as a molecule with two centers, i.e. an α particle and an antiproton (with the last one having negative charge though), bound with a single electron (see [18]).

[1]notice that in practice $\overline{p}\mathrm{He}^{++}$ is nothing else than an antiproton bound to an α particle

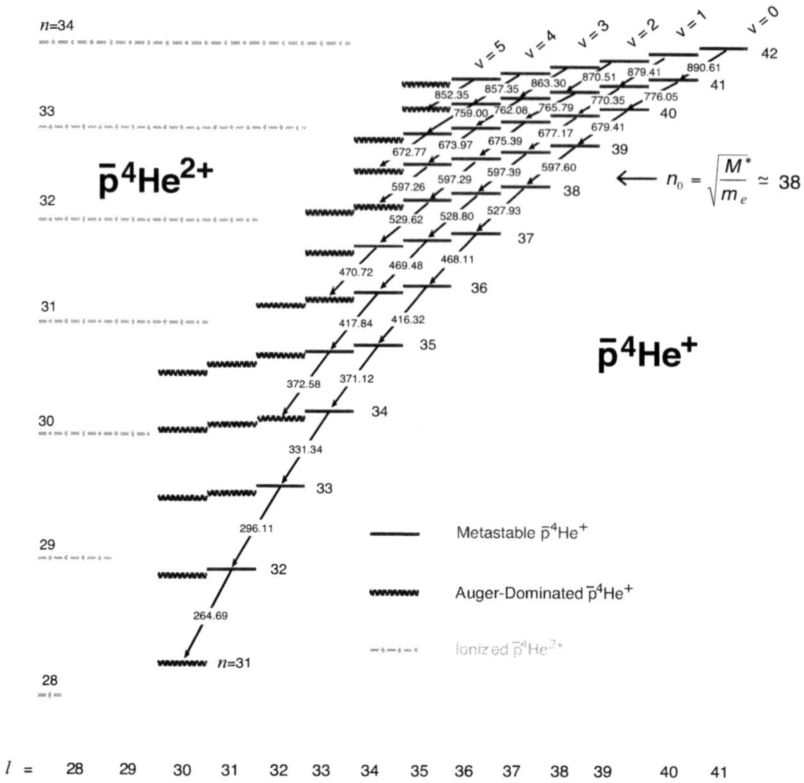

Figure 2. Levels scheme for $\overline{p}^4$He$^+$. Red, straight lines: metastable, radiative-dominated states; wavy, blue lines: short-lived, Auger-dominated states. The arrows indicate some possible transitions, while the figures near the arrows indicate their respective wavelength in nm. For completeness, also $\overline{p}^4$He^{++} levels are reported with dashed-dotted, green line. Reproduced with permission from [15].

This view is somehow dual to the one we have approached till now, showing some properties which are not at all evident in the other way: for example, the molecule can be described by its vibrational and rotational quantum numbers (v, J) which will be related to (n, l) via $J = l$ and $v = n - l - 1$. Within this description, the radiative transitions described before are characterized by the "propensity rule" $\Delta v = 0$ (other kinds of transitions will be discussed in 6.3.).

To stress the importance of both views, the antiprotonic helium is sometime referred to as an atom-molecule or "atomcule".

5. Other Important Effects

5.1. Isotope Effect

A helium nucleus is usually formed by two protons and two neutrons. But there is another kind of stable helium nucleus, namely formed by two protons and only one neutron.

This alternate helium form, ^{3}He, is actually extremely rare in nature, but it can be found as byproduct of tritium decay (which makes it possible to get it on Earth, in spite of his rareness, see e.g. [19]), and then be used to capture antiprotons instead of normal ^{4}He.

It is quite interesting that many of the properties of antiprotonic helium depend also on its reduced mass, so in this respect $\overline{p}^3\mathrm{He}^+$ slightly differs from $\overline{p}^4\mathrm{He}^+$. For example, equantion (8) applied to the case of ^{3}He gives a bit smaller, $n \simeq 37$, for the most probable populated level.

In fact, systematic investigations on $\overline{p}^3\mathrm{He}^+$ DATS [14, 20] gave quite different results from the corresponding ones obtained with $\overline{p}^4\mathrm{He}^+$: the overall lifetime is shorter, with the presence of a fast component with an "intermediate" lifetime of $\sim 0.5\mu$s, and also the trapping fraction is significantly smaller. Anyway, these observations cannot be explained in a simple way, because deeply involve the energy level structure and properties of both species [21].

In this work, actually, apart from the current section, we will almost completely neglect $\overline{p}^3\mathrm{He}^+$ and we focus on $\overline{p}^4\mathrm{He}^+$: more informations on this alternative subject can be found in the mentioned papers, e.g. [15]

5.2. Quenching by Contaminants

Quite from the very early experiments with gaseous helium, it was realized that a small contamination with hydrogen (H_2) resulted in an evident shortening of the metastable atoms lifetime. As a consequence, a systematic investigation on the effect of the presence of impurities has been performed [22].

Two kinds of admixtures were investigated: first, helium admixtures with other noble gases (Ne, Ar, Kr, Xe); second, helium admixtures with molecular gases (H_2, O_2, N_2). Experiments were performed with different concentration of contaminants.

For admixtures with other noble gases, the data indicate a linear increase of the rate (defined as the inverse of the mean lifetime of the metastable atom) with the density of the contaminant. At the same density, the effect increases with the atomic weight of the noble gas, being maximal for Xe. Apart from the latter case, the effect is rather weak, making the lifetime smaller by a factor 2 with an admixture fraction of 10%. In the case of Xe, a similar effect is caused by an admixture fraction 200 times smaller, i.e. 0.05% or 500 ppm.

The situation is completely different for diatomic molecular contaminants. For H_2 and O_2, the same concentration of 500 ppm reduces the lifetime by a factor of 10, while N_2 is slightly less effective in quenching the metastable atoms: its effect is about 5 times weaker, and more comparable to Xe than to the two other molecules.

Further studies, presented here in section 6.6., showed that the effect of H_2 contamination depends very strongly on the level occupied by the antiproton, being the lifetime of the states with $n = 39$ reduced much more than the states with $n \leq 38$ (at the same hydrogen concentration). This evidence stresses the fact that the overall observed shortening of the lifetime is actually a combination of effects depending also and the state of the metastable atom, weighted with the initial energy level occupancy.

6. Laser Spectroscopy

6.1. Experimental Method

The information about $\overline{p}$He^{+} collected with the DATS is extremely interesting, but its intrinsic limit lies in the fact that it refers to a set of different metastable states (all with $n \simeq 38$) which are populated simultaneously. Although the effect of the presence of other helium atoms and of contaminants shed more light on the properties of such a system, the problem was to find a method to study the properties of individual states of $\overline{p}$He^{+}.

This was eventually done by applying a new method of laser spectroscopy. The principle is relatively easy: if we stimulate the transfer of antiprotons between one long-lived, radiative-dominated level and one short-lived, Auger-dominated level by means of a resonant laser beam, almost all the antiprotons involved in the two states will annihilate on the short timescale, producing a sharp increment of the annihilation number in the DATS. The practical issue was to find a suitable laser system, but fortunately the required energies were about 2 eV, see equation(9), i.e. the photons are inside the visible light spectrum, making it possible to find plenty of laser systems to be utilized with this aim, with also a lot of know-how about their features.

The experiments on $\overline{p}$He^{+} laser spectroscopy started in 1993 at LEAR: a system of two tunable pulsed dye lasers (pumped by two Xe-Cl excimer lasers) with a bandwidth of 4–7 GHz (~ 0.007 nm at 600 nm), a pulse duration of ~ 40 ns and an energy of 2–5 mJ per pulse were fired into the helium target through two quartz windows attached to the container. The target was cooled down to cryogenic temperatures to limit the Doppler broadening of the transitions. The laser power density was roughly 1 MW/cm^{2}.

After scanning a large range of wavelengths near the expected value for $\overline{p}$He^{+} transitions between states with n=39 and states with n=38, in 1993 the first resonance was found at $\lambda = 597.26$ nm [23]. These results are reported in figure 3: it is very clear that the shape of the DATS, which is not different at all from usual (i.e. without any laser light) when the laser is off-resonance, shows a large peak simultaneous to the firing of the laser light, when the laser is on-resonance. In other words, the on-resonance laser light practically moves the antiprotons from the metastable state to the short-lived state, while the off-resonance laser light has almost no effect on the antiproton population. After more detailed investigations, this transition was identified as $(39,35) \rightarrow (38,34)$.

6.2. Improvements and Further Transitions Discovered

Quite soon it was realized that the limiting factor of this kind of experiment performed with the continuous $\overline{p}$ beam was the relatively long dead time of the laser, which could neither be fired sooner than 1.3 μs after the trigger, nor at a repetition frequency higher than a few hundred Hz.

To address this issue, new experiments were performed with a pulsed antiproton beam supplied by the same facility, LEAR; in this case, about 10^{7} antiprotons were extracted at once as a single bunch lasting ~ 200 ns. This permitted to illuminate all the formed metastable atoms ($\sim 3 \times 10^{5}$) at the same time, with a single laser pulse, which could be fired at whatever time after the arrival of the antiprotons in the helium target. The data now were collected by Lucite Cerenkov detectors, whose signal was proportional to the number

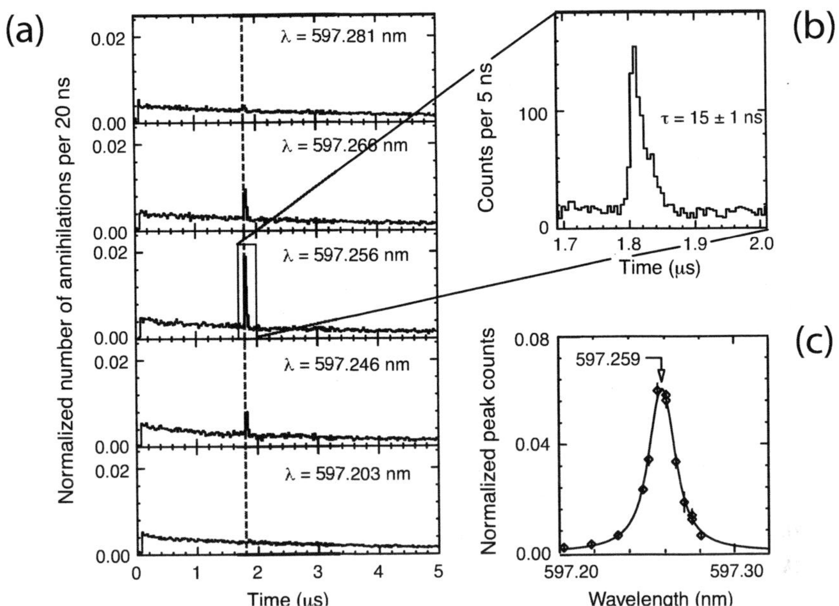

Figure 3. (a): DATS obtained with laser (having the wavelength indicated in each panel) firing at the time t=1.8 μs (dotted vertical line), being $t = 0$ the time of the antiproton arrival (given by a trigger); (b): zoom for the third case of (a), where the laser frequency is the closest to the resonance; (c) results of the peak height for different laser wavelengths. Reproduced with permission from [15].

of pions passing through, i.e. to the number of annihilating antiprotons per unit of time [2]. This kind of counters were preferred to scintillators, because the former provided better temporal response with respect to the latter, giving less background due to the preceding annihilations.

This new technique allowed to make faster wavelength scans, which resulted in the discovery of new resonances: for example, the second observed transition was $(37,34) \rightarrow (36,33)$ in 1994 [24]. Its measured wavelength was $\lambda = 470.72$ nm. Under the experimental conditions described till now, the main source of uncertainty was the laser light bandwidth, which was subsequently improved as well.

Another big improvement was due to the new theoretical calculations of Korobov [25], who was able to find the energy values of the $\overline{p}He^+$levels with an accuracy never reached before (see details in section 7.). This dramatically affected the work of the experimentalists, because it gave precise information on the frequencies/wavelengths to be scanned, and helped in saving most of the time to perform experiments.

In fact, already after the first stage of these calculations, some new transitions were discovered [26, 27], part of them being of a new type, called unfavoured, which we will describe now.

[2]The signal coming from these detectors is usually called "ADATS", Analog Delayed Time Spectrum, to be distinguished from the usual DATS obtained in a continuous beam

6.3. Favoured and Unfavoured Transitions

The transition discovered in 1993 is of the type $(n,l) \to (n-1,l-1)$, i.e. it is characterized by $\Delta v = 0$, and spontaneously occurs in antiprotonic helium, even though on the timescale of the metastable state, which in that case was $(n,l) = (39,35)$. More or less, that transition is just forced by the laser light. And the same thing holds also for the second discovered transition.

All the transitions of this kind are electric dipole transitions ($\Delta l = 1$), but one can find that also another type of electric dipole transition can lie in the visible light range: if we think of the transitions $(n,l) \to (n+1,l-1)$, they still are electric dipole transitions, but characterized by $\Delta v = 2$, and cannot spontaneously occur because the energy of the final state in higher than that of the initial one.

Actually, at the same laser power, and supposing to be exactly on resonance, the first kind of transition is more likely than the second kind: this is the reason why the former is called "favoured transition", while the latter is called "unfavoured transition".

The two transitions observed in [27] are of this new kind (see also figure 4a).

6.4. Double Resonance Method

Even with these improvements on the knowledge of the expected transition wavelength and consequently on the time needed to complete a full wavelength/frequency scan, the number of transitions observable by this method was quite limited: only the transitions with $\Delta v = 0$ or $\Delta v = 2$ having an Auger-dominated final state could be measured.

To extend the number of measurable transitions, the following idea was exploited. Having two independent laser system, they might be set to two different wavelengths and they can be fired simultaneously (within a few ns): the wavelength of the first is set to be on a known resonance $(n,l) \to (n-1,l-1)$ with a short-lived final state; the wavelength of the second laser being scanned, the annihilation peak will approximately increase by a factor of two when this second laser is on resonance with the transitions $(n+1,l+1) \to (n,l)$ because in this case both metastable (n,l) and $(n+1,l+1)$ levels are emptied at the same time feeding the $(n-1,l-1)$ Auger-dominated level.

This method was successfully applied to the chain $(38,35) \to (37,34) \to (36,33)$ [28] (reported in figure 4a).

6.5. Pressure Shift and Broadening

The continuous improvements on the accuracy of these measurements (with better frequency calibration etc.) revealed changes in the resonance wavelengths, depending on the physical conditions of the helium gas, which were sensibly larger than the uncertainty itself. Because the temperature was well controlled and kept low enough in order to limit the Doppler broadening of the resonance, it was soon found that there was a dependence on the pressure (or alternatively on the density) concerning not only the width of the resonance, but also (and much more important) the central value of the resonance itself [29].

The presence of pressure-depending broadenings and shifts in resonance profiles is a quite well known phenomenon in spectroscopy, but the resonance shift was not expected to be significant in $\overline{p}He^{+}$spectroscopy until then. Actually, its effect on some transitions was

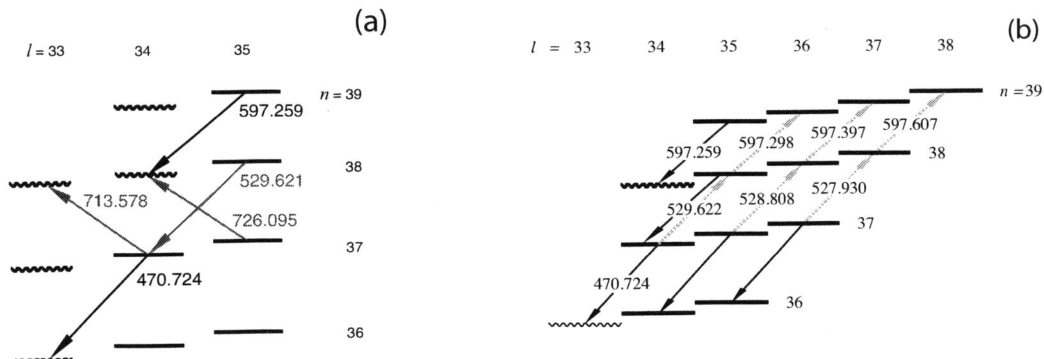

Figure 4. Summary of some transitions explored with different methods. (a): in black, favoured transitions; in red, unfavoured transitions; in blue: transitions explored with the double resonance method. (b) in green: HAIR transitions.

so large that some of the previously published results were affected by systematic errors and the "corrected" resonance wavelength had to be extrapolated to low pressures/densities, approaching as much as possible the "in vacuo" conditions featured by theoretical calculations.

After dedicated experiments, it was found that the wavelength shift scales with the gas density ρ better than with the pressure p:

$$\lambda = \lambda_0 + \left(\frac{\Delta\lambda}{\Delta\rho}\right)\rho \tag{10}$$

where $\frac{\Delta\lambda}{\Delta\rho}$ strongly depends on the considered transition, some transitions being much more sensitive to this density effect,

As a consequence, since then the resonance wavelengths have been generally measured after linear extrapolation to $\rho \to 0$, permitting to reduce the uncertainty on the resonance wavelength to about 1ppm or 1 part in 10^6.

6.6. HAIR Transitions

As discussed in 5.2., the admixture of a small amount of hydrogen to the helium target causes a dramatic change in the lifetime of metastable states.

If all the metastable states were equally quenched by H_2, the DATS would give approximately a single exponential distribution with lifetime governed by the common quenching rate. But this is not the case for H_2: as discussed in [30], from the DATS taken at hydrogen concentrations of 2–1000 ppm, it looks like some metastable states are violently quenched already at low H_2 concentrations, while some others remain almost unaffected even at higher concentrations.

Further studies (some of them performed with the ADATS method, which permitted to fire the laser at a very early time) showed that, increasing the H_2 concentration, the states with $n = 39$ are quenched first, followed by states with $n = 38$, and eventually by states with $n = 37$.

It was then realized that, by conveniently dosing the fraction of H_2, it was possible to stimulate with a laser inverse transitions of the kind $(n,l) \to (n+1,l+1)$, where $n = 38$ or $n = 37$ according to the H_2 concentration, and measure the corresponding wavelength. They are usually called HAIR (Hydrogen Assisted Inverse Transitions). Notice that they are all radiative dipole transitions: the only difference with respect to the favoured transition is that the short-lived state is now the one with higher energy, and its behaviour is not due to Auger decay but to Stark decay caused by collisional quenching with H_2.

This kind of experiments [31] successfully started with the observation of the transition $(38,35) \to (39,36)$, which is the transition adjacent to the first transition discovered in 1993. More than six new resonances were observed with this new method (see figure 4b).

7. Calculations of Transition Frequencies

7.1. First Calculations

Antiprotonic helium represents a three body quantum mechanical system, so the calculations of his energetic levels is not a trivial problem. Luckily, the fact that (for the energy levels we are interested in) the antiproton motion is much slower than electron motion allow to treat the problem adiabatically, i.e. as the electron was bound to a quasi-static system formed by the antiproton and the α particle.

Before 1994, calculations of the antiprotonic helium energy levels were done by different authors using various approaches, see e.g. the molecular approach of Shimamura [18], whose transition energies were found to be accurate within 0.1% (1000 ppm), which was a typical uncertainty of the results of the calculations a that stage.

The big step further was done in 1995 by Korobov [25] who applied the variational principle to the antiprotonic atom after introducing various excited electronic configurations in his so-called "molecular-expansion variational method". Using some thousands of function basis, he reached a convergence of the energy values which let him claim to have reached a precision of 1ppm.

These results were very useful to select the wavelength range to be scanned in the experiments, leading to the discovery of many new resonances (as explained in 6.2.) but anyway Korobov results looked to have a systematic shift when compared to the experimental measurements (see figure 5a). This shift amounted to $\sim$ 50 ppm (for favoured transitions) or $\sim$100 ppm (for unfavoured transitions).

After a few months Korobov and Bakalov [32] took into account the relativistic corrections to the electron energy, and surprisingly the gap between theory and experiment reduced dramatically to a few ppm (see again figure 5a) .

Soon later Elander and Yarevsky [34] pointed out the importance of the QED corrections (Lamb shift) in the theoretical calculations concerning antiprotonic helium. In fact, taking into account also these corrections, the agreement between theory and experiment became better than 1 ppm.

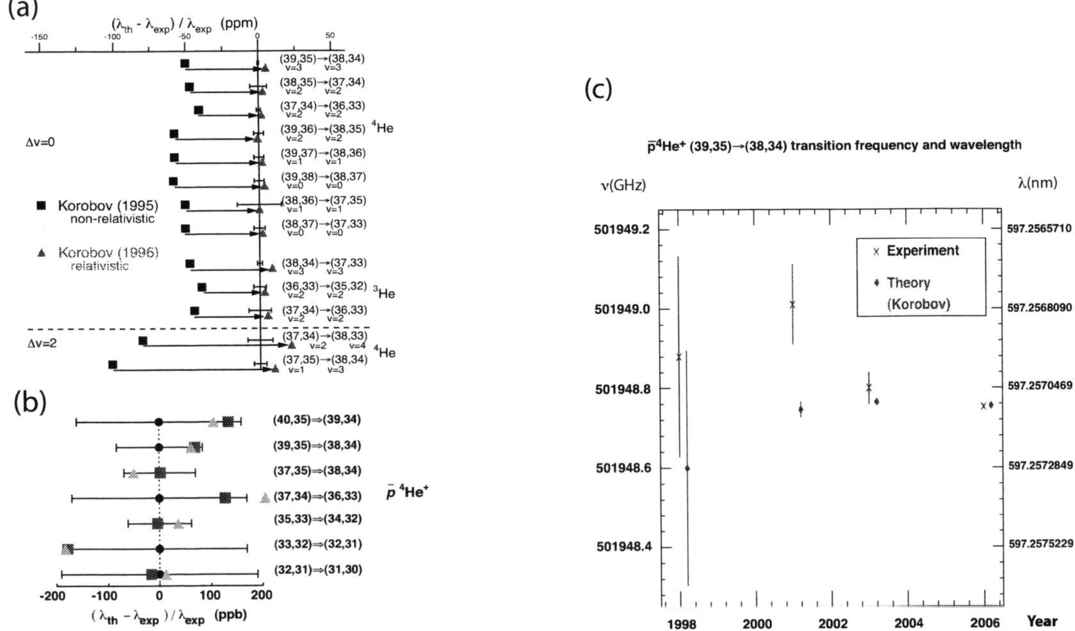

Figure 5. (a) Comparison between experimental and theoretical values for some transitions as of 1995 [25], black squares, and as of 1996 [32], i.e. after relativistic corrections, red triangles. (b) Analogue comparison, but using the data as of 2003; theoretical data are extracted from [33], red squares, and from [35], green triangles. Notice the scale in ppb while in (a) it was in ppm. (c) Comparison between experimental and theoretical values for the transition $(39,35) \rightarrow (38,34)$, one of the best-known transitions, as a function of the year. Reproduced with permission from [36].

7.2. Current Situation

Since then, some other refinements have been made in the calculations by Korobov as well as by some other groups (see e.g. [33,35]), that lead to take into account even smaller corrections in order to reach the same accuracy of the experimental results. For example, typical last-generation Korobov calculations [33] solve the non-relativistic Hamiltonian and perturbatively take into account relativistic [32] and quantum-electrodynamycal [34] corrections up to the order α^4. In this way, more than 11 terms contribute and must be calculated for each energy level of $\overline{p}He^+$.

This is a continuous improvement process which is still going on today; it is pushed by the need to compare more and more accurate experimental data, and in turn it pushes itself the experiments to reduce systematic and statistical errors in order to have better accuracy. For the sake of completeness, the experimental accuracy was brought from about 1 ppm after the extrapolation to zero density, introduced in 1996 (see section 6.5.), to about 100 ppb with the AD-RFQD advent a few years later (see figure 5b and the discussion reported in section 8.), but in the last years it has been further improved, as we will discuss later on, in the last mentioned section.

This produces an effect of convergence of both the experimental results and the theoretical results as a function of the elapsed time (see figure 5c). Till now, it is really pleasant and encouraging that both series are converging to a same value, within the uncertainty of the values coming from the two sides that have reached nowadays the extraordinary accuracy of less than 1 part in 10^8 or 10 ppb [36].

8. Further Experimental Improvements

8.1. New Antiproton Facility

After the end of LEAR activities in 1996, and after a stop of a few years, the experiments with antiprotonic helium were moved to the CERN new facility called AD (Antiproton Decelerator), which started working in 2000 [12].

This new facility, still in operation at the time being (2010), provides antiprotons at 5.3 MeV in bunches of $\sim 10^7$ $\overline{p}$s (and $\sim$ 100 ns duration) every 100 seconds, and consequntly all the experiments in this second phase were performed with the ADATS method.

Already before the restart of the $\overline{p}He^+$experiments at the AD, it was quite clear that the limiting factor for the achievable accuracy, after a substantial improvement of the laser system, was the density shift, due to collisional effects inside the gas target, already discussed in section 6.5.. Actually, the need to stop the antiprotons (with kinetic energy $E_k =$ 5.3 MeV) in a gas target with size of a few centimeters (which had to be totally irradiated by laser light) required a typical density of 10^{21} atoms per cm^3.

In order to address this issue, already since the experimental proposal [12], it was planned to build a further decelerator (called RadioFrequency Quadrupole Decelerator or in short RFQD) able to reduce the antiprotons kinetic energy down to ≈ 100 keV with relatively high efficiency. These antiprotons could be made to come to rest in a cm-sized experimental gas target having densities more than one thousand times lower than before, i. e. of the order of 10^{17} cm^{-3}, dramatically reducing the density shift.

Before putting into operation the RFQD, the first data taking performed with the AD beam [37,38] (at usual densities, temperatures $\sim$ 6 K and pressures $\sim$ 1 bar) confirmed the results taken at LEAR and permitted to test the new laser system (consisting in an improved pulsed dye laser coupled to a series of Fizeau interferometers [39], that permitted to select only the shots in which the effective laser bandwidth was relatively narrow ($\sim$ 0.6 GHz). Theoretically, the wavelength accuracy reachable with this laser system was about 1 part in 10^7, but at this stage it could not be fully exploited because of the systematic error induced by the density shift and its extrapolation to zero-density, so that the difference between experiment and theory was about five times larger.

8.2. RadioFrequency Quadrupole Decelerator

Soon after the AD startup, the RFQD was arranged and set up inside the antiproton beamline; this is in practice a decelerating cavity designed to decelerate part (about 40%) of the incoming antiprotons from 5.3 MeV to 65 keV, that could in turn be biased by applying to the whole cavity a suitable constant voltage; in this way, the energy of the delivered antiprotons could be chosen in the range 10–120 keV.

This low $\overline{p}$ energy permitted to start experiments at reduced target densities: the new target was only slightly larger than before (now about 15 cm $\times$ 30 cm), irradiated by a laser beam $\sim$ 5 cm in diameter, but its density was only $\sim 5 \times 10^{17}$ cm^{-3}, corresponding to a temperature of $\sim$ 10 K and to a pressure of $\sim$ 1mbar.

The use of helium at so low pressure not only permitted to measure the known resonances with an effective accuracy of 10^{-7} [40], but surprisingly led to the discovery of new metastable states which had never been observed before: for example, for the first time it was observed the transition $(32,31) \rightarrow (31,30)$, while the transition $(33,32) \rightarrow (32,31)$ was no longer detectable; this can be interpreted as an effect of collisional quenching of the state (32,31), which is metastable in near-vacuum conditions but became short-lived at the densities of the previous experiments[3].

8.3. Frequency Comb

After overcoming the pressure broadening issue with the RFQD and the low-pressure target, the limiting factor become the laser accuracy and bandwidth. As explained in 6., the use of pulsed lasers was necessary because only plused lasers are able to provide the high ($\gtrsim$ 1 MW/cm^2) intensities required to induce the $\overline{p}$He$^+$transitions. But such a kind of laser usually features relatively high fluctuations in the frequency and in the bandwidth: in a plused laser these fluctuations are of the order of 1 part in 10^7, to be compared the value of 1 part in 10^{10} reachable with a continuous-wave (cw) laser, whose bandwidth can be made as narrow as $\sim$ 100 kHz.

The essential idea for a large improvement in the laser system arose from the recent accomplishment of the optical frequency comb [41], a method which has permitted to extend microwave frequency counting techniques into the optical spectral region, making it possible to to calibrate a cw laser against a reference atomic clock with a precision never reached before.

As a consequence, a new generation laser system was arranged, consisting in a cw laser whose frequency was stabilized against an optical frequency comb generator with a precision better than 4×10^{-10}, measured by Doppler-free spectroscopy of Rb and I_2 in the seed beam [42]. Being the cw laser power (of the order of 1 watt) much lower than needed for the experiment, an subsequent amplifier was used to produce a pulsed beam with intensity 10^6 times larger than the one of the cw laser. This high amplification gain entailed also the so-called frequency chirp effect, due to nanosecond-scale changes in the refractive index of the dye during laser amplification, which causes small (but still appreciable) and non-reproducible deviations from the cw laser frequency, which needed to be compensated by a convenient electro-optic modulator.

The experiments described in [42] used a cw laser with a linewidth of $\sim$ 1 MHz, which was amplified to a pulse having a linewidth of about 60 MHz, 10–100 times better than the previously used pulsed dye lasers. However, the Doppler broadening caused by the motion of the antiprotonic helium atoms (corresponding to a line width of $\sim$ 400 MHz at 10 K) limited the accuracy of the $\overline{p}$He$^+$resonance measurements presented therein to about 10^{-8}.

[3]Another consequence of this effect is the increase of the average $\overline{p}$He$^+$lifetime, which at so low densities appears to be around 5 μs, instead of 3 μ s, as described in [40]

8.4. Two Photon Spectroscopy

Very recent developments on the laser system further improved the achievable spectral resolution: according to [43], the laser linewidth after amplification stage can be made as narrow as ~ 6 MHz.

As mentioned before, the limiting factor is currently the intrinsic Doppler broadening of the resonances line, even with a target at cryogenic temperatures. Recalling that Doppler broadening can be quantified as:

$$\Delta\nu = 2.35\,\nu\sqrt{\frac{kT}{Mc^2}} \tag{11}$$

where k is the Boltzmann constant, T the gas target temperature, M the mass of the antiprotonic atom and ν is the frequency of the resonance itself. This imply $\Delta\nu \simeq 400$ MHz for typical experimental conditions.

To avoid this kind of issue, laser spectroscopic measurements of normal atoms often exploit two laser beams of the same optical frequency disposed in a counter-propagating geometry, that effectively eliminates this Doppler broadening. But unfortunately, in our case, the calculations reveal that the laser intensities required to excite the $\overline{p}He^+$ transitions within the lifetime of the atom ($\sim 1\mu s$) are much greater than those available from any tunable laser available nowadays.

Although this fully Doppler-cancelling spectroscopy is impracticable with antiprotonic helium, still Doppler broadening can be sensitively reduced by utilizing two counter-propagating lasers with non-equal frequencies ν_1 and ν_2 so that:

- $\nu_1 + \nu_2$ is on resonance with a transition $(n,l) \rightarrow (n-2, l-2)$, having $\Delta l = 2$;
- the virtual intermediate state involved in the two-photon transition is tuned to within a few GHz of the real state $(n-1, l-1)$; this fact increases by orders of magnitude the two-photon transition probability with respect to the case $\nu_1 = \nu_2$

In this case, the Doppler broadening will be reduced to:

$$\Delta\nu' = \left|\frac{\nu_1 - \nu_2}{\nu_1 + \nu_2}\right| \Delta\nu \tag{12}$$

which implies, taking into account the energy levels shown in figure 2, a residual Doppler width narrower by about one order of magnitude.

The results of this near-resonant two-photon spectroscopy are still unpublished, although a preliminary analysis has been reported in [44, 45]. The estimated spectral resolution is of the order of 100 MHz, and the main source of uncertainty has been identified as the AC Stark effect, i.e. the interaction of the strong laser fields with the atom which can shift the measured transition frequencies even by several MHz.

One possibility to avoid the Doppler broadening and simultaneously also the AC Stark effect, which requires the use of smaller laser intensities/powers, has been attempted very recently and consists in the so-called two-color saturation spectroscopy [13]. At the moment it looks still difficult to achieve this kind of measurements with the current setup, so major changes in the apparatus are under way, but it is worthwhile going on because this method can allow to mesure all the transitions accessible till now with a precision determined only by the natural width of the short-lived daughter Auger state, which is $\Gamma_n \simeq 20$ MHz for the most favourable cases.

8.5. Fine and Hyperfine Structure

Till now, we have implicitly supposed that each $\overline{p}\mathrm{He}^+$state is fully characterized by the two quantum numbers (n,l). Actually, every (n,l) state has a hyperfine structure due to the coupling between the magnetic moment of the 1s electron (i.e. its spin) and the (total) magnetic moment of the antiproton:

$$\vec{F} = \vec{L} + \vec{S_e} \tag{13}$$

This interaction splits every (n,l) state into a doublet with quantum numbers $F_+ = L + \frac{1}{2}$ and $F_- = L - \frac{1}{2}$, as depicted in figure 6a. Following [15], we will call this splitting as hyperfine (HF) splitting[4] .

In the antiprotonic helium case we have an additional effect due to the spin-orbit coupling of the antiproton spin with its own orbital angular momentum (plus other corrective terms):

$$\vec{J} = \vec{F} + \vec{S_{\overline{p}}} \tag{14}$$

This further splits every level F_+ and F_- into two sublevels, originating four different levels from each (n,l) state , namely $J_{-+} = F_- + \frac{1}{2} = L$, $J_{--} = F_- - \frac{1}{2} = L - 1$, $J_{++} = F_+ + \frac{1}{2} = L + 1$, $J_{+-} = F_+ + \frac{1}{2} = L$ (listed in order of decreasing energy: see figure 6a). Although in a simple $\overline{p}$-nucleus two-body-system this spin-orbit coupling results in the well-known fine-structure doublet ($j = l \pm \frac{1}{2}$), in the context of the antiprotonic helium we will call this effect super hyperfine (SHF) splitting, as suggested in [15]. The motivation for this consists in avoiding confusion, because for antiprotonic helium in typical (n,l) states , the $\overline{p}$ spin-orbit splitting (of the order of 100 MHz) is expected to be much smaller than the hyperfine splitting (of the order 10 GHz).

As a consequence, each laser resonance corresponding to the transition $(n,l) \to (n',l')$ is actually split in four parts: first, because of the HF structure of the parent and daughter state (the cross-over transition $F_- \leftrightarrow F'_+$ and $F_+ \leftrightarrow F'_-$ are strongly suppressed); and second, because of the SHF structure (again the cross-over transitions are very unlikely to happen, i.e. the predominant laser transitions do not change neither the electron nor the antiproton spin direction).

Considering that the further line splitting (of ~ 10 MHz) due to the SHF structure cannot be easily resolved even today because of the Doppler/AC Stark broadening (see section 8.4.) conversely the HF splitting (of ~ 1 GHz) is large enough to be observed especially for unfavoured transitions, which feature the maximum value for the difference $\Delta\nu = \nu_+ - \nu_-$.

The first transition which revealed its twin structure was the transition $(37,35) \to (38,34)$ and this happened already at LEAR [46]; when explored with a relatively high resolution scan for that time (laser bandwidh of ~ 1.2 GHz), this resonance clearly showed two peaks centered in $\overline{\lambda} = (726.097 \pm 0.003)$ nm and separated by $\Delta\lambda = (2.98 \pm 0.09)$ pm, corresponding to $\Delta\nu = (1.70 \pm 0.05)$ GHz, see figure 6b.

[4]Notice that, differently from the hyperfine splitting in the hydrogen ground state case, for the antiprotonic helium case the F_- state has a higher energy than the F_+ state because of the $\overline{p}$ negative charge.

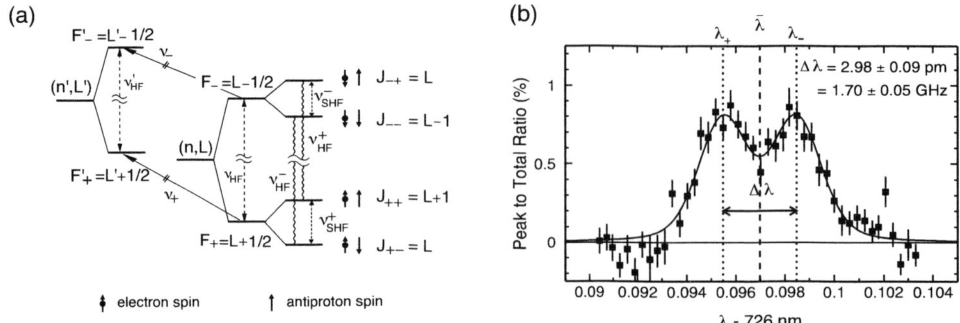

Figure 6. (a): Pictorial view of the splitting of a $\overline{p}He^+$ state caused by magnetic interactions. The straight arrows indicate the observable laser transitions $(n,l) \longrightarrow (n',l')$. Each state is split into two levels $F_{\pm}$,$F'_{\pm}$, each of them being split again into two sublevels. The wavy lines indicate the possible microwave transitions. (b): Observed hyperfine splitting of the unfavoured laser transition $(37,35) \longrightarrow (38,34)$. Reproduced with permission from [46].

This observation was subsequently confirmed and improved at the AD, with a better resolution laser system [37], and followed by the observation of the splitting in some more transitions (especially in $\overline{p}^3He^+$, [42])

Being the precision in measuring the HF splitting limited by the laser resolution, a clever method has been developed [47] in order to measure directly the HF splitting for the state $(37,35)$, stimulating the transitions $J_{-+} \leftrightarrow J_{++}$ (ν_{HF}^+) and $J_{--} \leftrightarrow J_{+-}$ (ν_{HF}^-), see figure 6b, the remaining ones being highly suppressed and very unlikely. Since these frequencies fall in the microwave (MW) range, for this experiment the gaseous helium target was arranged inside a tunable MW cavity. The method consists in three steps performed within ~ 100 ns:

1. depopulate the F_+ state with a laser pulse centered at the frequency ν_+;
2. transfer the population from F_- to F_+ by MW radiation;
3. probe the population of the state F_+ with a second laser pulse centered at the frequency ν_+.

By varying the MW frequency, the ADATS will show two peaks simultaneous to the firing of each laser, the signal in the second peak being maximal when the MW frequency corresponds to either ν_{HF}^+ or ν_{HF}^-. The results presented in [47] report ν_{HF}^+ =12.89596 GHz and ν_{HF}^- = 12.92467 GHz, both measured with a relative precision of $\sim$20 ppm (resolution $\lesssim 1$ MHz), much higher than the resolution achievable by the usual laser spectroscopy method.

9. Comparison between Data and Calculations: Antiproton Mass and CPT Test

The results of the antiprotonic helium laser spectroscopy have two main applications: the precise determination of the antiproton mass (with respect to the electron mass) and the comparison between the antiproton and the proton charge and mass.

First, from the comparison of the measured resonance frequencies with the corresponding theoretical values (see section 7.), it can be inferred the ratio between the antiproton mass and the electron mass. The most recent measurements and calculations give a relative precision of some parts in 10^8.

Compiling the 2006 self-consistent set of values of the basic constants and conversion factors of physics and chemistry for international use [48], the Committee on Data for Science and Technology (CODATA) has recently included the following value deduced from twelve transition frequencies in antiprotonic helium measured with both helium isotopes and their theoretical values,

$$\frac{M_{\overline{p}}}{m_e} = 1836.152674(5) \tag{15}$$

It is then clear how the a fractional precision is $\sim 10^{-8}$, competing therefore with the most precise determination of the ration between the *proton* mass and the electron mass:

$$\frac{M_p}{m_e} = 1836.15267261(85) \tag{16}$$

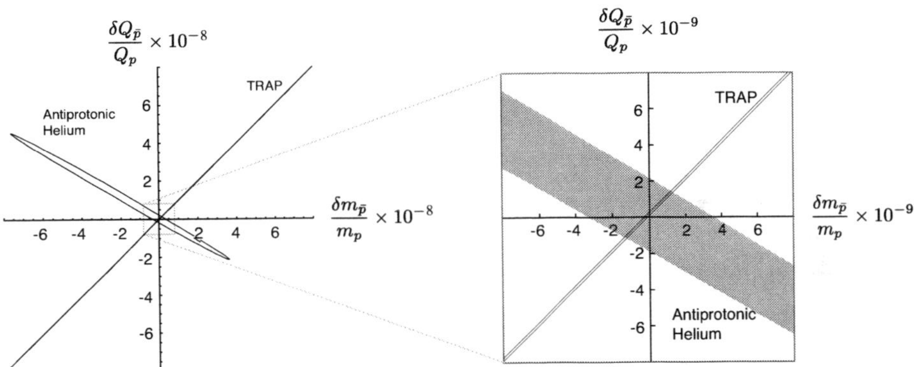

Figure 7. Possible difference between the proton and the antiproton charge versus the difference between the proton and the antiproton mass. A 90% confidence limit contour on the antiproton mass and charge deduced from the $\overline{p}He^+$laser spectroscopy is shown together with the results obtained from TRAP collaboration [49]. Reproduced with permission from [36].

This result can be subsequently combined with the cyclotron frequency of the antiproton measured in a Penning trap with the overwhelming precision of 9 parts in 10^{11} by the TRAP experiment at CERN [49]. This permits to look for any evidence of any difference between the proton/antiproton charge and mass, that would be a violation of the famous CPT invariance theorem. In fact, it can be shown (see figure 7) that combining the results of these two experiments the limit of the CPT violation is set by:

$$\delta_{\overline{p}} = \frac{Q_p + Q_{\overline{p}}}{Q_p} \sim \frac{M_p - M_{\overline{p}}}{M_p} = \frac{1}{f}\frac{\nu_{th} - \nu_{exp}}{\nu_{exp}} \tag{17}$$

with $f \simeq 2-6$ depending on the transition [37,40]

The most recent $\overline{p}He^+$ laser spectroscopy results indicate that this difference must be less than 2 parts in 10^9 [13].

10. Conclusions

This work is about the unique story of antiprotonic helium, a system made by matter and antimatter mixed together, which is here illustrated since its discovery to the most recent days. The extraordinary property of this atom (or molecule, depending on the point of view) is that to possess some long-lived metastable states that survive for a few microseconds before collapsing and annihilating. When compared to usual matter-antimatter interaction timescale, this is a fairly long time, and makes it possible to employ a new kind of laser spectroscopy, a sort of annihilation-inducing-spectroscopy, acting on single states of the atoms and which enables us to analyze transitions with perfect selectivity.

The accuracy of such laser spectroscopy has been continuously improved since the first experiments as of today, reaching the impressive level of a few parts in 10^9 in recent days. Theorists, on the other hand, have been simultaneously improving their models to calculate the energies involved in such transitions, with the results that the comparison of experimental results and theoretical calculations gives nowadays the most precise measurement of the antiproton mass or (alternatively) the most stringent CPT test with barions up to now.

Although at this stage it is hard to imagine whether the antiprotonic helium will ever have any practical application, or whether it will remain confined to a few leading-edge nuclear physics laboratories, particularly with regard to the high cost of the antiprotons production, it is nonetheless nice to regard antiprotonic helium t as a "naturally occurring trap for antiprotons", borrowing the expressions of J. Eades [50]. Actually antiprotonic helium has been the only known system of this kind till a few years ago, when a parallel experiment at CERN [51] discovered the possibility to produce protonium ($\overline{p}p$) in vacuo, with a lifetime of the order of 1 microsecond, through the interaction of antiprotons with hydrogen molecular ions (H_2^+). Yet, antiprotonic helium remains by far the most studied and best known system of this kind.

Acknowledgments

The author is grateful to all the ASACUSA laser spectroscopy group (especially R.S Hayno, E. Widmann and A. Dax) for collaborating together so friendly and so proficiently at the same time.

References

[1] Eisberg, R. and Resnick, R. Quantum Phiysics of Atoms, Molecules, Solids, Nuclei, and Particles; John Wiley & Sons, Inc: New York, NY, 1985; p 650.

[2] Fermi, E. and Teller, E. Phys. Rev. 1947, 72, 399-408.

[3] Fetkovich, J.G. and Pewitt, E.G. *Phys. Rev. Lett.* 1963, 11, 290-293.

[4] Condo, G.T. *Phys. Lett.* 1964, 9, 65-66.

[5] Russell, J.E. *Phys. Rev. Lett.* 1969, 23, 63-64.

[6] Chamberlain, O.; Segre, E.; Wiegand, C. and Ypsilantis, T. *Phys. Rev.* 1955, 100, 947-950.

[7] Yamazaki, T. et al. *Phys. Rev. Lett.* 1989, 63, 1590-1592.

[8] Nakamura, S.N. et al. *Phys. Rev. A* 1992, 45, 6202-6208.

[9] Iwasaki, M. et al. *Phys. Rev. Lett.* 1991, 67, 1246-1249.

[10] Yamazaki, T. et al. *Nature* 1993, 361, 238-240.

[11] Widmann, E. et al. *PS 205 Interim Report* (1994), CERN/SPSLC 94-4 (SPSLC/M 530)

[12] T. Azuma et al. *ASACUSA proposal* (1997), CERN/SPSC 97-19 (SPSC P-307).

[13] Hayano, R.S. *ASACUSA 2009 report* (2010), CERN-SPSC-2010-005 (SPSC-SR-056).

[14] Nakamura, S.N. et al. *Phys. Rev. A* 1994, 49, 4457-4465.

[15] Yamazaki T. et al., *Phys. Rep.* 2002, 366, 183-329.

[16] Mott, N.F. and Massey, H.S.W. The Theory of Atomic Collisions, 3rd Edition, Oxford University Press: Oxford, 1965.

[17] Korenman, G.Ya. *Hyperfine Interactions* 1996, 101/102, 463-469.

[18] Shimamura, I. *Phys. Rev. A* 1992, 46, 3776-3788.

[19] Millero, F.G *Chemical oceanography*, 3rd Edition, CRC Press - Taylor and Francis Group: Boca Raton, FL, 2006; p 44.

[20] Ketzer, B. et al. *Phys. Rev. A* 1996, 53, 2108-2117.

[21] Yamazaki, T. and Ohtsuki, K. *Phys. Rev. A* 1992, 45, 7782-7786.

[22] Widmann, E. et al. *Phys. Rev. A* 1996, 53, 3129-3139.

[23] Morita, N. et al. *Phys. Rev. Lett.* 1994, 72, 1180-1183

[24] Maas, F.E. et al. *Phys. Rev. A* 1995, 52, 4266-4269.

[25] Korobov, V.I. *Phys. Rev. A* 1996, 54 R1749-R1752.

[26] Torii, H.A. et al. *Phys. Rev. A* 1996, 53, R1931-R1934.

[27] Yamazaki, T. et al. *Phys. Rev. A* 1997, 55, R3295-R3298.

[28] Hayano, R.S. et al. *Phys. Rev. A* 1997, 55, R1-R4.

[29] Torii, H.A. et al. *Phys. Rev. A,* 1999, 59, 223-229.

[30] Yamazaki, T. et al. *Chem. Phys. Lett.*, 1997, 265, 137-144.

[31] Ketzer, B. et al. *Phys. Rev. Lett.*, 1997, 78, 1671-1674.

[32] Korobov, V.I. and Bakalov, D.D. *Phys. Rev. Lett.* 1997, 79, 3379-3382.

[33] Korobov V. I. *Phys. Rev. A* 2003, 67, 62501-7

[34] Elander, N. and Yarevsky, E. *Phys. Rev. A* 1997, 56 , 1855-1864

[35] Kino, Y.; Yamanaka, N.; Kamimura, M. and Kudo, H. *Hyperfine Interact.* 2003, 146-147, 331-336

[36] Hayano, R.S.; Hori, M.; Horváth, D.; Widmann, E. *Rep. Prog. Phys.* 2007, 70, 1995-2065

[37] Hori, M. et al. *Phys. Rev. Lett.* 2001, 87, 093401-4

[38] Hori, M. et al. *Phys. Rev. Lett.* 2002, 89, 093401-4

[39] Hori, M. Hayano, R. S. Widmann, E. and Torii, H. *Opt. Lett.* 2003, 28, 2479-2481

[40] Hori, M. et al. *Phys. Rev. Lett.* 2003, 91, 123401-4

[41] Udem, T.; Holzwarth, R. and Hänsch, T. W. *Nature* 2002, 416, 233237

[42] Hori, M. et al. *Phys. Rev. Lett.* 2006, 96, 243401-4

[43] Hori, M. and Dax, A. *Optics Letters* 2009, 34, 1273-1275

[44] Hayano, R.S. *ASACUSA 2007 report* (2008), CERN-SPSC-2008-002 (SPSC-SR-027).

[45] Hayano, R.S. *ASACUSA 2008 report* (2009), CERN-SPSC-2009-005 (SPSC-SR-040).

[46] Widmann, E. et al. *Physics Letters B* 1997, 404, 15-19.

[47] Widmann, E. et. al *Phys. Rev. Lett*. 2002, 89, 243402-4.

[48] Mohr, P.J.; Taylor, B.N. and Newell, D.B.; *Rev. Mod. Phys.* 2008 , 80(2), 633-730.

[49] Gabrielse, G.; Khabbaz, A.; Hall, D. S.; Heimann, C.; Kalinowsky, H. and Jhe, W. *Phys. Rev. Lett.* 1999, 82, 3198-3201.

[50] J. Eades et al., *Hyperfine Interact.* 1993, 81, 227-237

[51] N. Zurlo et al., *Phys. Rev. Lett.* 2006, 97, 153401-5

INDEX

A

B

C

D

E

F

G

H

I

J

K

L

M

N

O

P

Q

R

S

T

U

V

W

Z